International Press Series in Geometry and Topology

he Third Pacific Rim Geometry Conference	Jaigyoung Choe	25
urveys in Differential Geometry, vol. 3	Hsiung/ Yau	24
urveys in Differential Geometry, vol. 2	Hsiung/ Yau	23
urveys in Differential Geometry, vol. 1	S.-T. Yau/ H.B. Lawson	22
ectures on Low-Dimensional Topology	K. Johannson	21
ectures on Harmonic Maps	R. Schoen/ S.-T. Yau	20
ectures on Differential Geometry	R. Schoen / S.-T. Yau	19
oukova Geometry and Topology, v.4	S. Akbulut, et al.	18
oukova Geometry and Topology, v.3	S. Akbulut et al	17
oukova Geometry and Topology, v.2	S. Akbulut et al	16
oukova Geometry and Topology, v 1	S. Akbulut et al	15
eometric Topology, vol. 2	W.H. Kazez	14
eometric Topology, vol. 1	W.H. Kazez	13
hern: A Great Geometer, Expanded Edition	S.-T. Yau	12
First Course in Differential Geometry	C.C. Hsiung	11
lirror Symmetry III	A. Montpetit	10
nternational Press Lectures, vol. 2	Ronald Stern	9
ntl. Conference of Algebra and Geometry	M.-K. Kang	8
opics in Symplectic Manifolds	Ronald Stern	7
lirror Symmetry 1, 2nd ed.	S.-T. Yau	6
ingularities and Complex Geometry	Q._K. Lu, et. al.	5
eometry, Topology and Physics for Raoul Bott	S.-T. Yau	4
lgebraic Geometry and Related Topics	J.-H. Yang, et. al.	3
^{2} Moduli Spaces on 4-Manifolds with ylindrical Ends	C.H. Taubes	2
he L^2 Moduli Space and a Vanishing Theorem or Donaldson Polynomial Invariants	J. Morgan et. al.	1

The Third Pacific Rim Geometry Conference

Jaigyoung Choe, editor

Dec. 16--19, 1996
Seoul, Korea
sponsored by the Global Analysis Research Center
and the Topology and Geometry Research Center

International Press

International Press
P.O. Box 38-2872
Cambridge, MA 02238-2872
USA
www.intlpress.com
orders@descartes.intlpress.com

The Third Pacific Rim Geometry Conference,
Jaigyoung Choe, editor

ISBN 1-57146-064-0
Typeset using the TeX system
Printed in the U.S.A. on acid-free paper.

CONTENTS

PREFACE

The third Pacific Rim Geometry Conference was held December 16-19, 1996 at the Korea University in Seoul, Korea. The Pacific Rim Geometry Conference had been held twice before, at Hong Kong in 1992 and at Singapore in 1994. During the four-day conference there were 45 talks and a total of 147 participants, including mathematicians from China, Hong Kong, Japan, Singapore, Taiwan and the United States, as well as Korea.

This proceedings contains twenty articles by the participants of the conference which, after refereeing process, have been accepted for publication. The editor would like to thank the referees for their valuable jobs, and the authors for reformatting their manuscripts.

The Pacific Rim Geometry Conference was supported by the Global Analysis Research Center, the Topology and Geometry Research Center, the Korea Science and Engineering Foundation and the Korean Mathematical Society. The PRGC was organized by Hyeong In Choi, Jun-Muk Hwang, Hong-Jong Kim, Young Ho Kim and Young Wook Kim.

Jaigyoung Choe
Seoul, September 1998

THE EXISTENCE THEOREM OF HARMONIC OBJECTS VIA GREEN FUNCTION

S. BANDO
Mathematical Institute, Tohoku University,
Sendai, Japan

Let (M, g) be an m-dimensional $(m \geq 2)$ non-compact Riemannian manifold. We assume M admits the positive Green function $G(x, y)$. For any compact smooth domain $D \subset M$ admits the Green function $G_D(x, y)$ with the Dirichelt boundary condition. The assumption means that $\lim_{D \uparrow M} G_D(x, y) = G(x, y)$ exists for $x \neq y$.

The standard way to use the Green function G is as follows. Let f_0 be a smooth function M with suitable growth condition on $\triangle f_0$. Put

$$f(x) = f_0(x) - \int_{y \in M} G(x, y) \triangle f_0(y).$$

Then f is a harmonic function on M. If we assume $\lim_{x \to \infty} G(x, y) = 0$, then $\lim f - f_0 = 0$. In this sense f is a harmonic function with the same boundary value as f_0.

This simple idea can be used to prove the existence theorems of various "harmonic" objects in geometry.

1 Harmonic maps

Let (N, h) an n-dimensional complete Riemannian manifold of non-positive sectional curvature. We denote the distance function on N by d. For a mapping $f_0 : M \longrightarrow N$, we define

$$u(x) = \int_M G(x, y) |\triangle f_0|(y).$$

Theorem 1 *If the integral u converges and defines a continuous function on M, then f_0 can be deformed by the heat equation to a harmonic mapping h which satisfies*

$$d(h(x), f_0(x)) \leq u(x).$$

Proof. Let D be a compact domain in M with smooth boundary. By the result of Hamilton, we can solve the following heat equation on $(0,\infty)$.

$$\frac{df}{dt} = \triangle f,$$
$$f\big|_{t=0} = f_0,$$
$$f\big|_{\partial D} = f_0\big|_{\partial D}.$$

Since N has non-positve sectional curvature, differentiating the equation we get

$$(*) \qquad \frac{d}{dt}\left|\frac{df}{dt}\right| \leq \triangle\left|\frac{df}{dt}\right|.$$

Since $\frac{df}{dt}\big|_{\partial D} = 0$ and $\frac{df}{dt}\big|_{t=0} = \triangle f_0$, we have

$$\left|\frac{df}{dt}\right|(t,x) \leq \int_D H_D(t,x,y)|\triangle f_0(y)|dy,$$

where H_D is the heat kernel of D with the Dirichlet boundary condition. Thus for $0 \leq t_0 < t_1$

$$\begin{aligned} d(f(t_0,x), f(t_1,x)) &\leq \int_{t_0}^{t_1} dt \int_D H(t,x,y)|\triangle f_0(y)|dy, \\ &\leq \int_D G_D(x,y)|\triangle f_0(y)|dy. \end{aligned}$$

where $G_D(x,y) = \int_0^\infty H_D(t,x,y)\,dt$ is the Green function on D. Since $G_D \leq G$, the standard method says that letting D tends to large, we get the solution of the heat equation on the whole manifold M which satisfies

$$d(f(t_0,x), f(t_1,x)) \leq u(x).$$

Taking $t \to \infty$. one gets the desired harmonic map h as the limit of f_t. ∎

Note the crucial point of the proof is the inequality $(*)$ comming from non-positive curvature of N.

In the case that the manifolds M, N are simply connected hyperbolic spaces of constant curvature -1, we can identify the ideal boundaries ∂M and ∂N to the unit speres by the unit ball models. Li-Tam has shown the following,

Theorem 2 **([L-T])** *If a C^3 map $\phi : \partial M \longrightarrow \partial N$ satisfies the nondegeneracy condition $e(\phi) = |d\phi|^2 > 0$ everywhere, then we can construct a harmonic map $h : M \longrightarrow N$ which has ϕ as the boundary value.*

Here we give a simple alternate proof applying Theorem 1. We denote the distance function from the fixed origin o on M by ρ.

Lemma 1 **([L-T])** *Under the assumption of the Theorem 2, we can extend the map ϕ to a C^2 map f_0 defined on $M \cup \partial M$ such that $|\triangle f_0| = O(e^{-\rho})$.*

Let $(\tilde{\rho}, \theta) \in M$ and $(\tilde{r}, \phi) \in N$ be poler coordinates in the unit ball models. Near infinity, f_0 is given as follows.

$$\begin{aligned} \phi &= \phi(\theta), \qquad \text{(extends the given function radially constant)}, \\ \tilde{r} &= 1 - (1 - \tilde{\rho})\sqrt{\frac{e(\phi)}{m-1}}. \end{aligned}$$

Lemma 2

$$\int_M G(x,y)e^{-\rho(y)}dy = \begin{cases} O(\rho(x)e^{-\rho(x)}), & \textit{for } m = 2, \\ O(e^{-\rho(x)}), & \textit{for } m \geq 3. \end{cases}$$

Proof. Use the following facts.

1. G is a radial function such that $G(x, o) = O(e^{-(m-1)\rho(x)})$ for $\rho(x) \geq 1$.

2. Let r be the distance between x and y, θ the angle at x made by the geodesics to o and y. Then it holds that

$$\cosh(\rho(y)) = \cosh(\rho(x))\cosh(r) - \sinh(\rho(x))\sinh(r)\cos(\theta).$$

3. Denote the first coordinate of the unit sphere S^{m-1} by x^1. Then for a constant $0 < \epsilon < 1$, we have

$$\int_{S^{m-1}} \frac{1}{1 - \epsilon x^1} = \begin{cases} O((1-\epsilon)^{-1/2}) & \text{for } m = 2, \\ O(\log(1-\epsilon)^{-1}) & \text{for } m = 3, \\ O(1) & \text{for } m \geq 4. \end{cases}$$

∎

Combining Lemma 1 and Lemma 2, we get

Theorem 3 *Under the condition of Theorem 2, we have a harmonic map h such that*

$$d(h, f_0) = \begin{cases} O(\rho(x)e^{-\rho(x)}), & \textit{for } m = 2, \\ O(e^{-\rho(x)}), & \textit{for } m \geq 3. \end{cases}$$

Theorem 4 ([L-T]) *The harmonic map constructed in Theorem 3 defines a* $C^{1,\alpha}$ *(*$0 < \alpha < 1$*) map on the closed ball* $M \cup \partial M$.

Proof. We write down the harmonic map h using the exponential map based on f_0.

$$h = \exp_{f_0} X, \qquad X \in \Gamma(M, f_0^* TN).$$

Then we can easily see

$$|X| = \begin{cases} O(\rho e^{-\rho}), & m = 2, \\ O(e^{-\rho}), & m \geq 3, \end{cases}$$

$$|\Delta X| = \begin{cases} O(\rho e^{-\rho}), & m = 2, \\ O(e^{-\rho}), & m \geq 3. \end{cases}$$

Thus

$$|\nabla X| = \begin{cases} O(\rho e^{-\rho}), & m = 2, \\ O(e^{-\rho}), & m \geq 3. \end{cases}$$

Note that $|\nabla X|$ for hyperbolic metrics and Euclidian metrics are compatible. ∎

For recent development refer to the article by S. Nishikawa in this volume.

Remark. We used the notaion Δ to stand the trace of the second covariant derivatives.

2 Einstein-Hermitian metrics

We can apply the similar arguement for the Einstein-Hermitian metrics.

Definition 1 *Let* (E, h) *be a Hermitian holomorphic vector bundle over a Kähler manifold* (M, g). *Then* (E, h) *is called Einstein if the trace* ΛF *of the curvature* F *is a constant times identity endomorphism of* E.

Let (E, h_0) is a Hermitan holomorphic vector bundle on the complex Euclidean plane $\mathbf{C}^2$. We assume that its curvature F_0 satisfys the following decay condition.

$$|\Lambda F_0| = O(r^{-2-\epsilon}) \qquad (\epsilon > 0)$$

Then the integral

$$\int_{y \in \mathbf{C}^2} G(x, y) |\Lambda F_0|(y)$$

converges and gives a continuous function. We can obtain an Einstein-Hermitian metric on E by solving the heat equation on Hermitian metric h.

$$\frac{dh}{dt}h^{-1} = -\sqrt{-1}\Lambda F,$$
$$h\big|_{t=0} = h_0.$$

Differentiating the equation in t, we obtain the following crucial inequality.

$$(**) \qquad \frac{d}{dt}\left|\frac{dh}{dt}\right| \leq \frac{1}{2}\Delta\left|\frac{dh}{dt}\right|.$$

The same arguement as in the case of harmonic maps concludes the convergence of the solution metrics to an Einstein-Hermitian metrc in the limit of $t \longrightarrow \infty$.

In this way, we can give a PDE proof of the theorem of Donaldson.

Theorem 5 ([D]) *There is a natural correspondence between the moduli space of anti-self-dual connections with finite action on the complex Euclidean plane* $\mathbf{C}^2$ *and the moduli space of holomorphic vector bundles on the complex projective plane* $\mathbf{P}^2$ *whose restriction to the complex line at infinity are trivial.*

For a detail of the proof and generalization, refer to [B].

3 Scalar curvature

Let (M, g) be the m-dimensional ($m \geq 3$) Riemannian manifold and u be a positive function on M. The scalar curvature K of the conformally changed metric $\tilde{g} = u^{4/(n-4)}g$ is given by the following equation.

$$\frac{4(n-1)}{n-2}\Delta u - ku + Ku^{\frac{n+2}{n-2}} = 0,$$

where Δ and k are Laplacian and the scalar curvature for g, respectively.

In [N] Ni treated in the case of Euclidean space $\mathbf{R}^m$. Split $x \in \mathbf{R}^m$ as $x = (x_1, x_2) \in \mathbf{R}^3 \times \mathbf{R}^{m-3}$.

Theorem 6 ([N]) *If there exist constants* $C > 0$, $l > 2$ *such that* $|K(x)| \leq C/|x_1|^l$, *then the equation*

$$(***) \qquad \Delta u + Ku^p = 0 \qquad \text{with fixed } p > 1$$

admits infinitly many bounded positive solutions which are bounded away from zeros.

One can show this result by using heat equation as in the previous sections, but it would be easier to use Green function directly.

Theorem 7 *Assume the integral*

$$\int_{y \in M} G(x,y)|K(y)|$$

converges and gives a bounded countious function in x. Then the equation (***) admits infinitly many bounded positive solutions which are bounded away from zeros.

Proof. Define an operator T on the space C(M) of bounded continous functions on M by

$$\mathrm{T}u(x) = \int_{y \in M} G(x,y)K(y)u(y)^p.$$

We have $\|\mathrm{T}u\| \leq C\|u\|^p$ with a constant $C > 0$. One can easily solve the following equation for sufficiently small $\epsilon > 0$ with the uniform estimate independent of ϵ.

$$u = 1 + \epsilon \mathrm{T}u.$$

Then $\epsilon^{1/(p-1)}u$ gives the desired solution.

Acknowledgments

This work was partially supported by the Inamori Foundation.

References

[B] S. Bando, Einstein-Hermitian metrics on non-compact Kähler manifolds, in *Einstein metrics and Yang-Mills connections*, ed. T. Mabuchi and S. Mukai, Lect. Notes in Pure and Appl. Math., 27–33, Marcel Dekker, Inc., New York, 1993.

[D] S. K. Donaldson, Instanton and geometric invariant theory, Comm. Math. Phys. 93 (1984), 453–460.

[L-T] P. Li and L.-F. Tam, The heat equation and harmonic maps of complete manifolds, Invent. math. 105 (1991), 1–46.

[N] W.-M. Ni, On the elliptic equation $\Delta u + K(x)u^{(n+2)/(n-2)} = 0$, its generalization and applications in geometry, Indiana Univ. Math. J. 31 (1982), 493–529.

STRINGS OF RIEMANNIAN INVARIANTS, INEQUALITIES, IDEAL IMMERSIONS AND THEIR APPLICATIONS

BANG-YEN CHEN
Department of Mathematics, Michigan State University, East Lansing, Michigan 48824-1027, U. S. A.
E-mail: bychen@math.msu.edu

The main purpose of this paper is to explain the two strings of new type of Riemannian curvature invariants and the sharp inequalities, involving these invariants and the squared mean curvature, which are originally introduced and established in [9]. These invariants have several interesting connections to several areas of mathematics. For instance, they give rise to new obstructions to minimal, Lagrangian and slant isometric immersions. Moreover, these invariants relate closely to the notion of order and to the first nonzero eigenvalue λ_1 of the Laplacian Δ on a Riemannian manifold. These invariants together with the sharp inequalities gives rise naturally to the notion of "ideal immersions" or the notion of "the best ways of living".

After introducing the Riemannian invariants and inequalities, we explain the method to establish a new maximum principle by utilizing these invariants and these inequalities. Some sharp relationship between the eigenvalues of the Laplacian Δ and these new invariants are then presented. We also explain in details the physical meaning of the notion of "ideal immersions" for Riemannian manifolds in a Riemannian space form based again on the sharp inequalities. Many examples of ideal immersions are illustrated in this paper. Also classification, existence and non-existence theorems in this respect are given. Furthermore, we explain the method for applying the sharp inequalities to establish rigidity theorems for submanifolds in space forms without global assumption on the submanifolds and regardless of codimension. In this paper, we also present the sharp relationships between the k-Ricci curvatures, squared mean curvature, and the shape operator for an arbitrary submanifold in a Riemannian space form which were originally discovered in [11].

Contents

1 Introduction

One of the most fundamental problems in the theory of submanifolds is the immersibility (or non-immersibility) of a Riemannian manifold in a Euclidean space (or, more generally, in a space form). According to a well-known theorem of J. F. Nash, every Riemannian manifold can be isometrically immersed in some Euclidean spaces with sufficiently high codimension. In order to study this fundamental problem, in view of Nash's theorem, it is natural to impose a suitable condition on the immersions. For example, if one imposes the minimality condition on the immersions, it leads to

Problem 1. *Given a Riemannian manifold M, what are necessary conditions for M to admit a minimal isometric immersion in a Euclidean m-space E^m?*

It is well-known that for a minimal submanifold in E^m, the Ricci tensor satisfies $\text{Ric} \leq 0$. For many years this was the only known necessary Riemannian condition for a general Riemannian manifold to admit a minimal isometric immersion in a Euclidean space regardless of codimension (cf. [5,24]).

An immersion of a Riemannian n-manifold M in a Hermitian manifold $\tilde{M}$ is called *totally real* (or *isotropic* in symplectic geometry) if the almost complex structure J of $\tilde{M}$ maps each tangent space of M into its corresponding normal space. In particular, a totally real immersion is said to be *Lagrangian* if $\dim M = \dim_{\mathbf{C}} \tilde{M}$. For Lagrangian immersions in complex Euclidean n-space $\mathbf{C}^n$, a result of Gromov [36] states that a compact n-manifold M admits a Lagrangian immersion (not necessary isometric) into $\mathbf{C}^n$ if and only if the complexification $TM \otimes \mathbf{C}$ of the tangent bundle of M is trivial. Gromov's result implies that there is no topological obstruction to Lagrangian immersions for compact 3-manifolds in $\mathbf{C}^3$, because the tangent bundle of a 3-manifold is always trivial.

From the Riemannian point of view, it is natural to ask the following basic question.

Problem 2 *What are necessary conditions for a compact Riemannian manifold to admit a Lagrangian isometric immersion in $\mathbf{C}^n$?*

The class of Lagrangian immersions is included in a much larger class of immersions, namely, the class of slant immersions. A submanifold M in a Hermitian manifold is called *slant* if its Wirtinger angle (i.e., its slant angle) is constant. There exist abundant examples of slant submanifolds in $\mathbf{C}^n$ (cf. [4,19,46,47]).

There is a topological obstruction obtained in [4] for slant immersions; namely, a compact $2k$-manifold M with $H^{2i}(M;\mathbf{R}) = 0$ for some $1 \le i \le k$ admits no slant immersion in any Kaehlerian manifold $\tilde{M}^m$ unless it is totally real (or Lagrangian when $m = 2k$). Moreover, the author and Tazawa proved in [19] that there exist no slant immersions of a compact n-manifold in $\mathbf{C}^m$ unless it is totally real. On the other hand, there do exist compact slant submanifolds in a complex n-torus.

For slant immersions it is natural to ask the following.

Problem 3 *What are necessary conditions for a compact Riemannian n-manifold to admit a slant isometric immersion in a complex torus CT^n?*

In [5,6] the author introduced a Riemannian invariant and obtained a sharp inequality between this invariant and the squared mean curvature for arbitrary submanifolds in a Riemannian space form. By applying this sharp inequality, the author was able to obtain some sharp solutions to Problems 1, 2 and 3. Submanifolds which satisfy the equality case of this type of inequalities have been investigated recently by many geometers (cf. for instance, [2,5,6,9,14-16,21,22,27-30,32,33,38]).

In section 2, we explain the two strings of Riemannian curvature invariants $\delta(n_1,\ldots,n_k)$ and $\hat{\delta}(n_1,\ldots,n_k)$ first introduced in [9]. We simply called these invariants the δ-invariants. The first string of δ-invariants, $\delta(n_1,\ldots,n_k)$, extend naturally the Riemannian invariant introduced in [5,6]. In section 3 we present some new characterizations of Einstein and conformally flat manifolds in terms of δ-invariants and also in terms of the notion of the scalar curvature of k-plane sections [18]. In section 4, we present the general sharp inequalities of [9] which relate the δ-invariants to the squared mean curvature for submanifolds in both Riemannian and Kaehlerian space forms. By applying these sharp general inequalities, we establish a new maximum principle in section 5. This maximum principle allows us to obtain some relationships between the δ-invariants. In this section we also introduce the notion of "ideal immersions" (or "best ways of living") for Riemannian manifolds in a Riemannian space form. Roughly speaking, an isometric immersion of a Riemannian manifold M in a space form is called an ideal immersion (or a best way of living) if the tension the submanifold receives from the surrounding space at each point is equal to the least possible amount at that point. In section 6 we establish some sharp relationship between the eigenvalues of the Laplacian Δ and the δ-invariants. Moreover, we establish a very simple necessary and sufficient condition for a compact homogeneous Riemannian manifold with irreducible isotropy action to admit an ideal immersion in some Euclidean space. Several of their applications are presented. In section 7 we explain how to apply our general inequalities to provide sharp solutions to Problems 1, 2 and 3; thus extending the previous solutions obtained in [5,6]. In section 8 we present applications of our inequalities to rigidity problem for submanifolds, regardless of codimension. Many examples of ideal immersions are illustrated in section 9. Several classification theorems of tubular ideal hypersurfaces are given in section 10. In section 11 we present the classification of codimension one ideal immersions for conformally flat submanifolds. In section 12 we define complex and totally real δ-invariants and provide sharp inequalities involving such invariants for submanifolds in a Kaehlerian space form. We use the δ-invariants in section 13 to present simple new characterizations of Kaehlerian space forms, Bochner-Kaehler and Einstein-Kaehler manifolds. In section 14 we present some classification theorems for real hypersurfaces in a non-flat Kaehlerian space form. In the last section, we provide sharp relationship between the notion of k-Ricci curvature and the shape operator at the mean curvature vector for arbitrary submanifolds in a Riemannian space form which were obtained originally in [11].

2 Riemannian invariants (RNA): DNA of Riemannian manifolds

It is well-known that Riemannian invariants are the intrinsic characteristics of the Riemannian manifold. Riemannian invariants of a Riemannian manifold affect the behavior in general of the Riemannian manifold. Borrow a term from biology, Riemannian invariants are DNA of Riemannian manifolds. For this reason we simply call Riemannian invariants, RNA : the DNA of Riemannian manifolds.

Curvature invariants are the $N^o\,1$ Riemannian invariants and the most natural ones. Curvatures invariants also play key roles in physics. For instance, the magnitude of a force required to move an object at constant speed, according to Newton's laws, a constant multiple of the curvature of the trajectory. The motion of a body in a gravitational field is determined, according to Einstein, by the curvatures of space time. All sorts of shapes, from soap bubbles to red blood cells, seem to be determined by various curvatures (cf. [43]).

Classically, among the Riemannian curvature invariants, people have been studying sectional, scalar and Ricci curvatures. In this section we explain the two strings of new types of curvature invariants originally introduced and investigated by the author in [6,9]. These two strings of Riemannian curvature invariants seem to play significant roles in several areas of mathematics including submanifold theory and Riemannian, spectral and symplectic geometries.

Let M be a Riemannian n-manifold. Denote by $K(\pi)$ the sectional curvature of M associated with a plane section $\pi \subset T_pM$, $p \in M$. For any orthonormal basis $e_1, \dots, e_n$ of the tangent space T_pM, the scalar curvature τ at p is defined to be

$$\tau(p) = \sum_{i<j} K(e_i \wedge e_j). \tag{2.1}$$

Let L be a subspace of T_pM of dimension $r \geq 2$ and $\{e_1, \dots, e_r\}$ an orthonormal basis of L. We define the scalar curvature $\tau(L)$ of the r-plane section L by

$$\tau(L) = \sum_{\alpha<\beta} K(e_\alpha \wedge e_\beta), \quad 1 \leq \alpha, \beta \leq r. \tag{2.2}$$

Given an orthonormal basis $\{e_1, \dots, e_n\}$ of the tangent space T_pM, we simply denote by $\tau_{1\cdots r}$ the scalar curvature of the r-plane section spanned by $e_1, \dots, e_r$. The scalar curvature $\tau(p)$ of M at p is nothing but the scalar curvature of the tangent space of M at p; and if L is a 2-plane section, $\tau(L)$ is nothing but the sectional curvature $K(L)$ of L. Geometrically, $\tau(L)$ is nothing but the scalar curvature of the image $\exp_p(L)$ of L at p under the exponential map at p.

For an integer $k \geq 0$ denote by $\mathcal{S}(n, k)$ the finite set consisting of unordered k-tuples $(n_1, \dots, n_k)$ of integers ≥ 2 satisfying $n_1 < n$ and $n_1 + \cdots + n_k \leq n$. Denote by $\mathcal{S}(n)$ the set of unordered k-tuples with $k \geq 0$ for a fixed n.

Let $\#\mathcal{S}(n)$ denote the cardinal number of $\mathcal{S}(n)$. Then $\#\mathcal{S}(n)$ increases quite rapidly with n. For instance, for

$$n = 2, 3, 4, 5, 6, 7, 8, 9, 10, 11, 12, \cdots, 20, \dots,$$
$$50, \dots, 100, \dots, 200, \dots,$$

$\#\mathcal{S}(n)$ are given respectively by

$$1, 2, 4, 6, 10, 14, 21, 29, 41, 54, 76, \cdots, 626, \dots,$$
$$204225, \dots, 190569291, \dots, 3972999029387, \dots .$$

In general, the number $\#\mathcal{S}(n)$ is equal to $p(n) - 1$, where $p(n)$ denotes the partition function. According to a result of H. A. Radamacher, we have

$$\#\mathcal{S}(n) = \frac{1}{\pi\sqrt{2}} \sum_{k=1}^{\infty} A_k(n) k^{1/2} \left[\frac{d}{dx} \frac{\sinh((\pi/k)(\frac{2}{3}(x - 1/24))^{1/2})}{(x - 1/24)^{1/2}} \right]_{x=n} - 1,$$

where

$$A_k(n) = \sum_{h \bmod k,\,(h,k)=1} \omega_{h,k} e^{-2\pi i n h/k}$$

with $\omega_{h,k}$ a certain $24k$-th root of unity given by

$$\omega_{h,k} = -\tfrac{k}{h} \exp\Big(- \pi i \Big(\tfrac{1}{4}(2 - hk - h) + \tfrac{1}{12}(k - k^{-1})(2h - h' + h^2 h') \Big) \Big) \quad \text{if } h \text{ is odd},$$

and

$$\omega_{h,k} = -\tfrac{h}{k} \exp\Big(- \pi i \Big(\tfrac{1}{4}(k - 1) + \tfrac{1}{12}(k - k^{-1})(2h - h' + h^2 h') \Big) \Big) \quad \text{if } k \text{ is odd},$$

and h' is a solution of the congruence: $hh' \equiv -1 \pmod{k}$.

The asymptotic behavior of $\#\mathcal{S}(n)$ is given by

$$\#\mathcal{S}(n) \approx \frac{1}{4n\sqrt{3}} \exp\left[\pi \left(\frac{2n}{3} \right)^{1/2} \right] \quad \text{as } n \to \infty.$$

For each k-tuple $(n_1, \dots, n_k) \in \mathcal{S}(n)$ the two sequences of Riemannian invariants $S(n_1, \dots, n_k)(p)$ and $\hat{S}(n_1, \dots, n_k)(p)$ are defined respectively by

$$\begin{aligned} S(n_1, \dots, n_k)(p) &= \inf\{\tau(L_1) + \cdots + \tau(L_k)\}, \\ \hat{S}(n_1, \dots, n_k)(p) &= \sup\{\tau(L_1) + \cdots + \tau(L_k)\}, \end{aligned} \tag{2.3}$$

where $L_1, \dots, L_k$ run over all k mutually orthogonal subspaces of T_pM such that $\dim L_j = n_j$, $j = 1, \dots, k$. The two strings of Riemannian curvature invariants $\delta(n_1, \dots, n_k)(p)$ and $\hat{\delta}(n_1, \dots, n_k)(p)$ introduced in [9] are given by

$$\begin{aligned} \delta(n_1, \dots, n_k)(p) &= \tau(p) - S(n_1, \dots, n_k)(p), \\ \hat{\delta}(n_1, \dots, n_k)(p) &= \tau(p) - \hat{S}(n_1, \dots, n_k)(p). \end{aligned} \tag{2.4}$$

In terms of these δ-invariants, the scalar curvature τ is nothing but $\delta(\emptyset) = \hat{\delta}(\emptyset)$ (with $k = 0$); moreover, the invariant introduced in [5,6] is nothing but the invariant $\delta(2)$ (with $k = 1$ and $n_1 = 2$). Obviously, one has $\delta(n_1, \dots, n_k) \geq \hat{\delta}(n_1, \dots, n_k)$ for any k-tuple $(n_1, n_2, \dots, n_k)$ in $\mathcal{S}(n)$.

I shall point out that our δ-invariants $\delta(n_1, \dots, n_k)$, $\hat{\delta}(n_1, \dots, n_k)$ with $k > 0$ and the scalar curvature are very much different in nature.

A Riemannian manifold M is simply called an $S(n_1, \dots, n_k)$*-space* if it satisfies $S(n_1, \dots, n_k) = \hat{S}(n_1, \dots, n_k)$ identically. Such spaces are completely determined by the following two propositions.

Proposition 2.1 [18] *Let M be a Riemannian n-manifold with $n > 2$. Then we have*

(1) *For any integer j with $2 \leq j \leq n-2$, M is an $S(j)$-space if and only if M is a Riemannian space form.*

(2) *M is an $S(n-1)$-space if and only if M is an Einstein space.*

Proposition 2.2 [18] *Let M be a Riemannian n-manifold such that n is not a prime and k an integer ≥ 2. Then*

(1) *if M is an $S(n_1, \ldots, n_k)$-space, then M is a Riemannian space form unless $n_1 = \ldots = n_k$ and $n_1 + \cdots + n_k = n$, and*

(2) *M is an $S(n_1, \ldots, n_k)$-space with $n_1 = \ldots = n_k$ and $n_1 + \cdots + n_k = n$ if and only if M is a conformally flat space.*

Statement (2) of Proposition 2.2 gives rise to a simple new characterization of conformally flat manifolds. In particular, statement (2) of Proposition 2.2 yields the following.

Corollary 2.3 *Let M be a Riemannian $2r$-manifold with $r > 1$. Then M is a conformally flat manifold with vanishing scalar curvature if and only if*

$$\tau(L) = -\tau(L^{\perp}) \tag{2.5}$$

for any r-plane section $L \subset T_pM$, $p \in M$, where $L^{\perp}$ denotes the orthogonal complement of L in T_pM.

Remark 2.1 For a fixed k-tuple $(n_1, \ldots, n_k) \in \mathcal{S}(n)$, we simply call a Riemannian n-manifold M a *$\delta(n_1, \ldots, n_k)$-space* (respectively, a *$\hat{\delta}(n_1, \ldots, n_k)$-space*) if its δ-invariant $\delta(n_1, \ldots, n_k)$ (respectively, its $\hat{\delta}(n_1, \ldots, n_k)$) is constant on M.

It is clear that a homogeneous Riemannian manifold M is a $\delta(n_1, \ldots, n_k)$-space and also a $\hat{\delta}(n_1, \ldots, n_k)$-space for each fixed $(n_1, \ldots, n_k) \in \mathcal{S}(n)$. In general, the class of $\delta(n_1, \ldots, n_k)$-spaces and the class of $\hat{\delta}(n_1, \ldots, n_k)$-spaces are much bigger than the class of homogeneous Riemannian manifolds. For instance, there exist many 3-dimensional $\delta(2)$-spaces which are non-homogeneous (cf. for instance, [3,5,14-17,27-30,32]).

In views of Propositions 2.1 and 2.2, it is interesting to investigate the geometries of $\delta(n_1, \ldots, n_k)$-spaces and of $\hat{\delta}(n_1, \ldots, n_k)$-spaces.

Remark 2.2 It was pointed out by R. Osserman (personal communication) that $\tau = \delta(\emptyset)$ and $\delta(2)$ satisfy

$$\frac{\delta(2)}{N-1} \geq \frac{\tau}{N}, \quad N = \binom{n}{2}, \tag{2.6}$$

for a Riemannian n-manifold with $n > 2$, with equality holding only for Riemannian space forms.

3 Characterizations of Einstein and conformally flat spaces

By using the notion of the scalar curvature of r-plane sections, we have the following simple characterization of Einstein spaces which generalizes the well-known characterization of Einstein 4-manifolds given by I. M. Singer and J. A. Thorpe obtained in [45].

Theorem 3.1 [18] *Let M be a Riemannian $2r$-manifold. Then M is an Einstein space if and only if*

$$\tau(L) = \tau(L^{\perp}) \tag{3.1}$$

for any r-plane section $L \subset T_pM$, $p \in M$.

Moreover, also by using the notion of the scalar curvature of r-plane sections, we have the following simple characterization of conformally flat spaces which generalizes a well-known result of R. S. Kulkarni obtained in [37].

Theorem 3.2 [18] *Let M^n be a Riemannian manifold with $n \geq 4$, and let s be an integer satisfying $2 < 2s \leq n$. Then M is a conformally flat manifold if and only if, for any orthonormal set $\{e_1, \ldots, e_{2s}\}$ of vectors, one has*

$$\tau_{1\cdots s} + \tau_{s+1\cdots 2s} = \tau_{1\cdots s-1\, s+1} + \tau_{s\, s+2\cdots 2s}. \tag{3.2}$$

Remark 3.1 In general, the δ-invariants $\delta(n_1, \ldots, n_k)$ are independent invariants. However, Theorem 3.1 implies that, for a $2r$-dimensional Einstein manifold, we have the following relations:

$$\delta(\emptyset) + \delta(r,r) = 2\delta(r), \quad \hat{\delta}(\emptyset) + \hat{\delta}(r,r) = 2\hat{\delta}(r). \tag{3.3}$$

4 Sharp inequalities involving δ-invariants

For each $(n_1, \ldots, n_k) \in \mathcal{S}(n)$, let $c(n_1, \ldots, n_k)$ and $b(n_1, \ldots, n_k)$ denote the constants given by

$$c(n_1, \ldots, n_k) = \frac{n^2(n + k - 1 - \sum n_j)}{2(n + k - \sum n_j)}, \tag{4.1}$$

$$b(n_1, \ldots, n_k) = \frac{1}{2}\Big(n(n-1) - \sum_{j=1}^{k} n_j(n_j - 1)\Big). \tag{4.2}$$

The following sharp inequalities involving the δ-invariants and the squared mean curvature obtained in [9] play the most fundamental role in this paper.

Theorem 4.1 (Fundamental Inequalities) *For each $(n_1, \ldots, n_k) \in \mathcal{S}(n)$ and each n-dimensional submanifold M in a Riemannian space form $R^m(\epsilon)$ of constant sectional curvature ϵ, we have*

$$\delta(n_1, \ldots, n_k) \leq c(n_1, \ldots, n_k)H^2 + b(n_1, \ldots, n_k)\epsilon. \tag{4.3}$$

The equality case of inequality (4.3) *holds at a point $p \in M$ if and only if there exists an orthonormal basis $e_1, \ldots, e_m$ at p such that the shape operators of M in $R^m(\epsilon)$ at p take the following forms:*

$$A_r = \begin{bmatrix} A_1^r & \cdots & 0 & 0 & \cdots & 0 \\ \vdots & \ddots & \vdots & \vdots & & \vdots \\ 0 & \cdots & A_k^r & 0 & \cdots & 0 \\ 0 & \cdots & 0 & \mu_r & \cdots & 0 \\ \vdots & \ddots & \vdots & \vdots & \ddots & \vdots \\ 0 & \cdots & 0 & 0 & \cdots & \mu_r \end{bmatrix}, \tag{4.4}$$

$$r = n+1, \ldots, m,$$

where each A_j^r are symmetric $n_j \times n_j$ submatrices satisfying

$$\operatorname{trace}(A_1^r) = \cdots = \operatorname{trace}(A_k^r) = \mu_r. \tag{4.5}$$

The exact proof of Theorems 4.1 given in [9] also yields the following.

Theorem 4.2 *Let M be an n-dimensional totally real (in particular, Lagrangian) submanifold in a Kaehlerian (respectively, quaternionic) space form $\tilde{M}^m(4\epsilon)$ of constant holomorphic (respectively, quaternionic) sectional curvature 4ϵ. Then, for any k-tuple $(n_1, \ldots, n_k) \in \mathcal{S}(n)$, we have*

$$\delta(n_1, \ldots, n_k) \leq c(n_1, \ldots, n_k)H^2 + b(n_1, \ldots, n_k)\epsilon. \tag{4.6}$$

The equality case of (4.6) *holds at a point $p \in M$ if and only if there exists an orthonormal basis $e_1, \ldots, e_m$ at p such that the shape operators of M in $\tilde{M}^m(\epsilon)$ at p take the forms of* (4.4) *which satisfy* (4.5).

Remark 4.1 Inequality (4.3) implies that for arbitrary k mutually orthogonal subspaces $L_1, \ldots, L_k$ of T_pM with $\dim L_j = n_j$, $j = 1, \ldots, k$, one has

$$\tau(p) - \tau(L_1) - \cdots - \tau(L_k) \leq c(n_1, \ldots, n_k)H^2(p) + b(n_1, \ldots, n_k)\epsilon. \tag{4.7}$$

In particular, we have

$$\hat{\delta}(n_1,\dots,n_k) \le c(n_1,\dots,n_k)H^2 + b(n_1,\dots,n_k)\epsilon. \tag{4.8}$$

Remark 4.2 On the other hand, for a given arbitrary intrinsic invariant, say ρ, of a Riemannian n-manifold M, there doesn't exist in general a positive constant, say C, such that

$$H^2(p) \ge C\rho(p). \tag{4.9}$$

This fact can be seen from the following simple example.

Example 4.1 Let $x : M \to E^{n+1}$, $n > 2$, be a minimal hypersurface whose shape operator is non-singular at some point $p \in M$. Then by the minimality there exist two principal directions at p, say e_1, e_2, such that their corresponding principal curvatures κ_1, κ_2 are of the same sign. This implies that the sectional curvature K_{12} at p is positive. Since $H = 0$, this shows that there does not exist any positive constant C such that $H^2(p) \ge C\mu(p)$, where μ is the Riemannian invariant defined by $\mu = \max K$.

Example 4.2 For any k-tuple $(n_1,\dots,n_k) \in \mathcal{S}(n)$, we have

$$c(n_1,\dots,n_k) \ge b(n_1,\dots,n_k), \tag{4.10}$$

with equality holding if and only if either (1) $k = 0$ or (2) $n_1 = \cdots = n_k$ and $kn_1 = n$. This fact follows from Theorem 4.1 and that $\delta(n_1,\dots,n_k) = b(n_1,\dots,n_k)$ for the unit n-sphere.

5 A maximum principle, ideal immersions or best ways of living

In general there doesn't exist direct relationship between δ-invariants. However, Theorem 4.1 gives rise to the following maximum principle.

A Maximum Principle *Let M be an n-dimensional submanifold of a Riemannian space form $R^m(\epsilon)$. If it satisfies the equality case of* (4.3), *i.e.,*

$$\delta(n_1,\dots,n_k) = c(n_1,\dots,n_k)H^2 + b(n_1,\dots,n_k)\epsilon, \tag{5.1}$$

for a k-tuple $(n_1,\dots,n_k) \in \mathcal{S}(n)$, then

$$\Delta_\epsilon(n_1,\dots,n_k) \ge \Delta_\epsilon(m_1,\dots,m_j) \tag{5.2}$$

for any $(m_1,\dots,m_j) \in \mathcal{S}(n)$, where

$$\Delta_\epsilon(n_1,\dots,n_k) = \frac{\delta(n_1,\dots,n_k) - b(n_1,\dots,n_k)\epsilon}{c(n_1,\dots,n_k)}. \tag{5.3}$$

Let $x : M \to R^m(\epsilon)$ be an isometric immersion of a Riemannian n-manifold M into $R^m(\epsilon)$. Theorem 4.1 implies

$$H^2(p) \geq \hat{\Delta}_\epsilon(p), \quad p \in M, \tag{5.4}$$

where $\hat{\Delta}_\epsilon$ is the Riemannian invariant on M defined by

$$\hat{\Delta}_\epsilon = \max \Delta_\epsilon(n_1, \ldots, n_k), \tag{5.5}$$

where $(n_1, \ldots, n_k)$ runs over all elements in $\mathcal{S}(n)$.

Using inequality (5.4) we introduce the notion of ideal immersions.

Definition 5.1 An isometric immersion $x : M \to R^m(\epsilon)$ of a Riemannian n-manifold M into a Riemannian space form $R^m(\epsilon)$ is called an *ideal immersion* if equality case of (5.4) holds identically on M.

Remark 5.1 (*Physical Interpretation of Ideal Immersions*) "An isometric immersion $x : M \to R^m(\epsilon)$ is an ideal immersion" means that *"M lives in $R^m(\epsilon)$ in a best way"*; in the sense that M receives the least possible amount of tension (given by $\hat{\Delta}_\epsilon(p)$) from the surrounding space at each point p on M. This is due to (5.4) and the fact that the mean curvature vector field is exactly the tension field for an isometric immersion of a Riemannian manifold in another Riemannian manifold (a well-known fact since the time of Laplace). Therefore, the squared mean curvature at a point on the submanifold simply measures the amount of tension the submanifold receives from the surrounding space at that point.

If one defines a "best world" to be a complete Riemannian space with the highest degree of homogeneity, then, according to work of Lie, Klein and Killing, the family of best worlds consists of Euclidean spaces, Riemannian spheres, real projective spaces, and real hyperbolic spaces. These spaces have the highest degree of homogeneity, since their groups of isometries have the maximal possible dimension. In this sense, a best world in the terminology of differential geometry is nothing but a Riemannian space form $R^m(\epsilon)$.

Remark 5.2 (*Physical Interpretation of Theorem 4.1*) In terms of tension and RNA, Theorem 4.1 establishes a simple, sharp and direct relationship between the RNA string $\delta(n_1, \ldots, n_k)$ of M and the amount of tension M receiving from its surrounding space $R^m(\epsilon)$, regardless of codimension.

These show that each member in the string of invariants $\delta(n_1, \ldots, n_k)$ plays some significant roles in submanifold theory as well as in mathematical physics.

It is interesting to point out that, in terms of tension and RNA, Remark 4.1, Remark 4.2 and Example 4.1 can be interpreted as the following

Observation *Not every RNA of a Riemannian manifold M relates directly to the amount of tension M receiving from its surrounding space when M lives in a best world.*

Remark 5.3 Our maximum principle yields the following important fact.

Fact *If an isometric immersion $x : M \to R^m(\epsilon)$ satisfies equality* (5.1) *for some k-tuple $(n_1, \ldots, n_k) \in \mathcal{S}(n)$, then it is an ideal immersion automatically.*

From the definition of ideal immersions, we have

Proposition 5.1 *Let $x : M^2 \to R^m(\epsilon)$ be an isometric immersion of a Riemannian 2-manifold into $R^m(\epsilon)$. Then it is an ideal immersion if and only if it is a totally umbilical immersion.*

This proposition follows from Theorem 4.1 and the fact that the only δ-invariant for a Riemannian 2-manifold is the scalar curvature.

If the submanifold itself is a Riemannian space form, then we have the following result from [18].

Proposition 5.2 *Let $x : M \to R^m(\epsilon)$ be an isometric immersion of a Riemannian space form into $R^m(\epsilon)$. Then x is an ideal immersion if and only if it is a totally umbilical immersion.*

Remark 5.4 Theorem 4.1 and statement (2) of Theorem 2.1, yield the following.

Proposition 5.3 *The only ideal immersion of an Einstein n-manifold M into a Riemannian space form associated with the 1-tuple $(n-1) \in \mathcal{S}(n)$ is the totally geodesic one.*

Similarly, by applying Theorem 4.1 and statement (2) of Theorem 2.2 we have the following.

Proposition 5.4 *The only ideal immersion of a conformally flat kn_1-manifold into a Riemannian space form associated with the k-tuple $(n_1, \ldots, n_1) \in \mathcal{S}(kn_1)$ is the totally umbilical one.*

Remark 5.5 (*Size of the Smallest Ball Containing an Ideal Submanifold*) According to Nash's embedding theorem, every compact Riemannian n-manifold can be isometrically embedded in any small portion of Euclidean space if the codimension is large. In contrast, the following theorem states that an ideal compact submanifold in a Euclidean space cannot be contained in a very small ball of the Euclidean space, regardless of codimension.

Theorem 5.5 *Suppose that $x : M \to E^m$ is an ideal immersion of a compact Riemannian n-manifold M in Euclidean m-space with arbitrary codimension. Let B(R) denote the smallest ball which contains the image of M. Then the radius R of B(R) satisfies*

$$R^2 \geq \frac{vol(M)}{\int_M \hat{\Delta}_0 dV}, \tag{5.6}$$

where $\hat{\Delta}_0$ is the Riemannian invariant on M defined by (5.5) *and $vol(M)$ is the volume of M.*

The equality sign of (5.6) *holds if and only if x is a 1-type ideal immersion.*

This Theorem follows from Theorem 4.1 and Theorem 6.28 of [8, p.150].

Remark 5.6 The estimate of the radius of the smallest ball containing a compact ideal submanifold given in Theorem 5.5 is sharp. For example, let M be the unit n-sphere and let x be the ideal immersion which is the standard embedding of M in E^{n+1}. Then the intrinsic invariant given in the right hand side of (5.6) is equal to one. On the other hand, a ball of radius one is clearly the smallest ball which can contain the image of M.

Remark 5.7 The ideal immersions of a Riemannian manifold in a given Riemannian space form are not unique in general. For instance, the compact homogeneous Riemannian 24-manifold $F_4/\text{Spin}(8)$ admits more than one ideal immersions in Euclidean spaces associated with 2-tuple $(12, 12) \in \mathcal{S}(24)$.

Remark 5.8 From (5.3) it follows that if $\bar{\epsilon} > \epsilon$, we have $\hat{\Delta}_{\bar{\epsilon}} < \hat{\Delta}_{\epsilon}$ for each Riemannian manifold M. Thus, if M is a Riemannian manifold with $\hat{\Delta}_{\epsilon} < 0$ at some point in M, then M admits no ideal immersion in any Riemannian space form $R^m(\bar{\epsilon})$ with $\bar{\epsilon} > \epsilon$, according to (5.4).

Remark 5.9 A Riemannian n-sphere of constant sectional curvature K admits ideal immersions in every complete, simply-connected Riemannian space form $R^m(\epsilon)$ with $\epsilon \leq K$ and $m > n$.

Remark 5.10 A Riemannian manifold may admit an ideal immersion in one Riemannian space form but does not admit any ideal immersion in another Riemannian space form (cf. Remark 5.9).

6 Relations between eigenvalues of Laplacian and δ-invariants

Let M be a compact Riemannian n-manifold and Δ denote the Laplacian operator of M. Denote by λ_j the j-th nonzero eigenvalue of Δ on M. The main purpose of this section is to explain the close relationship between the eigenvalues of the Laplacian Δ and the δ-invariants $\delta(n_1, \ldots, n_k)$ of M. In order to do so, we need to recall the notions of order and type.

For an isometric immersion $x : M \to E^m$ of M in E^m, let

$$x = x_0 + \sum_{t=p}^{q} x_t, \quad \Delta x_t = \lambda_t x_t, \tag{6.1}$$

denote the spectral resolution of x, where x_0 is center of mass of M in E^m.

The set

$$T(x) = \{t \in \mathbf{Z} : x_t \neq \text{constant map}\} \tag{6.2}$$

is called the order of the submanifold. The smallest element p in $T(x)$ is called the *lower order* of x and the supremum q of $T(x)$ is called the *upper order* of x. The immersion is said to be *of finite type* if the upper order q is finite; and it is said to be of infinite type if the upper order q is infinite. Moreover, the immersion is said to be *of k-type* if $T(x)$ contains exactly k elements.

Clearly, the immersion is of 1-type if and only if $p = q$. In this case, the immersion is called a *1-type immersion of order* $\{p\}$ (see [8] for a recent survey on submanifolds of finite type).

For any isometric immersion of a compact Riemannian manifold M in a Euclidean space, one has (cf. [8])

$$\left(\frac{\lambda_q}{n}\right) vol(M) \geq \int_M H^2 dV \geq \left(\frac{\lambda_p}{n}\right) vol(M), \tag{6.3}$$

where p and q are the upper and the lower order of M, respectively. Either equality sign of (6.3) holds if and only if the immersion is of 1-type with order $\{q\}$ or order $\{p\}$, respectively.

Since $p \geq 1$, (6.3) implies in particular the following result of [44]:

$$\int_M H^2 dV \geq \left(\frac{\lambda_1}{n}\right) vol(M). \tag{6.4}$$

When a compact submanifold M of a Euclidean m-space E^m is contained in a ball $B(R)$ of radius R, we have the following sharp relationship between the total mean curvature and the first two nonzero eigenvalues λ_1, λ_2 of the Laplacian on M (cf. Theorem 6.34 of [8]):

$$\int_M H^2 dV \geq \frac{1}{n^2}\{n(\lambda_1 + \lambda_2) - R^2 \lambda_1 \lambda_2\} vol(M), \tag{6.5}$$

with equality holding if and only if M is contained in the boundary $S(R)$ of the ball $B(R)$ and the immersion is of order $\{1\}, \{2\}$ or $\{1, 2\}$. Moreover, if the equality sign of (6.5) holds, M has constant squared mean curvature.

By combining Theorem 4.1 with inequality (6.3), we obtain the following sharp relationships between the order, the eigenvalues of the Laplacian, and the δ-invariants.

Theorem 6.1 *Let $x : M \to E^m$ be an isometric immersion of a compact Riemannian n-manifold. Then*

$$\lambda_q \geq \left(\frac{n}{vol(M)}\right) \int_M \Delta_0(n_1, \ldots, n_k) dV, \tag{6.6}$$

for each k-tuple $(n_1, \ldots, n_k) \in \mathcal{S}(n)$, where q denotes the upper order of the immersion.

The equality case of (6.6) holds for a k-tuple $(n_1, \ldots, n_k) \in \mathcal{S}(n)$ if and only if x is a 1-type ideal immersion of order $\{q\}$ associated with $(n_1, \ldots, n_k)$, i.e., x is a 1-type immersion of order $\{q\}$ satisfying

$$\Delta_0(n_1, \ldots, n_k) = H^2 \tag{6.7}$$

identically.

In particular, Theorem 6.1 yields the following.

Theorem 6.2 *Let M be a compact Riemannian n-manifold. If M admits a 1-type isometric immersion into a Euclidean space, then the p-th nonzero eigenvalue λ_p of the Laplacian of M satisfies*

$$\lambda_p \geq \left(\frac{n}{vol(M)}\right) \int_M \Delta_0(n_1, \ldots, n_k) dV, \tag{6.8}$$

for each k-tuple $(n_1, \ldots, n_k) \in \mathcal{S}(n)$, where p denotes the order of the 1-type immersion.

The equality case of (6.8) holds for a k-tuple $(n_1, \ldots, n_k) \in \mathcal{S}(n)$ if and only if the 1-type immersion is an ideal immersion associated with $(n_1, \ldots, n_k)$.

Remark 6.1 Theorem 6.1 provide a simple way to estimate the upper order q for an isometric immersion of a compact Riemannian manifold in a Euclidean space in terms of the δ-invariants. For instance, let $M := S^2(1/a^2) \times S^2(1/b^2)$ denote the Riemannian product of two 2-spheres with curvatures $1/a^2, 1/b^2$, respectively. Assume that

$$a^2 + b^2 = 1, \quad 0 < b < \frac{1}{\sqrt{s(s+1)}}, \tag{6.9}$$

for some natural number $s \in N_0$. Then the first s nonzero eigenvalues $\lambda_1, \lambda_2, \ldots, \lambda_s$ of M are given by $2/a^2, 6/a^2, \ldots, s(s+1)/a^2$, respectively. Moreover, the invariant $\hat{\Delta}_0$ of M is equal to $1/(4a^2b^2)$. Since $4\hat{\Delta}_0 = 1/(a^2b^2) > \lambda_s$, we conclude from Theorem 6.1 that the upper order of any isometric immersion of $S^2(1/a^2) \times S^2(1/b^2)$ in any Euclidean space is at least $s+1$.

Similarly, if $b^2 = 1/s(s+1)$, then the upper order q of any isometric immersion of $S^2(1/a^2) \times S^2(1/b^2)$ in any Euclidean space is at least s. In particular, the upper order is equal to s if and only if $s = 1$; in this case, $a = b = 1/\sqrt{2}$.

Consequently, regardless of codimension, when b is a small number, the upper order q of any isometric immersion of $S^2(1/a^2) \times S^2(1/b^2)$ must be large.

We remark that the standard embedding of $S^2(1/a^2) \times S^2(1/b^2)$, $a > b$, in E^6 is a 2-type ideal embedding of order $\{1, q\}$ whose upper order q satisfies $\lambda_q = 2/b^2$.

Since every compact homogeneous Riemannian manifold with irreducible isotropy action does admit a 1-type isometric immersion of order $\{1\}$ and all of the δ-invariants of a compact homogeneous Riemannian manifold are constant, Theorem 6.2 implies the following sharp relationship between the first nonzero eigenvalue λ_1 of the Laplacian Δ and the δ-invariants $\delta(n_1, \ldots, n_k)$ for compact homogeneous Riemannian manifolds.

Theorem 6.3 *If M is a compact homogeneous Riemannian n-manifold with irreducible isotropy action, then the first nonzero eigenvalue λ_1 of the Laplacian on M satisfies*

$$\lambda_1 \geq n\, \Delta_0(n_1, \ldots, n_k) \tag{6.10}$$

for any k-tuple $(n_1, \ldots, n_k) \in \mathcal{S}(n)$. In particular, we have

$$\lambda_1 \geq n\, \hat{\Delta}_0. \tag{6.11}$$

The equality sign of (6.11) *holds if and only if M admits a 1-type ideal immersion in a Euclidean space.*

Remark 6.2 The example in Remark 6.1 shows that Theorem 6.3 is false if M is reducible.

Remark 6.3 When $k = 0$, inequality (6.10) reduces to $\lambda_1 \geq n\rho$, where ρ is the normalized scalar curvature of M. In this special case, the result is due to T. Nagano [39]. Nagano's inequality was later extended to compact Einstein n-manifolds in [40].

Remark 6.4 A Riemannian n-manifold with $n \geq 3$ satisfies inequality $\Delta_0(2) > \Delta_0(\emptyset) = \tau$ if and only if $\inf K < \tau/(n-1)^2$. Therefore, a Riemannian n-

manifold ($n \geq 3$) with vanishing scalar curvature satisfies $\Delta_0(2) > \Delta_0(\emptyset)$ automatically, unless M is flat.

For a r-flat, $r \geq 2$, homogeneous Riemannian manifold with scalar curvature $\tau \geq 0$ (respectively, a symmetric space of compact type with rank $r \geq 2$), one has

$$\tau = \delta(\emptyset) \leq \delta(2) \leq \cdots \leq \delta(r)$$
$$(respectively, \delta(\emptyset) = \delta(2) = \cdots = \delta(r) > \delta(r+1)),$$
$$\rho = \Delta_0(\emptyset) < \Delta_0(2) < \cdots < \Delta_0(r).$$

By applying Theorem 6.3, we have the following.

Proposition 6.4 *Let M be a compact homogeneous Einstein Kaehler manifold with positive scalar curvature. Then, for each $(n_1, \ldots, n_k) \in \mathcal{S}(n)$, we have*

$$\Delta_0(n_1, \ldots, n_k) \leq \left(2 - \frac{2}{n}\right)\Delta_0(\emptyset), \tag{6.12}$$

where n denotes the real dimension of M.

This Proposition follows from Theorem 6.3 and the fact that τ and λ_1 are related by $n\lambda_1 = 4\tau$ for compact homogeneous Einstein Kaehler manifolds with positive scalar curvature [41].

Let M be a compact Riemannian manifold. It has a unique kernel of the heat equation:

$$K : M \times M \times \mathbf{R}_+^* \to \mathbf{R}. \tag{H}$$

If there exists a function $\Psi : \mathbf{R}_+ \times \mathbf{R}_+^* \to \mathbf{R}$ such that $K(u, v, t) = \Psi(d(u, v), t)$ for every $u, v \in M$ and $r \in \mathbf{R}_+^*$, then M is said to be *strongly harmonic*.

Since every strongly harmonic manifold admits a 1-type immersion of order $\{1\}$, Theorem 6.2 yields immediately the following.

Corollary 6.5 *Let M be a compact strongly harmonic manifold. Then the first nonzero eigenvalue λ_1 of M satisfies*

$$\lambda_1 \geq \left(\frac{n}{vol(M)}\right) \int_M \Delta_0(n_1, \ldots, n_k) dV, \tag{6.13}$$

for any k-tuple $(n_1, \ldots, n_k) \in \mathcal{S}(n)$.

The equality case of (6.13) *holds for a k-tuple $(n_1, \ldots, n_k) \in \mathcal{S}(n)$ if and only if M admits a 1-type ideal immersion in a Euclidean space associated with the k-tuple $(n_1, \ldots, n_k)$.*

Conversely, Theorem 4.1 also yields the following necessary intrinsic condition for the existence of ideal immersions.

Theorem 6.6 *Let M be a compact Riemannian n-manifold. We have*

(1) *If M admits an ideal immersion in a Euclidean space associated with a k-tuple $(n_1, \ldots, n_k)$, then*

$$\lambda_1 \leq \left(\frac{n}{vol(M)}\right) \int_M \Delta_0(n_1, \ldots, n_k) dV. \tag{6.14}$$

(2) *If M satisfies*

$$\lambda_p \leq \left(\frac{n}{vol(M)}\right) \int_M \hat{\Delta}_0 dV,$$

then every order $\{p\}$, 1-type isometric immersion of M in a Euclidean space is an ideal immersion.

(3) *An ideal immersion of M satisfies the equality case of* (6.14) *if and only if the immersion is a 1-type ideal immersion of order* $\{1\}$.

Combining Corollary 6.5, Theorem 6.6 and our maximum principle, we obtain the following.

Proposition 6.7 *Let M be a compact Riemannian manifold which admits a 1-type isometric immersion of order $\{1\}$ in a Euclidean space. If M admits an ideal immersion in a Euclidean space, then*

$$\lambda_1 vol(M) = n \int_M \hat{\Delta}_0 dV, \tag{6.15}$$

In particular, if a compact strongly harmonic n-manifold admits an ideal immersion in a Euclidean space, then (6.15) *holds.*

By applying Theorem 6.6 we obtain the following simple Riemannian obstruction to ideal immersions for compact Riemannian manifolds in terms of the δ-invariant $\hat{\Delta}_0$.

Theorem 6.8 *Let M be a compact Riemannian n-manifold. If M satisfies*

$$\lambda_1 > \left(\frac{n}{vol(M)}\right) \int_M \hat{\Delta}_0 dV, \tag{6.16}$$

then M admits no ideal immersion in a Euclidean space for any codimension.

In particular, every compact Riemannian manifold with non-positive sectional curvatures admits no ideal immersion in a Euclidean space for any codimension.

By combining Theorems 6.3 and 6.6, we have the following "simple" necessary and sufficient condition for a compact homogeneous Riemannian manifold with irreducible isotropy action to admit an ideal immersion in Euclidean spaces with arbitrary codimension.

Theorem 6.9 *The following statements hold.*

(1) *A compact homogeneous Riemannian n-manifold M with irreducible isotropy action admits an ideal immersion in a Euclidean space if and only if it satisfies* $\lambda_1 = n\,\hat{\Delta}_0$.

(2) *Every ideal immersion of a compact homogeneous Riemannian manifold with irreducible isotropy action in a Euclidean space is a 1-type immersion of order* $\{1\}$.

(3) *If a compact homogeneous Riemannian n-manifold M with irreducible isotropy action admits an ideal immersion in a Euclidean space, then* $\hat{\Delta}_0 = \Delta_0(n_1,\ldots,n_1)$, *for some* $(n_1,\ldots,n_1) \in \mathcal{S}(n,k)$ *with* $kn_1 = n$.

(4)*If a compact homogeneous Riemannian n-manifold M with irreducible isotropy action admits an ideal immersion in a Euclidean m-space such that the image of M is contained in a hypersphere with radius r, then* $\lambda_1 = n/r^2$ *and* $\hat{\Delta}_0 = 1/r^2$.

Remark 6.5 Besides Riemannian spheres, there do exist other compact homogeneous Riemannian manifolds which admit ideal immersions in the Euclidean space. For instance, the following three compact homogeneous Riemannian manifolds:

$$SU(3)/T^2, \quad Sp(3)/Sp(1)^3, \quad \text{and} \quad F_4/\mathrm{Spin}(8)$$

admit ideal immersions in E^8, E^{14}, and E^{26} of codimension 2 associated with $(3,3) \in \mathcal{S}(6)$, $(3,3,3,3) \in \mathcal{S}(12)$, and $(12,12) \in \mathcal{S}(24)$, respectively. These ideal immersions of $SU(3)/T^2$, $Sp(3)/Sp(1)^3$, and $F_4/\mathrm{Spin}(8)$ in E^8, E^{14} and E^{26} arise from their isometric immersions in S^7, S^{13} and S^{25} respectively as minimal isoparametric hypersurfaces.

Remark 6.6 The assumption on the irreducibility in Theorem 6.9 cannot be omitted. For instance, a standard embedding of $S^2(a) \times S^2(b)$, $a \neq b$, in E^6 is an ideal embedding of the compact homogeneous 4-manifold $S^2(a) \times S^2(b)$ which is not of 1-type.

For a given compact irreducible homogeneous n-manifold M, Theorem 6.9 can be applied to determine whether M admits an ideal immersion. In principle, λ_1 (using Freudenthal's formula for Casimir's operator) and the invariant $\hat{\Delta}_0$ are both "computable" for every compact irreducible homogeneous Riemannian manifold. For many compact irreducible symmetric spaces $M = G/H$

with G being classical groups, λ_1 of M has been computed by T. Nagano in [39].

Although S^n does admit an ideal immersion in E^{n+1}, Theorem 6.9 also yields the following non-existence results for other compact rank one symmetric spaces in Euclidean space.

Corollary 6.10 *Let FP^n $(n > 1)$ denote a projective space over real, complex, or quaternion field equipped with a standard Riemannian metric, where the real dimension of FP^n is $n, 2n$ or $4n$, according to $F = \mathbf{R}, \mathbf{C}$ or $\mathbf{H}$. Then FP^n doesn't admit ideal immersions in a Euclidean space, regardless of codimension.*

Corollary 6.11 *The Cayley plane $\mathcal{O}P^2$ with a standard Riemannian metric does not admit ideal immersions in a Euclidean space, regardless of codimension.*

Corollaries 6.10 and 6.11 were proved by comparing λ_1 with $\hat{\Delta}_0$. In fact, we have

$$\lambda_1 = 2(n+1)\epsilon,\ \hat{\Delta}_0 = \epsilon \quad \text{for} \quad RP^n(\epsilon);$$
$$\lambda_1 = 4(n+1)\epsilon,\ \hat{\Delta}_0 = \Delta_0(n,n) = (n+3)\epsilon/n \quad \text{for} \quad CP^n(4\epsilon),\ \ (n>1);$$
$$\lambda_1 = 8(n+1)\epsilon,\ \hat{\Delta}_0 = \Delta_0(n,n,n,n) = (n+3)\epsilon/n \quad \text{for} \quad HP^n(4\epsilon),\ \ (n>1);$$
$$\lambda_1 = 48\epsilon,\ \hat{\Delta}_0 = \Delta_0(2,2,2,2,2,2,2,2) = 145\epsilon/56 \quad \text{for} \quad \mathcal{O}P^2(4\epsilon).$$

Since $\lambda_1 = 8\tau/s^2$ for the compact symmetric space $SU(s)/SO(s)$ [39], by applying Theorem 6.9 we may prove the following.

Corollary 6.12 *The compact symmetric spaces $SU(s)/SO(s)$, $s > 2$, admit no ideal immersions in a Euclidean space, regardless of codimension.*

By applying Nagano's table in [39], similar arguments also apply to other compact symmetric spaces:

$$SU(2h)/Sp(h), h > 2,\ SO(r+s)/SO(r) \times SO(s),$$
$$Sp(r+s)/Sp(r) \times Sp(s),\ r, s \geq 2$$

associated with classical Lie groups.

For an n-dimensional compact homogeneous Einstein Kaehler manifold M with positive scalar curvature, one has $\lambda_1 = 4\tau/n$ [39,41]. Thus, $\lambda_1 > n\Delta_0(n_1, \ldots, n_k)$ if and only if

$$\left(1 + \frac{1}{n+k-1-\sum n_j}\right)\delta(n_1, \ldots, n_k) < 2\tau. \tag{6.17}$$

It is obvious that the coefficient of $\delta(n_1, \ldots, n_k)$ in (6.17) is equal to two if and only if one of the three cases occurs: (1) $k = 0, n = 2$, (2) $k = 1, n = n_1 + 1$, or (3) $k = 2, n_1 + n_2 = n$. In particular, when M is an irreducible Hermitian symmetric space of compact type, the above observations together with some geometric properties of irreducible symmetric spaces of compact type imply that (6.17) holds unless $n = 2, k = 0$. Consequently, we have the following.

Theorem 6.13 *CP^1 is the only irreducible Hermitian symmetric space which admits an ideal immersion in a Euclidean space, regardless of codimension.*

Remark 6.7 By applying an argument similar to the proof of Theorem 6.13, we also have the following.

Proposition 6.14 *Let M be an irreducible symmetric n-space of compact type. If M satisfies $\lambda_1 \geq 4\tau/n$, then M admits no ideal immersion in a Euclidean space unless $n = 2$.*

For a general ideal immersion of a compact Riemannian manifold in a Euclidean space with arbitrary codimension, we have the following sharp relationship between $\hat{\Delta}_0$, λ_1 and λ_2.

Theorem 6.15 *Let $x : M \to E^m$ be an ideal immersion of a compact Riemannian n-manifold in Euclidean m-space E^m. Then*

$$\int_M \hat{\Delta}_0 dV \geq \frac{1}{n^2}\{n(\lambda_1 + \lambda_2) - R^2\lambda_1\lambda_2\} vol(M), \tag{6.18}$$

where R denotes the radius of the smallest ball $B(R)$ in E^m which contains the image of M.

The equality sign of (6.18) holds if and only if the image of M is contained in the boundary S^{m-1} of the ball $B(R)$ and the immersion x is a 1-type ideal immersion of order $\{1\}$, or a 1-type ideal immersion of order $\{2\}$, or a 2-type ideal immersion of order $\{1, 2\}$; moreover, if the equality case of (6.18) holds, then M is mass-symmetric in S^{m-1} and $\hat{\Delta}_0$ is a constant on M.

Remark 6.8 The relationship between the intrinsic invariants $\hat{\Delta}_0, \lambda_1, \lambda_2$ given in (6.18) is sharp. For instance, the example mentioned in Remark 6.1 also satisfies the equality case of (6.18).

Remark 6.9 Similar to Theorem 6.8, one has the following obstruction to the ideal immersions in Riemannian spheres and in real hyperbolic spaces.

Theorem 6.16 *Let M be a compact Riemannian n-manifold. If M satisfies*

$$\lambda_1 > \left(\frac{n}{vol(M)}\right) \int_M \left(\hat{\Delta}_\epsilon + \epsilon\right) dV, \tag{6.19}$$

then M admits no ideal immersion in any complete, simply-connected, Riemannian space form $R^m(\epsilon)$, regardless of codimension.

Theorem 6.16 implies that the real, complex, and quaternionic projective spaces $RP^n(1), CP^n(4)$ and $HP^n(4)$ with $n > 1$, and also the Cayley plane $OP^2(4)$ admit no ideal immersions in a Riemannian m-sphere $S^m(\epsilon)$ with $\epsilon \leq 1$. Also they admit no ideal immersions in a real hyperbolic space, regardless of codimension.

Remark 6.10 For compact irreducible homogeneous Riemannian manifolds in a Riemannian unit sphere, we have the following.

Theorem 6.17 *Every compact homogeneous Riemannian n-manifold with irreducible isotropy action admits an isometric immersion in the m-sphere $S^m(\epsilon)$ of constant sectional curvature ϵ for some $\epsilon > 0$ and, moreover, it satisfies $\lambda_1 \geq n(\epsilon + \hat{\Delta}_\epsilon)$.*

In particular, if a compact homogeneous Riemannian n-manifold with irreducible isotropy action admits an isometric immersion in $S^m(\epsilon)$ and if it satisfies $\lambda_1 = n(\epsilon + \hat{\Delta}_\epsilon)$, then the immersion is an ideal immersion in $S^m(\epsilon)$.

Remark 6.11 It should be pointed out that $\epsilon + \hat{\Delta}_\epsilon \neq \hat{\Delta}_0$ in general for Riemannian n-manifolds with $n > 2$. For instance, the compact irreducible homogeneous Riemannian 3-manifold $SO(3)/\mathbf{Z}_2 \times \mathbf{Z}_2$ satisfies

$$\lambda_1 = 3, \quad \hat{\Delta}_0 = \tfrac{8}{9}, \quad \hat{\Delta}_1 = 0, \tag{6.20}$$

up to a homothetic factor.

From (6.20) and Theorem 6.9, it follows that $SO(3)/\mathbf{Z}_2 \times \mathbf{Z}_2$ does not admit ideal immersions in a Euclidean space. In contrast, $SO(3)/\mathbf{Z}_2 \times \mathbf{Z}_2$ does admit an ideal immersion in $S^4(1)$ as a Cartan hypersurface.

Remark 6.12 The estimates given in this section are all sharp. For instance, if M is the Riemannian product $S^2(2) \times S^2(2)$, then

$$\begin{aligned} &\delta(\emptyset) = \delta(2) = \delta(2,2) = 4, \quad \delta(3) = 2, \\ &c(\emptyset) = 6, \quad c(2) = \tfrac{16}{3}, \quad c(2,2) = c(3) = 4, \quad \hat{\Delta}_0 = 1. \end{aligned} \tag{6.21}$$

On the other hand, let $x : S^2(2) \times S^2(2) \to E^6$ be the standard embedding of $S^2(2) \times S^2(2)$ in $S^5(1) \subset E^6$. Then x is a 1-type ideal immersion of order $\{1\}$.

Since the first two positive eigenvalues λ_1, λ_2 of the Laplacian on M are 4 and 8, respectively, we obtain the equality cases of (6.6) and (6.8). This shows that the relationships given in (6.6) and (6.8) are both sharp.

Also, if M is $S^n(1)$, then $\lambda_1 = n$ and $\hat{\Delta}_0 = 1$. Since $S^n(1)$ admits an ideal immersion in E^{n+1}, the estimate given in Theorem 6.9 is also sharp.

For the detailed proofs of results presented in this section, see [9].

7 New obstructions to minimal, Lagrangian and slant immersions

In this section, we report applications of the δ-invariants to the theory of minimal, Lagrangian and slant immersions. The results presented in this section illustrate again the significance of the δ-invariants.

First we mention the following new obstructions to minimal isometric immersions.

Theorem 7.1 *Let M be a Riemannian n-manifold. If there exists a point $p \in M$ and a k-tuple $(n_1, \ldots, n_k) \in \mathcal{S}(n)$ such that*

$$\delta(n_1, \ldots, n_k)(p) > \tfrac{1}{2}\left(n(n-1) - \sum n_j(n_j - 1)\right)\epsilon, \qquad (7.1)$$

then M admits no minimal isometric immersion into a Riemannian space form $R^m(\epsilon)$, regardless of codimension.

In particular, if $\delta(n_1, \ldots, n_k)(p) > 0$ at some point $p \in M$, then M admits no minimal isometric immersion into Euclidean space for any codimension.

Remark 7.1 For each integer $n \geq 2$ and each k-tuple $(n_1, \ldots, n_k) \in \mathcal{S}(n)$, the condition on $\delta(n_1, \ldots, n_k)$ given in Theorem 7.1 is sharp. This can be seen as follows. Let $f_j : M_j^{n_j} \to E^{m_j}$, $j = 1, \ldots, k$, be k minimal submanifolds and ι a totally geodesic immersion of a Euclidean $(n - \sum n_j)$-space into another Euclidean space. Then $\delta(n_1, \ldots, n_k)$ of $M_1^{n_1} \times \cdots \times M_k^{n_k} \times E^{n - \sum n_j}$ vanishes identically. Clearly, the product immersion $f_1 \times \cdots f_k \times \iota$ is a minimal immersion.

The following result provides new solutions to Problems 2 and 3; thus extending results obtained in [6].

Theorem 7.2 *Let M be a compact Riemannian n-manifold with null first Betti number b_1 or finite fundamental group π_1. If there is a k-tuple $(n_1, \ldots, n_k) \in \mathcal{S}(n)$ such that*

$$\delta(n_1, \ldots, n_k) > 0, \qquad (7.2)$$

then M admits no slant isometric immersion in a complex n-torus CT^n or in $\mathbf{C}^n$.

In particular, M admits no Lagrangian isometric immersion in a complex n-torus or in complex Euclidean n-space.

Remark 7.2 An immediate important consequence of Theorem 7.2 is the first known necessary Riemannian condition for compact Lagrangian submanifolds in $\mathbf{C}^n$; namely, the Ricci curvature of every compact Lagrangian submanifold M in $\mathbf{C}^n$ must satisfies

$$\inf_u Ric(u) \leq 0, \tag{7.3}$$

where u runs over all unit tangent vectors of M. For Lagrangian surfaces, this consequence implies that the Gaussian curvature of every compact Lagrangian surface M in $\mathbf{C}^2$ must be nonpositive at some points on M. Another immediate consequence is that every irreducible symmetric space of compact type cannot be isometrically immersed in a complex Euclidean space as a Lagrangian submanifold.

Remark 7.3 The assumptions on the finiteness of $\pi_1(M)$ and vanishing of $b_1(M)$ given in Theorem 7.2 are both necessary for $n \geq 3$. This can be seen from the following example:

Let $F : S^1 \to \mathbf{C}$ be the unit circle in the complex plane given by $F(s) = e^{is}$ and let $\iota : S^{n-1} \to E^n$ $(n \geq 3)$ be the unit hypersphere in E^n centered at the origin. Denote by $f : S^1 \times S^{n-1} \to \mathbf{C}^n$ the complex extensor defined by $f(s,p) = F(s) \otimes \iota(p)$, $p \in S^{n-1}$. Then f is an isometric Lagrangian immersion of $M =: S^1 \times S^{n-1}$ into C^n which carries each pair $\{(u,p), (-u,-p)\}$ of points in $S^1 \times S^{n-1}$ to a point in $\mathbf{C}^n$ (cf. [10]). Clearly, $\pi_1(M) = \mathbf{Z}$ and $b_1(M) = 1$, and moreover, for each k-tuple $(n_1, \ldots, n_k) \in \mathcal{S}(n)$, the δ-invariant $\delta(n_1, \ldots, n_k)$ on M is a positive constant. This example shows that both the conditions on $\pi_1(M)$ and $b_1(M)$ cannot be removed.

For Lagrangian immersions in complex space forms, we have the following.

Theorem 7.3 *Let M be a compact Riemannian n-manifold either with finite fundamental group or with null first Betti number. If*

$$\delta(n_1, \ldots, n_k) > \tfrac{1}{2}\left(n(n-1) - \textstyle\sum n_j(n_j - 1)\right)$$

for some k-tuple $(n_1, \ldots, n_k) \in \mathcal{S}(n)$, then M admits no Lagrangian isometric immersion into the complex projective n-space $CP^n(4)$.

Theorem 7.4 *Let M be a compact Riemannian n-manifold either with finite fundamental group or with null first Betti number. If*

$$\delta(n_1, \ldots, n_k) > \tfrac{1}{2}\left(\textstyle\sum n_j(n_j - 1) - n(n-1)\right)$$

for some k-tuple $(n_1, \ldots, n_k) \in \mathcal{S}(n)$, then M admits no Lagrangian isometric immersion into the complex hyperbolic n-space $CH^n(-4)$.

These results follow from Theorem 4.2 and the following vanishing theorem of [12].

Theorem 7.5 *Let M be a compact manifold with finite fundamental group $\pi_1(M)$ or vanishing first Betti number $b_1(M)$. Then every Lagrangian immersion from M into any Einstein-Kaehler manifold must have some minimal points.*

This vanishing theorem has several interesting geometric consequences:

(1) There do not exist Lagrangian isometric immersions from a compact Riemannian n-manifold with positive Ricci curvature into any flat Kaehler n-manifold or into any complex hyperbolic n-space;

(2) Every Lagrangian isometric immersion of constant mean curvature from a compact Riemannian manifold with positive Ricci curvature into any Einstein-Kaehler manifold is a minimal immersion; and

(3) Every Lagrangian isometric immersion of constant mean curvature from a spherical space form into a complex projective n-space CP^n is a totally geodesic immersion.

This vanishing theorem is sharp in the following sense:

(a) The conditions on b_1 and π_1 given in the vanishing theorem cannot be removed, since the standard Lagrangian embedding of $T^n = S^1 \times \cdots \times S^1$ into $\mathbf{C}^n = \mathbf{C}^1 \times \cdots \times \mathbf{C}^1$ is a Lagrangian embedding with nonzero constant mean curvature; and

(b) "Lagrangian immersion" in the theorem cannot be replaced by the weaker condition "totally real immersion", since S^n has both trivial first Betti number and trivial fundamental group; and also the standard totally real embedding of S^n in $E^{n+1} \subset \mathbf{C}^{n+1}$ is a totally real submanifold with nonzero constant mean curvature.

An n-dimensional submanifold M in the complex Euclidean n-space $\mathbf{C}^n$ is called *purely real* (or totally real in symplectic geometry) if $J(T_pM)$ is transversal to T_pM for every $p \in M$.

The δ-invariants via Theorem 4.1 also provide the following obstructions to purely real immersions in $\mathbf{C}^n$.

Proposition 7.6 *Let M be a compact Riemannian n-manifold with non-vanishing Euler characteristic, i.e., $\mathcal{X}(M) \neq 0$. If there is a k-tuple $(n_1, \ldots, n_k)$ $\in \mathcal{S}(n)$ such that $\delta(n_1, \ldots, n_k) > 0$, then M admits no purely real isometric immersions in $\mathbf{C}^n$.*

Remark 7.4 The conditions on $\delta(n_1, \ldots, n_k)$ given in Theorems 7.2, 7.3, 7.4

and Proposition 7.6 are all sharp. For instance, the following example of Whitney's sphere shows that the conditions on $\delta(n_1,\ldots,n_k)$ given in Theorem 7.2 and Proposition 7.6 are sharp.

Example 7.1 There is a well-known (non-isometric) Lagrangian immersion of the unit n-sphere S^n into complex Euclidean n-space $\mathbf{C}^n$, known as the Whitney immersion, which is defined by

$$w(y_0,y_1,\ldots,y_n)=\frac{1+\mathrm{i}\,y_0}{1+y_0^2}(y_1,\ldots,y_n)$$

with $y_0^2+y_1^2+\cdots+y_n^2=1$. The Whitney immersion has a unique self-intersection point

$$w(-1,0,\ldots,0)=w(1,0,\ldots,0).$$

The S^n together with the metric induced from Whitney's immersion w is called a *Whitney n-sphere.* Obviously, the Whitney immersion is also a purely real immersion of S^n in $\mathbf{C}^n$. Moreover, $\mathcal{X}(S^n)=2\neq 0$ when n is even. For each n-tuple $(n_1,\ldots,n_k)\in\mathcal{S}(n)$, the invariant $\delta(n_1,\ldots,n_k)$ of the Whitney n-sphere satisfies $\delta(n_1,\ldots,n_k)\geq 0$ and $\delta(n_1,\ldots,n_k)=0$ only at the unique point of self-intersection.

8 Applications to rigidity problems

Although the ideal immersions of a given Riemannian manifold in a Euclidean space is not necessary unique, very often the δ-invariants and Theorem 4.1 can be applied to obtain the rigidity for isometric immersion of arbitrary codimension from a Riemannian manifold into a Riemannian space form; in particular, to obtain a rigidity theorem for open portions of a homogeneous Riemannian manifold isometrically immersed in a Euclidean space, regardless of codimension.

The philosophy of the rigidity comes from the fact that, for a given Riemannian manifold M, inequality (4.1) provides us a lower bound of the squared mean curvature. When the inequality is actually an equality, the submanifold is an ideal submanifold according to our maximum principle. In this case Theorem 4.1 shows that the shape operators of the submanifold must take special simple form. In many cases, those informations on the Riemannian structure of M and on the shape operators are sufficient to conclude the rigidity of the submanifold *without global assumption and regardless of codimension.*

Here we provide three of many such applications.

Proposition 8.1 *Let M be an open portion of a unit n-sphere $S^n(1)$. Then, for any isometric immersion of M into E^m, we have*

$$H^2 \geq 1, \tag{8.1}$$

regardless of codimension.

The equality case of (8.1) holds identically if and only if M is immersed as an open portion of an ordinary hypersphere in an affine $(n+1)$-subspace E^{n+1} of E^m.

Proposition 8.2 *Let M be an open portion of $S^k(1) \times E^{n-k}, k > 1$. Then, for any isometric immersion of M into E^m, we have*

$$H^2 \geq \left(\frac{k}{n}\right)^2 \tag{8.2}$$

regardless of codimension.

The equality case of (8.2) holds identically if and only if M is immersed as an open portion of an ordinary spherical hypercylinder in an affine $(n+1)$-subspace E^{n+1} of E^m.

Proposition 8.3 *Let M be an open portion of $S^k(1) \times S^{n-k}(1), 1 < k < n-1$. Then for any isometric immersion of M into E^m, we have*

$$H^2 \geq \left(\frac{k}{n}\right)^2 + \left(\frac{n-k}{n}\right)^2, \tag{8.3}$$

regardless of codimension.

The equality case of (8.3) holds identically if and only if M is embedded in a standard way in an affine $(n+2)$-subspace of E^m.

Theorem 4.1 and Moore's lemma can be applied to provide us some decomposition results. For instance, we have

Proposition 8.4 *Let $M_1^{n_1}, \ldots, M_k^{n_k}$ $(k \geq 2)$ be k Riemannian manifolds satisfying $n_1 + \cdots + n_k \leq n$. Then, regardless of codimension, we have*

(1) Every isometric immersion of $M_1^{n_1} \times \cdots \times M_k^{n_k} \times H^{n-\sum n_j}(-\epsilon)$ into the hyperbolic m-space $H^m(-\epsilon)$, $\epsilon > 0$, satisfies

$$H^2 \geq \frac{b(n_1, \ldots, n_k)\epsilon}{c(n_1, \ldots, n_k)}.$$

In particular, if $H^2 = b(n_1, \ldots, n_k)\epsilon / c(n_1, \ldots, n_k)$ identically, then the immersion is a product immersion.

(2) *Every minimal isometric immersion from* $M_1^{n_1} \times \cdots \times M_k^{n_k} \times E^{n-\sum n_j}$ *into a Euclidean space is a product immersion.*

Statement (2) of Proposition 8.4 was due to N. Ejiri [35].

9 Examples of ideal submanifolds

Recall that an isometric immersion of a Riemannian n-manifold M into a Riemannian space form is called an ideal immersion i.e., a best way of living, if it achieves the least possible squared mean curvature or the least possible amount of tension at every point on the submanifold.

Applying our maximum principle, we know that an ideal immersion of a Riemannian n-manifold in a Riemannian space form is exactly a submanifold which satisfies the equality case of inequality (4.1) identically for some k-tuple $(n_1, \ldots, n_k)$.

In this section, we illustrate many examples of ideal submanifolds, that is, submanifolds satisfying the basic equality:

$$\delta(n_1, \ldots, n_k) = c(n_1, \ldots, n_k)H^2 + b(n_1, \ldots, n_k)\epsilon. \tag{9.1}$$

for some n-tuple $(n_1, \ldots, n_k) \in \mathcal{S}(n)$.

Example 9.1 For any integer $k \in \{0, 1, \ldots, [\frac{n}{2}]\}$, a spherical cylinder $E^k \times S^{n-k}$ is an ideal hypersurface of E^{n+1} which satisfies (9.1) for $(n_1, \ldots, n_k) = (2, \ldots, 2)$.

Example 9.2 For each positive even integer $n = 2k$, an n-dimensional totally umbilical submanifold in $R^m(\epsilon)$ is an ideal submanifold which satisfies (9.1) for $(n_1, \ldots, n_k) = (2, \ldots, 2)$.

In particular, a horosphere in a hyperbolic space H^{2k+1} is an ideal submanifold satisfies equality (9.1) for $(n_1, \ldots, n_k) = (2, \ldots, 2)$.

A totally umbilical submanifold of $R^m(\epsilon)$ also satisfies (9.1) with $k = 0$.

We call an n-dimensional submanifold M of a Riemannian m-manifold N^m *ultra-minimal* if, with respect to some suitable locally orthonormal frame fields, $e_1, \ldots, e_n, e_{n+1}, \ldots, e_m$, the shape operators A_r take the form:

$$A_r = \begin{bmatrix} A_1^r & \cdots 0 & 0 \cdots 0 \\ \vdots & \ddots \vdots & \vdots \quad \vdots \\ 0 & \cdots A_k^r & 0 \cdots 0 \\ 0 & \cdots 0 & 0 \cdots 0 \\ \vdots & \ddots \vdots & \vdots \ddots \vdots \\ 0 & \cdots 0 & 0 \cdots 0 \end{bmatrix}, \qquad \operatorname{trace} A_j^r = 0, \tag{9.2}$$

where $A_j^r, n+1 \leq r \leq m, j = 1, \ldots, k$, are $n_j \times n_j$ symmetric submatrices.

Obviously, *every ultra-minimal submanifold is an ideal submanifold,* according to Theorem 4.1.

A submanifold is called *austere* in the sense of Harvey and Lawson, if the set of eigenvalues of each shape operator A_ξ is invariant under multiplication by -1.

An ultra-minimal submanifold is minimal; and an ultra-minimal submanifold is austere if the submatrices $\{A_j^r\}_{r,j}$ in (9.2) are 2×2 submatrices. An ultra-minimal austere submanifold is called an *ultra-austere submanifold.*

Example 9.3 Every compact homogeneous minimal hypersurface of S^{n+1} with three distinct principal curvatures is ultra-austere in S^{n+1}. In particular, such hypersurfaces are ultra-minimal in S^{n+1}.

Example 9.4 Every austere hypersurface of $R^{n+1}(\epsilon)$, in particular, every minimal real Kaehler hypersurface, is ultra-austere; hence it is an ideal hypersurface which satisfies equality (9.1) with $(n_1, \ldots, n_k) = (2, \ldots, 2) \in \mathcal{S}(n)$ for some k.

By a real Kaehler hypersurface we mean a (real) codimension one isometric immersion of a Kaehler manifold. According to a result of [26], locally, a minimal hypersurface M of E^{2k+1} is Kaehler if and only if its Gauss map is of rank 2 and its spherical image is a pseudo-holomorphic surface.

The class of austere hypersurfaces of $R^{n+1}(\epsilon)$ is larger than the class of real minimal Kaehler hypersurfaces.

Example 9.5 The product of an austere submanifold of a Euclidean space and a totally geodesic submanifold of a Euclidean space is ultra-minimal. The products of austere submanifolds of Euclidean spaces are also ultra-minimal. Such submanifolds are ideal submanifolds which satisfy (9.1) for some k-tuple $(n_1, \ldots, n_k)$.

Example 9.6 Let $\iota_a : S^k(a^2) \to E^{k+1}$ be the standard embedding of a round k-sphere with constant curvature a^2. For any two positive numbers a, b with $a^{-2} + b^{-2} = 1$, the product embedding

$$\iota_a \times \iota_b : S^k\left(a^2\right) \times S^k\left(b^2\right) \to S^{2k+1}(1) \subset E^{2k+2} \tag{9.3}$$

is an ideal embedding which satisfies (9.1) with $(n_1, \ldots, n_k) = (2, \ldots, 2)$. In particular, if $a = b$, it gives rise to an ultra-minimal hypersurface of $S^{2k+1}(1)$ satisfying (9.1) with $(n_1, \ldots, n_k) = (2, \ldots, 2)$.

More generally, if ℓ and $n - \ell$ are two positive integers such that $\ell = kt$, $n - \ell = ks$ for some integers $k > 1, t, s$, then the standard embedding of the Riemannian product $S^\ell(a^2) \times S^{n-\ell}(b^2)$ in E^{n+2} is an ideal embedding associated with the k-tuple $(n_1, \ldots, n_k)$, where $n_1 = \cdots = n_k = t + s$.

Example 9.7 Let N be an ℓ-dimensional ultra-minimal submanifold of a hypersphere $S^{m-1} \subset E^m$ and CN the $(\ell+1)$-dimensional cone in E^m over N with vertex at the center of S^{m-1}. Then, for any $n \geq \ell$, the product submanifold $M = CN \times E^{n-\ell-1}$ is ultra-minimal in a Euclidean space which satisfies (9.1) for some k-tuple. Thus, M is an ideal submanifold of E^m.

Example 9.8 Let B^{2k} be an austere submanifold of $S^{n+1}(1)$ with zero relative nullity. For each point $p \in B^{2k}$, we put

$$\{\xi \in T_pS^{n+1} \,|\, \langle \xi, \xi \rangle = 1 \text{ and } \langle\, \xi, T_pB^{2k}\, \rangle = 0\}. \tag{9.4}$$

Then NB^{2k} lies in $S^{n+1}(1)$ and the identity map from NB^{2k} into $S^{n+1}(1)$ is an austere immersion on an open dense subset U of NB^{2k}. Moreover, U is an austere hypersurface of $S^{n+1}(1)$ with relative nullity $n-2k$. The NB^{2k} satisfies (9.1) with $(n_1, \ldots, n_k) = (2, \ldots, 2)$.

Conversely, according to a result of [25], every austere hypersurface of $S^{n+1}(1)$ with relative nullity $n-2k$ can be obtained in this way.

If $n = 2\ell$ is even, NB^{2k} gives rise to an ideal immersion in E^{n+2} which satisfies (9.1) with $(n_1, \ldots, n_\ell) = (2, \ldots, 2)$.

Let E_1^{n+2} denote the $(n+2)$-dimensional Minkowski space-time endowed with the Lorentzian metric

$$g = -dx_1^2 + dx_2^2 + \cdots + dx_{n+2}^2.$$

Recall that the hyperbolic space $H^{n+1}(-1)$ and the de Sitter space-time S_1^{n+1} can be isometrically embedded in E_1^{n+2} as

$$H^{n+1}(-1) = \{x \in E_1^{n+2} \mid \langle x, x \rangle = -1\},$$

$$S_1^{n+1} = \{x \in E_1^{n+2} | \, \langle x, x \rangle = 1\}.$$

Example 9.9 Let k be a natural number less than $\frac{n}{2}$. Put

$$N_{2k,n-2k} = \big\{x \in E_1^{n+2} : x_1^2 - x_2^2 - \cdots - x_{2k+1}^2 = 2,$$

$$x_{2k+2}^2 + \cdots + x_n^2 = 1\big\}.$$

Then $N_{2k,n-2k}$ is an ideal hypersurface of $H^{n+1}(-1)$ satisfying equality (9.1) with $(n_1, \ldots, n_k) = (2, \ldots, 2)$.

Example 9.10 Let B^{2k} be a $2k$-dimensional space-like austere submanifold of the de Sitter space-time S_1^{n+1} with zero relative nullity. Let NB^{2k} denote the tube over B^{2k} defined by

$$\{\xi \in T_p S_1^{n+1} : \langle \xi, \xi \rangle = -1 \text{ and } \langle\, \xi, T_p B^{2k}\, \rangle = 0\}. \tag{9.5}$$

Then NB^{2k} lies in the hyperbolic space $H^{n+1}(-1)$ and it gives rise to an austere hypersurface of $H^{n+1}(-1)$ with relative nullity equal to $n - 2k$ which satisfies (9.1) with $(n_1, \ldots, n_k) = (2, \ldots, 2)$.

Conversely, every austere hypersurface of $H^{n+1}(-1)$ with relative nullity $n - 2k$ can be obtained in this way.

Example 9.11 Let $x : M \to E^m$ be an ideal immersion of a Riemannian n-manifold M in E^m associated with the k-tuple $(n_1, \ldots, n_k) \in \mathcal{S}(n)$. Then the product immersion

$$x \times x : M \times M \to E^m \times E^m = E^{2m} \tag{9.6}$$

is an ideal immersion of the Riemannian product manifold $M \times M$ associated with $(2n_1, \ldots, 2n_k, 2, \ldots, 2) \in \mathcal{S}(2n)$ with $2n_1 + \cdots + 2n_k + 2 + \cdots + 2 = 2n$.

10 Tubular ideal hypersurfaces

It is an interesting problem to classify ideal submanifolds. In this section we provide several solutions to this problem when the submanifolds are tubular hypersurfaces which satisfy (9.1) for some k-tuple $(n_1, \ldots, n_k) = (2, \ldots, 2)$ in $\mathcal{S}(n)$.

Let B^ℓ be a topologically embedding ℓ-dimensional submanifold in $R^m(\epsilon)$ with $\ell < m-1$. Denote by $\nu_1(B^\ell)$ the unit normal bundle of B^ℓ in $R^m(\epsilon)$. For a sufficiently small $r > 0$, the map

$$f : \nu_1(B^\ell) \to R^m(\epsilon) : (p, e) \mapsto \exp_\nu(re) \tag{10.1}$$

is an immersion which is called the *tubular hypersurface* with radius r about B^ℓ; we denote it by $T_r(B^\ell)$.

For tubular hypersurfaces in Riemannian space forms we have the following classification results.

Theorem 10.1 *A tubular hypersurface $T_r(B^\ell)$ in E^{n+1} satisfies equality* (9.1) *for a k-tuple $(n_1, \ldots, n_k) = (2, \ldots, 2)$ if and only if one of the following two cases occurs:*

(1) $\ell = 0$ *and the tubular hypersurface is a hypersphere.*

(2) $\ell = k \in \{1, \dots, [\frac{n}{2}]\}$ *and the tubular hypersurface is an open portion of a spherical hypercylinder:* $E^\ell \times S^{n-\ell}(r)$.

Theorem 10.2 *Let* B^ℓ *be a topologically embedding* ℓ*-dimensional submanifold in* $S^{n+1}(1)$. *Then the tubular hypersurface* $M = T_r(B^\ell)$ *satisfies equality* (9.1) *for some* k*-tuple* $(n_1, \dots, n_k) = (2, \dots, 2)$ *if and only if one of the following three cases occurs:*

(1) $\ell = 0$ *and* M *is a geodesic sphere with radius* $r \in (0, \pi)$.

(2) $n > \ell \geq n/2$, $k = n - \ell$, $r = \pi/2$, *and* B^ℓ *is a totally umbilical submanifold of* $S^{n+1}(1)$.

(3) $\ell = 2k < n$, $r = \pi/2$, *and* B^ℓ *is an austere submanifold of* $S^{n+1}(1)$.

Theorem 10.3 *Let* $T_r(B^\ell)$ *be a tubular hypersurface about a submanifold* B^ℓ *in* $H^{n+1}(-1)$. *Then* $T_r(B^\ell)$ *satisfies equality* (9.1) *for some* k*-tuple* $(n_1, \dots, n_k) = (2, \dots, 2)$ *if and only if one of following two cases occurs:*

(1) $\ell = 0$ *and* M *is a geodesic sphere with radius* $r > 0$.

(2) $\ell = 2k$ *is even,* B^ℓ *is totally geodesic, and* $r = \coth^{-1}(\sqrt{2})$.

For ideal tubular hypersurfaces about a curve in a Riemannian space forms, we have the following.

Proposition 10.4 *A tubular hypersurface* $T_r(B^1)$ *in* E^{n+1} *satisfies equality* (9.1) *for some* k*-tuple* $(n_1, \dots, n_k) \in \mathcal{S}(n)$ *if and only if* $T_r(B^1)$ *is an open portion of a spherical hypercylinder* $E^1 \times S^{n-1}(r)$.

Proposition 10.5 *A tubular hypersurface* $T_r(B^1)$ *in* $S^{n+1}(1)$ *satisfies equality* (9.1) *for some* k*-tuple* $(n_1, \dots, n_k) \in \mathcal{S}(n)$ *if and only if* B^1 *is a geodesic of* $S^{n+1}(1)$ *and* $r = \tan^{-1}(\sqrt{\ell})$ *for some integer* ℓ *satisfying* $1 \leq \ell \leq n-3$.

Proposition 10.6 *A tubular hypersurface* $T_r(B^1)$ *in* $H^{n+1}(-1)$ *satisfies equality* (9.1) *for some* k*-tuple* $(n_1, \dots, n_k) \in \mathcal{S}(n)$ *if and only if* B^1 *is a geodesic of* $H^{n+1}(-1)$ *and* $r = \tanh^{-1}(\sqrt{\ell})$ *for some integer* ℓ *satisfying* $1 \leq \ell \leq n-3$.

11 Best ways to live in $R^4(\epsilon)$

Concerning ideal immersions the final goal is to solve the following.

Problem 11.1 *Determine ideal immersions of all Riemannian manifolds in a Riemannian space form.*

In terms of best ways of living and best worlds (cf. Remarks 5.1 and 5.2), Problem 11.1 can be rephrased as the following.

Problem 11.2 ("World Problem") *Determine the best ways of living for all individual Riemannian manifolds which live in a best world.*

Proposition 5.1, Proposition 5.2 and Theorem 6.9 solve this "World Problem" for the family of Riemannian 2-manifolds, the family of Riemannian space forms and the family of compact irreducible homogeneous Riemannian manifolds. In this section, we discuss this problem for another important family of Riemannian manifolds; namely, the family of conformally flat manifolds.

In order to do so, we recall some special families of conformally flat manifolds introduced in [7,22] (see also [20]).

Let $\mathrm{cn}(u,k), \mathrm{dn}(u,k)$ and $\mathrm{sn}(u,k)$ denote the three main Jacobi elliptic functions with modulus k. The nine other elliptic functions

$$\begin{gathered}\mathrm{nd}(u,k), \mathrm{nc}(u,k), \mathrm{ns}(u,k), \mathrm{sc}(u,k), \mathrm{cd}(u,k),\\ \mathrm{ds}(u,k), \mathrm{cs}(u,k), \mathrm{dc}(u,k), \mathrm{sd}(u,k)\end{gathered}$$

are defined by taking reciprocals and quotients; for examples,

$$\mathrm{sd}(u,k) = \mathrm{sn}(u,k)/\mathrm{dn}(u,k), \quad \mathrm{nd}(u,k) = 1/\mathrm{dn}(u,k).$$

We put

$$\mu_a = ak\,\mathrm{cn}(ax,k), \quad k = \frac{\sqrt{a^2-1}}{\sqrt{2}a}, \quad a > 1, \tag{11.1}$$

$$\eta_a = \frac{a}{k}\,\mathrm{dn}(\frac{a}{k}x,k), \quad k = \frac{\sqrt{2}a}{\sqrt{a^2+1}}, \quad 0 < a < 1, \tag{11.2}$$

$$\rho_a = ak\,\mathrm{cn}(ax,k), \quad k = \frac{\sqrt{a^2+1}}{\sqrt{2}a}, \quad a > 1. \tag{11.3}$$

Let $S^n(c)$ and $H^n(-c)$ denote the n-sphere with constant sectional curvature c and the hyperbolic n-space with constant sectional curvature $-c$, respectively.

For $n > 2$, P_a^n, D_a^n and C_a^n originally introduced in [7] are the Riemannian n-manifolds given by the warped product manifolds $I \times_{\mu_a} S^{n-1}(\frac{a^4-1}{4})$, $\mathbf{R} \times_{\eta_a} H^{n-1}(\frac{a^4-1}{4})$ and $I \times_{\rho_a} S^{n-1}(\frac{a^4-1}{4})$ with warped functions μ_a, η_a and ρ_a, respectively, where I are the open intervals on which the corresponding warped functions are positive. The two exceptional spaces F^n and L^n defined in [7] are the warped product manifolds $\mathbf{R} \times_{1/\sqrt{2}} H^{n-1}(-\frac{1}{4})$ and $\mathbf{R} \times_{\mathrm{sech}(x)} E^{n-1}$, respectively.

The D_a^n, F^n, L^n are complete Riemannian n-manifolds, but P_a^n and C_a^n are not complete. Topologically, S^n is the two point compactification of both P_a^n and C_a^n. The Riemannian metrics defined on P_a^n and C_a^n can be extended smoothly to their two point compactification S^n. We denote by $\hat{P}_a^n$ and $\hat{C}_a^n$ the manifold S^n together with the Riemannian metrics given by the (smooth) extensions of the metrics on P_a^n and C_a^n to S^n, respectively. We remark that P_a^n, D_a^n, C_a^n are indeed isometric to the warped product n-manifolds $I \times_{f_1} S^{n-1}(1), I \times_{f_2} H^{n-1}(-1), I \times_{f_3} S^{n-1}(1)$ with warped functions $2\rho_a/(a^2+1), 2\mu_a/(a^2-1), 2\eta_a/(a^2+1)$, respectively.

Let $A_a^n\,(a > 1)$, $B_a^n\,(0 < a < 1)$, G^n, $H_a^n\,(a > 0)$ and $Y_a^n\,(0 < a < 1)$ denote the warped product manifolds:

$$\mathbf{R} \times_{\sqrt{a^2-1}\cosh x} S^{n-1}(1), \quad (-\tfrac{\pi}{2}, \tfrac{\pi}{2}) \times_{\sqrt{1-a^2}\cos x} S^{n-1}(1),$$

$$\mathbf{R} \times_{\cosh x} E^{n-1}, \quad \mathbf{R} \times_{\sqrt{a^2+1}\cosh x} H^{n-1}(-1), \quad \mathbf{R} \times_{\sqrt{1-a^2}\sinh x} S^{n-1}(1),$$

respectively. When $n = 2$, the second factor S^{n-1} or H^{n-1} in each of the warped product manifolds shall be replaced either by $S^1(1)$ or by $\mathbf{R}$. The geometries of A_a^2, G^2 and H_a^2 are similar in the sense that one can be obtained from the others by applying some suitable scalings on the first factor $\mathbf{R}$.

Clearly, A_a^n, G^n, H_a^n are complete Riemannian manifolds. Topologically, S^n is the two point compactification of B_a^n. As for P_a^n and C_a^n, the warped metric on B_n^a can be extended smoothly to its two point compactification. We denote by $\hat{B}_a^n$ the S^n together with the Riemannian metrics on S^n extended from the metric on B_a^n.

For $n \geq 2$ and a real number $a > 0$, there is a Lagrangian immersion from the unit n-sphere S^n into a complex Euclidean n-space $\mathbf{C}^n$ defined by (cf. [6])

$$w_a(y_0, y_1, \ldots, y_n) = \frac{1 + iay_0}{1 + y_0^2}(y_1, \ldots, y_n), \tag{11.4}$$

where $y_0^2 + y_1^2 + \cdots + y_n^2 = 1$. This Whitney's immersion w_a has a unique self-intersection point $w_a(-1, 0, \ldots, 0) = w_a(1, 0, \ldots, 0)$. The S^n together with the metric induced from Whitney's immersion w_a, denoted by W_a^n, is also called a Whitney n-sphere.

For conformally flat 3-manifolds we have the following results by combining Theorem 4.1, the maximum principle, Proposition 5.4 and Theorems 4, 5, and 6 of [22].

Theorem 11.1 *The only 3-dimensional conformally flat manifolds which can achieve best ways of living in E^4 are open parts of E^3, $S^3(r)\,(r > 0)$, $S^2(r) \times E^1\,(r > 0)$, a round hypercone, and a Whitney 3-sphere W_a^3, $a > 0$.*

Moreover, up to rigid motions, their best ways of living are achieved by

(1) *living in the "most natural way" for open parts of* E^3, $S^3(r)$, $S^2(r)\times E^1$, *and a round hypercone;*

(2) *for an open part of* W_a^3 *it is*

$$\left(\frac{1}{2}\int_0^x \operatorname{sd}^2\left(\tfrac{\sqrt{2}}{a}x\right)dx,\, ay_1 \operatorname{sd}\left(\tfrac{\sqrt{2}}{a}x\right), ay_2 \operatorname{sd}\left(\tfrac{\sqrt{2}}{a}x\right), ay_3 \operatorname{sd}\left(\tfrac{\sqrt{2}}{a}x\right)\right)$$

where $y_1^2 + y_2^2 + y_3^2 = \frac{1}{2}$ *and* $k = \frac{1}{\sqrt{2}}$ *is the modulus of the Jacobi elliptic functions.*

Theorem 11.2 *The only 3-dimensional conformally flat manifolds which can achieve best ways of living in* $S^4(1)\,(\subset E^5)$ *are open parts of* $S^3(r)\,(r > 0)$, $\hat{B}_a^3\,(0 < a < 1)$, $\hat{P}_a^3\,(a > 1)$.

Moreover, up to rigid motions, their best ways of living are achieved by

(1) *living totally umbilically for* $S^3(r)$*;*

(2) *for an open part of* $\hat{B}_a^3$ *it is*

$$\left(\sin x, a\cos x, \sqrt{1-a^2}\,y_1 \cos x, \sqrt{1-a^2}\sin\theta\cos x\right)$$

(3) *for an open portion of* $\hat{P}_a^3$ *it is*

$$\frac{1}{ak'}\Bigg(\cos\theta\operatorname{cn}(ax), y_2\operatorname{cn}(ax), y_3\operatorname{cn}(ax),$$

$$\sqrt{a^2k'^2 - \operatorname{cn}^2(ax)}\cos\left(\frac{k'}{k}x + \frac{\mathrm{i}}{2}\left(\ln\frac{\Theta(ax-\gamma)}{\Theta(ax+\gamma)} + 2a\mathrm{Z}(\gamma)x\right)\right),$$

$$\sqrt{a^2k'^2 - \operatorname{cn}^2(ax)}\sin\left(\frac{k'}{k}x + \frac{\mathrm{i}}{2}\left(\ln\frac{\Theta(ax-\gamma)}{\Theta(ax+\gamma)} + 2a\mathrm{Z}(\gamma)x\right)\right)\Bigg),$$

$$y_1^2 + y_2^2 + y_3^2 = 1, \quad \gamma = \operatorname{sn}^{-1}\left(\frac{\mathrm{i}}{ak^2}\right), \quad \mathrm{i} = \sqrt{-1},$$

where $k = \sqrt{a^2-1}/(\sqrt{2}a)$, $k' = \sqrt{a^2+1}/(\sqrt{2}a)$ *are the modulus and the complementary modulus of Jacobi's elliptic functions respectively,* $\Theta(u) = \Theta(u,k)$ *is the Theta function and* $Z(u) = Z(u,k)$ *the Zeta function.*

Theorem 11.3 *The only 3-dimensional conformally flat manifolds which can achieve best ways of living in* $H^4(-1)\,(\subset E_1^5)$ *are open parts of* $S^3(r)\,(r > 0)$, E^3, $H^3(-r)\,(0 < r \leq 1)$, $A_a^3\,(a > 1)$, G^3, $H_a^3\,(a > 0)$, $Y_a^3\,(0 < a < 1)$, F^3, L^3, $\hat{C}_a^3\,(a > 1)$ *and* $D_a^3\,(0 < a < 1)$.

Moreover, up to rigid motions, their best ways of living are achieved by

(1) *living totally umbilically for open parts of* $S^3(r), E^3$, *and* $H^3(-r)$,

(2) *for an open portion of* A^3_a *it is*

$$\left(a\cosh x, \sinh x, \sqrt{a^2-1}\, y_1 \cosh x, \sqrt{a^2-1}\, y_2 \cosh x, \sqrt{a^2-1}\, y_3 \cosh x\right)$$

with $y_1^2 + y_2^2 + y_3^2 = 1$.

(3) *for an open portion of* G^3 *it is*

$$\left(\left(1 + \frac{u_2^2 + u_3^2}{2}\right)\cosh x, \frac{u_2^2 + u_3^2}{2}\cosh x, \sinh x, u_2 \cosh x, u_3 \cosh x\right),$$

(4) *for an open portion of* H^3_a *it is*

$$\left(\sqrt{a^2+1}\, y_1 \cosh x, \sqrt{a^2+1}\, y_2 \cosh x, \sqrt{a^2+1}\, y_3 \cosh x, a\cosh x, \sinh x\right)$$

with $y_1^2 - y_2^2 - y_3^2 = 1$.

(5) *for an open portion of* Y^3_a *it is*

$$\left(\cosh x, a\sinh x, \sqrt{1-a^2}\, y_1 \sinh x, \sqrt{1-a^2}\, y_2 \sinh x, \sqrt{1-a^2}\, y_3 \sinh x\right)$$

with $y_1^2 + y_2^2 + y_3^2 = 1$.

(6) *for an open portion of* F^3 it is

$$\left(\sqrt{2}\cosh u \cosh v, \sqrt{2}\cosh u \sinh v, \sqrt{2}\sinh u, \cos x, \sin x\right).$$

(7) *for an open portion of* L^3 *it is*

$$\operatorname{sech} x\Big(x^2 + u^2 + v^2 + \cosh^2 x + \tfrac{1}{4}, \\ x^2 + u^2 + v^2 + \cosh^2 x - \tfrac{1}{4}, x, u, v\Big).$$

(8)*for an open portion of* $\hat{C}^3_a$ *it is*

$$\frac{1}{ak'}\left(\sqrt{a^2k'^2 + \mathrm{cn}^2(ax)}\cosh\left(\frac{k'}{k}x - \frac{1}{2}\ln\frac{\Theta(ax-\gamma)}{\Theta(ax+\gamma)} - a\mathrm{Z}(\gamma)x\right),\right.$$

$$\sqrt{a^2k'^2 + \mathrm{cn}^2(ax)}\sinh\left(\frac{k'}{k}x - \frac{1}{2}\ln\frac{\Theta(ax-\gamma)}{\Theta(ax+\gamma)} - a\mathrm{Z}(\gamma)x\right),$$

$$\left. y_1\mathrm{cn}(ax), y_2\mathrm{cn}(ax), y_3\mathrm{cn}(ax)\right), \qquad y_1^2 + y_2^2 + y_3^2 = 1,$$

where $\gamma = \mathrm{sn}^{-1}\left(1/(ak^2)\right)$, $k = \sqrt{a^2+1}/(\sqrt{2}a)$ *and* $k' = \sqrt{a^2-1}/(\sqrt{2}a)$.

(9) *for an open portion of* D_a^3 *it is*

$$\frac{1}{ak'}\Big(k\,\mathrm{dn}(\tfrac{a}{k}x)\cosh u\cosh v, k\,\mathrm{dn}(\tfrac{a}{k}x)\cosh u\sinh v, k\,\mathrm{dn}(\tfrac{a}{k}x)\sinh u,$$

$$\sqrt{k^2\mathrm{dn}^2(\tfrac{a}{k}x) - a^2k'^2}\cos\left(k'x - \frac{\mathrm{i}}{2}\ln\frac{\Theta(\frac{a}{k}x-\gamma)}{\Theta(\frac{a}{k}x+\gamma)} - \mathrm{i}\frac{a}{k}\mathrm{Z}(\gamma)x\right),$$

$$\sqrt{k^2\mathrm{dn}^2(\tfrac{a}{k}x) - a^2k'^2}\sin\left(k'x - \frac{\mathrm{i}}{2}\ln\frac{\Theta(\frac{a}{k}x-\gamma)}{\Theta(\frac{a}{k}x+\gamma)} - \mathrm{i}\frac{a}{k}\mathrm{Z}(\gamma)x\right)\Big)$$

where $\gamma = \mathrm{sn}^{-1}(k/a)$, $k = \sqrt{2}a/\sqrt{1+a^2}$ *and* $k' = \sqrt{1-a^2}/\sqrt{1+a^2}$.

Remark 11.1 The immersion of the Whitney 3-sphere W_a^3 in E^4 given in Theorem 11.1 is of class C^∞ except at the two points corresponding to $\mathrm{sd}(\sqrt{2}x/2) = 0$. At these two points the immersion is of class C^2 instead.

Remark 11.2 If M is a conformally flat n-manifolds with $n \geq 4$ isometrically immersed in an $(n+1)$-dimensional Riemannian space form $R^{n+1}(\epsilon)$, then according to a well-know result of E. Cartan and J. A. Schouten, M has a principal curvature, say μ, with multiplicity at least $n-1$. Denote by λ the remaining principal curvature, where we put $\lambda(p) = \mu(p)$ when the multiplicity of μ is n at a point $p \in M$.

Assume that the immersion of M in $R^{n+1}(\epsilon)$ is ideal and non-totally umbilical. Then, by applying Theorem 4.1, we may prove that there exist three integers $t \geq 0, r \geq 1, k \geq 1$ such that

$$\lambda = (r-t)\mu, \quad n = t + 1 + (k-1)r. \tag{11.5}$$

Hence, by applying a result of [34,31], we conclude that M is a rotation hypersurface of $R^{n+1}(\epsilon)$. Therefore, every ideal immersion of a conformally flat n-manifold in a Riemannian space form $R^{n+1}(\epsilon)$ is a rotation hypersurface for $n \geq 4$. Moreover, up to rigid motions, the profile curve of the rotation hypersurface is completely determined by condition (11.5). In particular, if $R^{n+1}(\epsilon) = E^{n+1}$, the profile curve is congruent to the graph of a function ϕ of one variable which satisfies the following differential equation:

$$\phi\phi'' + (r-t)(1+\phi'^2) = 0, \tag{11.6}$$

where $r-t$ is the difference of two integers satisfying $n = t+1+(k-1)r$ for some $t \geq 0, r \geq 1$, and $k \geq 2$.

Remark 11.3 The proofs of Theorems 11.1, 11.2 and 11.3 rely strongly on Theorem 4.1 and the exact solutions of the following differential equations of Picard type discovered in [22].

Proposition 11.4 *For real numbers $a > 1$ and $\beta > 0$, the general solution of the second order differential equation:*

$$y''(x) + 2a\,\mathrm{sc}(ax)\mathrm{dn}(ax)y'(x) - a^2\beta^2 y(x) = 0 \tag{11.7}$$

is given by $y(x) = c_1 y_1(x) + c_2 y_2(x)$ with

$$y_1 = \sqrt{k'^2 - \beta^2\mathrm{cn}^2(ax)}\cos\left(\left(\frac{a\beta\sqrt{k^2+\beta^2}}{\sqrt{k'^2-\beta^2}}\right)x + \frac{\mathrm{i}}{2}\ln\frac{\Theta(ax-\gamma)}{\Theta(ax+\gamma)} + \mathrm{i}\,a\mathrm{Z}(\gamma)x\right)$$

and

$$y_2 = \sqrt{k'^2 - \beta^2\mathrm{cn}^2(ax)}\sin\left(\left(\frac{a\beta\sqrt{k^2+\beta^2}}{\sqrt{k'^2-\beta^2}}\right)x + \frac{\mathrm{i}}{2}\ln\frac{\Theta(ax-\gamma)}{\Theta(ax+\gamma)} + \mathrm{i}\,a\mathrm{Z}(\gamma)x\right),$$

where $k = \sqrt{a^2-1}/(\sqrt{2}a)$, $k' = \sqrt{a^2+1}/(\sqrt{2}a)$ are the modulus and the complementary modules of the elliptic functions, and $\gamma = \mathrm{sn}^{-1}(\mathrm{i}\,\beta/(k\sqrt{k'^2-\beta^2}))$, respectively.

Proposition 11.5 *For real numbers $a > 1$ and $\beta > 0$, the general solution of*

$$y''(x) + 2a\,\mathrm{sc}(ax)\mathrm{dn}(ax)y'(x) + a^2\beta^2 y(x) = 0 \tag{11.8}$$

is given by $y(x) = c_1 y_1(x) + c_2 y_2(x)$ with

$$y_1 = \sqrt{k'^2 + \beta^2\,\mathrm{cn}^2(ax)}\cosh\left(-\frac{a\beta\sqrt{k^2-\beta^2}}{\sqrt{k'^2+\beta^2}}x + \frac{1}{2}\ln\frac{\Theta(ax-\gamma)}{\Theta(ax+\gamma)} + a\mathrm{Z}(\gamma)x\right)$$

and

$$y_2 = \sqrt{k'^2 + \beta^2\,\mathrm{cn}^2(ax)}\sinh\left(\frac{a\beta\sqrt{k^2-\beta^2}}{\sqrt{k'^2+\beta^2}}x - \frac{1}{2}\ln\frac{\Theta(ax-\gamma)}{\Theta(ax+\gamma)} - a\mathrm{Z}(\gamma)x\right),$$

where $k = \sqrt{a^2+1}/(\sqrt{2}a)$, $k' = \sqrt{a^2-1}/(\sqrt{2}a)$ are the modulus and the complementary modules of the elliptic functions and $\gamma = \mathrm{sn}^{-1}(\beta/(k\sqrt{k'^2+\beta^2}))$.

Proposition 11.6 *For real numbers $0 < a < 1$ and $\beta > 0$, the general solution of*

$$y''(x) + 2ak\,\mathrm{cn}(\tfrac{a}{k}x)\mathrm{sd}(\tfrac{a}{k}x)y'(x) + \tfrac{a^2\beta^2}{k^2}y(x) = 0 \tag{11.9}$$

is given by $y(x) = c_1y_1(x) + c_2y_2(x)$ *with*

$$y_1 = \sqrt{\beta^2\mathrm{dn}^2(\tfrac{a}{k}x) - k'^2}\,\cos\left(\frac{a\beta\sqrt{\beta^2-1}}{k\sqrt{\beta^2-k'^2}}x - \frac{\mathrm{i}}{2}\ln\frac{\Theta(\frac{a}{k}x-\gamma)}{\Theta(\frac{a}{k}x+\gamma)} - \mathrm{i}\,\frac{a}{k}\mathrm{Z}(\gamma)x\right)$$

and

$$y_2 = \sqrt{\beta^2\mathrm{dn}^2(\tfrac{a}{k}x) - k'^2}\,\sin\left(\frac{a\beta\sqrt{\beta^2-1}}{k\sqrt{\beta^2-k'^2}}x - \frac{\mathrm{i}}{2}\ln\frac{\Theta(\frac{a}{k}x-\gamma)}{\Theta(\frac{a}{k}x+\gamma)} - \mathrm{i}\,\frac{a}{k}\mathrm{Z}(\gamma)x\right),$$

where $k = \sqrt{2}a/\sqrt{1+a^2}$ *and* $k' = \sqrt{1-a^2}/\sqrt{1+a^2}$ *are the modulus and the complementary modules of the elliptic functions and* $\gamma = \mathrm{sn}^{-1}(\beta/\sqrt{\beta^2-k'^2})$.

12 Complex δ-invariants and further inequalities for submanifolds in Kaehlerian space forms

Inequality (4.1) holds for arbitrary submanifolds in Kaehlerian and quaternionic space forms $\tilde{M}^m(4\epsilon)$ with $\epsilon < 0$ as well. More precisely, we have the following theorem from [9].

Theorem 12.1 *Let M be an n-dimensional submanifold in a Kaehlerian (or quaternionic) space form $\tilde{M}^m(4\epsilon)$ with $\epsilon < 0$. Then, for each k-tuple $(n_1, \ldots, n_k) \in \mathcal{S}(n)$, we have*

$$\delta(n_1, \ldots, n_k) \le c(n_1, \ldots, n_k)H^2 + b(n_1, \ldots, n_k)\epsilon. \tag{12.1}$$

Equality in (12.1) *holds at p if and only if there exists an orthonormal basis $e_1, \ldots, e_n, e_{n+1}, \ldots, e_{2m}$ at p such that*

(a) $S(n_1, \ldots, n_k)(p) = \tau(L_1) + \cdots + \tau(L_k)$, where, for each $j \in \{1, \ldots, k\}$, L_j is spanned by $e_{n_1+\cdots+n_{j-1}+1}, \ldots, e_{n_1+\cdots+n_j}$;

(b) the shape operators at p satisfy (4.4)–(4.5); *and*

(c) the subspace W of T_pM spanned by $e_{n_1+\cdots+n_k+1}, \cdots, e_n$ is totally real, i.e, $J(W) \subset T_p^{\perp}M$, and moreover, $P(L_j) \subset L_j$ for each $j = 1, \ldots, k$.

Furthermore, the equality sign of (12.1) *holds only when $m \ge 2n - \sum n_j$.*

For submanifolds in complex projective spaces, we have the following sharp inequality.

Theorem 12.2 *Let M be an n-dimensional submanifold in a Kaehlerian space form $\tilde{M}^m(4\epsilon)$ with $\epsilon > 0$. Then, for each k-tuple $(n_1, \ldots, n_k) \in \mathcal{S}(n)$, we have*

$$\delta(n_1, \ldots, n_k) \le c(n_1, \ldots, n_k)H^2 + \frac{\epsilon}{2}\Big(n^2 + 2n - \sum_{j=1}^{k} n_j(n_j - 1)\Big). \tag{12.2}$$

The equality of (12.2) *holds identical only when M is a holomorphic, totally geodesic submanifold.*

Both Theorem 12.1 and Theorem 12.2 follow from a general inequality for arbitrary submanifolds in a Kaehlerian space form $\tilde{M}^m(4\epsilon)$ involving the δ-invariants proved in [9].

Remark 12.1 Theorems 12.1 and 12.2 give rise to sharp Riemannian obstructions for a Riemannian n-manifold to admit a minimal isometric immersion in a non-flat Kaehlerian space form.

For Kaehler submanifolds of non-flat Kaehlerian space forms, Theorems 12.1 and 12.2 yield the following.

Corollary 12.3 *If M is a (real) n-dimensional Kaehler submanifold of the complex projective m-space $CP^m(4\epsilon)$. Then*

$$\delta(n_1,\ldots,n_k) \le \frac{\epsilon}{2}\Big(n^2+2n-\sum_{j=1}^{k} n_j(n_j-1)\Big), \tag{12.3}$$

for each k-tuple $(n_1,\ldots,n_k)\in\mathcal{S}(n)$. The equality of (12.3) *holds identical only when M is a totally geodesic Kaehler submanifold.*

Corollary 12.4 *If M is a (real) n-dimensional Kaehler submanifold in the complex hyperbolic m-space $CH^m(4\epsilon)$. Then*

$$\delta(n_1,\ldots,n_k) \le b(n_1,\ldots,n_k)\epsilon, \tag{12.4}$$

for each k-tuple $(n_1,\ldots,n_k)\in\mathcal{S}(n)$.

If the equality sign of (12.4) *holds at a point, then $n = n_1+\cdots+n_k$ and $n_1,\ldots,n_k$ are even integers.*

Let M be a real $2n$-dimensional Kaehler manifold. We introduce the *complex δ-invariants*:

$$\delta^c(2n_1,\ldots,2n_k), \quad \text{and} \quad \hat{\delta}^c(2n_1,\ldots,2n_k)$$

for each k-tuple $(2n_1,\ldots,2n_k)\in\mathcal{S}(2n)$ by

$$\delta^c(2n_1,\ldots,2n_k) = \tau - S^c(2n_1,\ldots,2n_k), \tag{12.5}$$

$$\hat{\delta}^c(2n_1,\ldots,2n_k) = \tau - \hat{S}^c(2n_1,\ldots,2n_k), \tag{12.6}$$

$$S^c(2n_1,\ldots,2n_k) = \inf\{\tau(L_1^c)+\cdots+\tau(L_k^c)\}, \tag{12.7}$$

$$\hat{S}^c(2n_1,\dots,2n_k)=\sup\{\tau(L_1^c)+\cdots+\tau(L_k^c)\}, \tag{12.8}$$

where $L_1^c,\dots,L_k^c$ run over all k mutually orthogonal complex subspaces of T_pM, $p\in M$, with real dimensions $2n_1,\dots,2n_k$, respectively.

For a (real) $2n$-dimensional Kaehler submanifold M of a Kaehlerian space form $\tilde{M}^m(4\epsilon)$ with constant holomorphic sectional curvature 4ϵ, the scalar curvature τ of M satisfies (cf. [42, pp.78-82])

$$\tau\le 2n(n+1)\epsilon, \tag{12.9}$$

with equality holding if and only if M is a totally geodesic Kaehler submanifold.

For the complex δ-invariants $\delta^c(2n_1,\dots,2n_k)$ of a Kaehler submanifold in a Kaehlerian space form, we have the following general result which extends (12.9). This result can be regarded as the complex version of Theorem 4.1.

Proposition 12.5 *Let M be a (real) $2n$-dimensional Kaehler submanifold of a Kaehlerian space form $\tilde{M}^m(4\epsilon)$. Then, for each k-tuple $(2n_1,\dots,2n_k)\in\mathcal{S}(2n)$, the complex δ-invariant $\delta^c(2n_1,\dots,2n_k)$ satisfies*

$$\delta^c(2n_1,\dots,2n_k)\le 2\Big(n(n+1)-\sum_{j=1}^k n_j(n_j+1)\Big)\epsilon. \tag{12.10}$$

The equality case of inequality (12.10) *holds at a point $p\in M$ if and only if, there exists an orthonormal basis*

$$e_1,\dots,e_{n_1},Je_1,\dots,Je_{n_1},\dots,e_{2(n_1+\cdots+n_{k-1})+1},\dots,e_{2(n_1+\cdots+n_{k_1})+n_k},\\ Je_{2(n_1+\cdots+n_{k-1})+1},\dots,Je_{2(n_1+\cdots+n_{k-1})+n_k},e_{2n+1},\dots,e_{2m}$$

at p, such that the shape operators of M in $\tilde{M}^m(4\epsilon)$ at p take the following form:

$$A_r=\begin{bmatrix} A_1^r & \cdots & 0 & 0 & \cdots & 0\\ \vdots & \ddots & \vdots & \vdots & & \vdots\\ 0 & \cdots & A_k^r & 0 & \cdots & 0\\ 0 & \cdots & 0 & 0 & \cdots & 0\\ \vdots & \ddots & \vdots & \vdots & \ddots & \vdots\\ 0 & \cdots & 0 & 0 & \cdots & 0\end{bmatrix}, \tag{12.11}$$

$$r=2n+1,\dots,2m,$$

where each A_j^r is a symmetric $(2n_j)\times(2n_j)$ submatrix with zero trace.

Similarly, for each k-tuple $(n_1, \ldots, n_k) \in \mathcal{S}(2n)$, we introduce the *totally real δ-invariants*:

$$\delta^r(n_1, \ldots, n_k), \quad \text{and} \quad \hat{\delta}^r(n_1, \ldots, n_k)$$

by

$$\delta^r(n_1, \ldots, n_k) = \tau - \inf\{\tau(L_1^r) + \cdots + \tau(L_k^r)\}, \tag{12.12}$$

$$\hat{\delta}^r(n_1, \ldots, n_k) = \tau - \sup\{\tau(L_1^r) + \cdots + \tau(L_k^r)\}, \tag{12.13}$$

where $L_1^r, \ldots, L_k^r$ run over all k mutually orthogonal totally real subspaces of T_pM, $p \in M$, with dimensions $n_1, \ldots, n_k$, respectively.

For totally real δ-invariants $\delta^r(n_1, \ldots, n_k)$ of a Kaehler submanifold in a Kaehlerian space form, we have the following.

Proposition 12.6 *Let M be a (real) $2n$-dimensional Kaehler submanifold of a Kaehlerian space form $\tilde{M}^m(4\epsilon)$. Then, for each k-tuple $(n_1, \ldots, n_k) \in \mathcal{S}(2n)$, the totally real δ-invariant $\delta^r(n_1, \ldots, n_k)$ satisfies*

$$\delta^r(n_1, \ldots, n_k) \leq \Big(2n(n+1) - \frac{1}{2}\sum_{j=1}^{k} n_j(n_j+1)\Big)\epsilon. \tag{12.14}$$

The equality case of inequality (12.14) holds at a point $p \in M$ if and only if, there exists an orthonormal basis $e_1, \ldots, e_{2n}, e_{2n+1}, \ldots, e_{2m}$ at p, such that $Span\{e_1, \ldots, e_{n_1}\}, \ldots, Span\{e_{n_1+\ldots+n_{k-1}+1}, \ldots, e_{n_1+\ldots+n_k}\}$ are totally real subspaces of T_pM and the shape operators of M in $\tilde{M}^m(4\epsilon)$ at p take the following form:

$$A_r = \begin{bmatrix} A_1^r & \cdots & 0 & 0 & \cdots & 0 \\ \vdots & \ddots & \vdots & \vdots & & \vdots \\ 0 & \cdots & A_k^r & 0 & \cdots & 0 \\ 0 & \cdots & 0 & 0 & \cdots & 0 \\ \vdots & \ddots & \vdots & \vdots & \ddots & \vdots \\ 0 & \cdots & 0 & 0 & \cdots & 0 \end{bmatrix},$$

$$r = n+1, \ldots, m,$$

where each A_j^r is a symmetric $n_j \times n_j$ submatrix with zero trace.

The proof of Propositions 12.5 and 12.6 are similar to that of Theorem 4.1.

Remark 12.2 Contrast to inequality (12.9), besides totally geodesic submanifolds there do exist Kaehler submanifolds of $\tilde{M}^m(4\epsilon)$ which satisfy the equality case of (12.10) identically. For instance, let Q_n denote the complex hyperquadric in $CP^{n+1}(4)$ defined by

$$\{(z_0, z_1, \dots, z_{n+1}) \in CP^{n+1}(4) : z_0^2 + z_1^2 + \cdots + z_{n+1}^2 = 0\}. \tag{12.15}$$

For Q_n we have

$$\tau = 2n^2, \quad \delta_k^c := \delta^c(2, \dots, 2) = 2n(n-1), \tag{12.16}$$

where 2 in $\delta^c(2, \dots, 2)$ repeats n times. The complex quadric Q_n in $CP^{n+1}(4\epsilon)$ satisfies the equality case of (12.10) identically for the n-tuple $(2, \dots, 2) \in \mathcal{S}(2n)$.

Also, direct products of Kaehler submanifolds of complex Euclidean spaces are Kaehler submanifolds of complex Euclidean spaces which satisfy the equality case of (12.10) for some suitable k-tuples.

In views of the above facts, it is an interesting problem to classify all Kaehler submanifolds of Kaehlerian space forms which satisfy either the equality case of inequality (12.10) or the equality case of inequality (12.14).

13 δ-invariants, Bochner-Kaehler and Einstein-Kaehler manifolds

A Kaehler manifold M is simply called an $S^c(2n_1, \dots, 2n_k)$-space if it satisfies $\delta^c(2n_1, \dots, 2n_k) = \hat{\delta}^c(2n_1, \dots, 2n_k)$ identically.

In terms of complex δ-invariants, we have the following.

Proposition 13.1 *Let M be a Kaehler $2n$-manifold $(n > 1)$ and k an integer greater than one.*

(1) *If M is an $S^c(2n_1, \dots, 2n_k)$-space, then M is a Kaehlerian space form unless $n_1 = \dots = n_k$ and $n_1 + \cdots + n_k = n$.*

(2) *M is an $S^c(2n_1, \dots, 2n_k)$-space with $n_1 = \dots = n_k$ and $n_1 + \cdots + n_k = n$ if and only if M is a Bochner-Kaehler manifold, i.e., M is a Kaehler manifold with vanishing Bochner curvature tensor.*

An r-plane π tangent to a Kaehler manifold M is called totally real if $J(\pi)$ is perpendicular to π. For any orthonormal basis $\{e_1, e_2\}$ of a totally real 2-plane, the sum $H(e_1, e_2) := K(e_1, e_2) + K(e_1, Je_2)$ of the two sectional curvatures $K(e_1, e_2)$ and $K(e_1, Je_2)$ is called a *totally real bisectional curvature*. For any unit tangent vector X of M, we denote by $H(X)$ the holomorphic sectional curvature at X.

Bochner-Kaehler manifolds, in particular, self-dual Kaehler surfaces, can be characterized in terms of totally real bisectional curvatures in a very simple way.

Theorem 13.2 [13] *Let M be a Kaehlerian $2n$-manifold with $n > 1$. Then M is a Bochner-Kaehler manifold if and only if every totally real bisectional curvature $H(X,Y)$ depends only only the totally real plane section spanned by X, Y and not on the choice of orthonormal basis X, Y.*

In particular, a Kaehler surface is self-dual if and only if every totally real bisectional curvature $H(X,Y)$ depends only only the totally real plane section spanned by X, Y and not on the choice of orthonormal basis X, Y.

The following theorem provides many new characterizations of Bochner-Kaehler manifolds.

Theorem 13.3 [13] *Let M be a Kaehlerian $2n$-manifold $(n > 1)$. Then the following statements are equivalent.*

(1) *M is a Bochner-Kaehler manifold.*

(2) *For every totally real 2-plane section π spanned by orthonormal vectors X, Y, $H(X)+H(Y)$ depends only only the plane section π and not on the choice of X, Y.*

(3) *For every totally real 2-plane section π spanned by orthonormal vectors X, Y, one has $K(X,Y) = \frac{1}{8}(H(X) + H(Y))$.*

(4) *For every totally real 2-plane section π spanned by orthonormal vectors X, Y, one has $K(X,Y) = K(X, JY)$.*

(5) *For every totally real 2-plane section π spanned by orthonormal vectors X, Y, one has $H(X) + H(Y) = H(X,Y) + 2K(X,Y)$.*

(6) *For any totally real 4-plane spanned by orthonormal vectors e_i, e_j, e_k, e_ℓ, one has*

$$H(e_i, e_j) + H(e_k, e_\ell) = H(e_i, e_k) + H(e_j, e_\ell). \tag{13.1}$$

(7) *For any totally real 4-plane spanned by orthonormal vectors e_i, e_j, e_k, e_ℓ, one has*

$$K(e_i, e_j) + K(e_k, e_\ell) = K(e_i, e_k) + K(e_j, e_\ell). \tag{13.2}$$

(8) *For any holomorphic $4k$-plane section spanned by orthonormal basis $e_1, \ldots, e_{2k}, Je_1, \ldots, Je_{2k}$, one has*

$$\tau^c_{1\cdots k} + \tau^c_{k+1\cdots 2k} = \tau^c_{1\cdots k-1 k+1} + \tau^c_{k\, k+2\cdots 2k}, \tag{13.3}$$

where $\tau^c_{1\cdots k}$ denotes the scalar curvature of the holomorphic $2k$-plane section spanned by $e_1, \ldots, e_k, Je_1, \ldots, Je_k$.

(9) *Given a fixed integer r, $4 \leq r \leq n$, for each point $p \in M$ and each totally real r-plane section L at p, $\exp(L)$ has vanishing Weyl conformal curvature tensor at p.*

The proofs of Theorems 13.2 and 13.3 base on the following result of [23].

Proposition 13.4 *Let M be a Kaehlerian $2n$-manifold ($n > 1$). Then M is a Bochner-Kaehler manifold if and only if there is a Hermitian quadratic form Q on M such that the sectional curvature with respect to a holomorphic section π is the trace of the restriction of Q to π.*

Similar to Theorem 3.1, we have the following

Proposition 13.5 [13] *Let M be a (real) $4n$-dimensional Kaehler manifold. Then M is Einstein-Kaehler if and only if for any holomorphic $2n$-plane section π, we have $\tau(\pi) = \tau(\pi^{\perp})$, where $\pi^{\perp}$ is the orthogonal complement of π.*

For Kaehlerian space forms, we have the following result from [13].

Proposition 13.6 *A Kaehler immersion of Kaehlerian space form M into another Kaehlerian space form $\tilde{M}^m(4\epsilon)$ is totally geodesic if and only if it satisfies the equality case of* (12.10) *identically for some k-tuple* $(2n_1, \ldots, 2n_k) \in \mathcal{S}(2n)$.

Theorem 13.3 implies the following.

Corollary 13.7 *Every r-dimensional ($r \geq 4$) totally real, totally umbilical submanifold B of a Bochner-Kaehler $2n$-manifold is conformally flat.*

A special case of Corollary 13.7 has been proved in [1].

14 δ-invariants and real hypersurfaces in Kaehlerian space forms

A real hypersurface M in a Kaehlerian space form is called a *Hopf hypersurface* if $J\xi$ is a principal curvature vector, where ξ is a unit normal vector of M.

The following result provides a very simple characterization of a horosphere in a complex hyperbolic space.

Proposition 14.1 *Let M be a real hypersurface in a complex hyperbolic m-space $CH^m(4\epsilon)$. Then*

$$\delta_k \leq \frac{(2m-1)^2(2m-k-2)}{2(2m-k-1)} H^2 + (2m^2 - 4k - 2)\epsilon, \tag{14.1}$$

for any natural number $k \leq m-1$, where δ_k is defined to be the δ-invariant $\delta(2, \ldots, 2)$ with 2 repeats k times.

Equality sign of (14.1) *holds for some k if and only if one of the two following cases occurs:*

(1) *m is odd, $k = m-1$, and M is an open portion of a tubular hypersurface of radius $r \in \mathbf{R}_+$ over a totally geodesic $CH^{(m-1)/2}(4\epsilon)$.*

(2) *M is an open portion of a horosphere in* $CH^m(4\epsilon)$.

For real hypersurfaces in complex projective spaces, we have the following.

Proposition 14.2 *Let M be a real hypersurface in* $CP^m(4)$ $(m \geq 2)$. *Then*

$$\delta_k \leq \frac{(2m-1)^2(2m-k-2)}{2(2m-k-1)} H^2 + 2m^2 - k - 2 \tag{14.2}$$

for any natural number $k \leq m-1$.

If M is a Hopf hypersurface, then equality sign of (14.2) *for some k if and only if one of the following three cases occurs:*

(1) $k = m-1$ *and M is an open portion of a geodesic sphere with radius* $\pi/4$.

(2) *m is odd,* $k = m-1$, *and M is an open portion of a tubular hypersurface with radius* $r \in (0, \frac{\pi}{2})$ *over a totally geodesic* $CP^{(m-1)/2}(4c)$.

(3) $m = 2$, $k = 1$, *and M is an open part of a tubular hypersurface over a complex quadric curve* Q_1 *with radius* $r = \arctan\left(\frac{1}{2}\left(1+\sqrt{5}-\sqrt{2+2\sqrt{5}}\right)\right)$ $= 0.33311971\cdots$.

15 k-Ricci curvature and shape operator

In this section we explain a sharp relationship between the k-Ricci curvature and the shape operator for a submanifold in a Riemannian space form with arbitrary codimension obtained in [11]. In order to do so we recall the notion of k-Ricci curvature.

For a Riemannian n-manifold M, denote by $K(\pi)$ the sectional curvature of a 2-plane section $\pi \subset T_pM$, $p \in M$. Suppose L^k is a k-plane section of T_pM and X a unit vector in L^k. We choose an orthonormal basis $\{e_1, \ldots, e_k\}$ of L^k such that $e_1 = X$. Define the Ricci curvature Ric_{L^k} of L^k at X by

$$Ric_{L^k}(X) = K_{12} + \cdots + K_{1k}, \tag{15.1}$$

where K_{ij} is the sectional curvature of the 2-plane section spanned by e_i, e_j. We simply called such a curvature a *k-Ricci curvature.*

Let V^ℓ be an ℓ-plane section in a tangent space T_pM of a Riemannian n-manifold M. Then V^ℓ is said to be *k-Einsteinian* if the k-Ricci curvatures of all k-plane sections in V^ℓ are equal. In particular, if V^ℓ is the whole tangent space T_pM at p, then M^n is said to be *k-Einsteinian* at p.

An ℓ-plane section V^ℓ is said to have constant sectional curvature if it is 2-Einsteinian; in this case, sectional curvatures of all 2-plane sections in V^ℓ

are equal. It follows from (15.1) that an ℓ-plane section is 2-Einsteinian if and only if it is k-Einsteinian for some $k < \ell$.

A Riemannian manifold is called *k-Einsteinian* if it is *k-Einsteinian* at every point. Obviously, a Riemannian n-manifold is an Einstein space if it is n-Einsteinian. On the other hand, a Riemannian n-manifold is a Riemannian space form if it is k-Einsteinian for some $k < n$.

For a given integer k, $2 \leq k \leq n$, a Riemannian invariant θ_k on a Riemannian n-manifold M was introduced in [11]. This invariant is defined by

$$\theta_k(p) = \left(\frac{1}{k-1}\right) \inf_{L^k, X} Ric_{L^k}(X), \quad p \in T_pM, \tag{15.2}$$

where L^k runs over all k-plane sections in T_pM and X runs over all unit vectors in L^k.

Recall that for a submanifold M in a Riemannian manifold, the *relative null space* of M at a point $p \in M$ is defined by

$$N_p = \{X \in T_pM : h(X,Y) = 0 \text{ for all } Y \in T_pM\}. \tag{15.3}$$

The following result from [11] establishes a sharp relationship between the k-Ricci curvature and the shape operator A_H, regardless of codimension.

Theorem 15.1 *Let $x : M \to R^m(\epsilon)$ be an isometric immersion of a Riemannian n-manifold M into a Riemannian space form $R^m(\epsilon)$ of constant sectional curvature ϵ. Then, regardless of codimension, for any integer k, $2 \leq k \leq n$, and any point $p \in M$ we have*

(1) *If $\theta_k(p) \neq \epsilon$, then the shape operator at the mean curvature vector H satisfies*

$$A_H > \frac{n-1}{n}(\theta_k(p) - \epsilon)I \quad at\ p, \tag{15.4}$$

where I denotes the identity map of T_pM.

(2) *If $\theta_k(p) = \epsilon$, then $A_H \geq 0$ at p.*

(3) *A unit vector $X \in T_pM$ satisfies $A_H X = \frac{n-1}{n}(\theta_k(p) - \epsilon)X$ if and only if $\theta_k(p) = \epsilon$ and X lies in the relative null space at p.*

(4) *$A_H \equiv \frac{n-1}{n}(\theta_k - \epsilon)I$ at p if and only if p is a totally geodesic point, i.e., the second fundamental form vanishes identically at p.*

Remark 15.1 Clearly the estimate of A_H given in statement (2) of Theorem 15.1 is sharp. Here we provide an example to illustrate that statement (1) of Theorem 15.1 is also sharp.

Consider a hyperellipsoid in E^{n+1} defined by

$$ax_1^2 + x_2^2 + \cdots + x_{n+1}^2 = 1, \tag{15.5}$$

where $0 < a < 1$. The principal curvatures $a_1, \ldots, a_n$ of the hyperellipsoid are given by

$$a_1 = \frac{a}{(1 + a(a-1)x_1^2)^{3/2}}, \quad a_2 = \cdots = a_n = \frac{1}{(1 + a(a-1)x_1^2)^{1/2}}. \tag{15.6}$$

Therefore, for any k, $2 \leq k \leq n$, the k-Ricci curvatures at a point p satisfies

$$Ric_{L^k}(X) \geq (k-1)\theta_k(p) := \frac{(k-1)a}{(1 + a(a-1)x_1^2)^2} > 0 \tag{15.7}$$

for any k-plane section L^k and any unit vector X in L^k and, moreover, the eigenvalues $\kappa_1, \ldots, \kappa_n$ of the shape operator A_H are given by

$$\kappa_1 = \cdots = \kappa_{n-1} = \frac{a + (n-1)(1 + a(a-1)x_1^2)}{n(1 + a(a-1)x_1^2)^2}, \tag{15.8}$$

$$\kappa_n = \frac{a(a + (n-1)(1 + a(a-1)x_1^2))}{n(1 + a(a-1)x_1^2)^3}. \tag{15.9}$$

From (15.7), (15.8) and (15.9) it follows that $A_H > \left(\frac{n-1}{n}\right)\theta_k(p)I_n$ and

$$\kappa_1 - \frac{n-1}{n}\theta_k(p) = \frac{a^2}{n(1 + a(a-1)x_1^2)^3} \longrightarrow 0$$

as $a \to 0$.

This example shows that our estimate of A_H in statement (1) is also sharp.

One may apply Theorem 15.1 to provide a lower bound for every eigenvalue of the shape operator A_H for all isometric immersions of a given Riemannian n-manifold, regardless of codimension. As the simplest example, Theorem 15.1 implies immediately the following.

Corollary 15.2 *Let M be a Riemannian n-manifold. If there is a point $p \in M$ such that every sectional curvature of M at p is equal to 1, then, for any isometric immersion of M in a Euclidean m-space with arbitrary codimension, each eigenvalue of the shape operator A_H at p is greater than $(n-1)/n$.*

Remark 15.2 The estimate of the eigenvalues of A_H given in Corollary 15.2 is sharp. For instance, assume that M is a surface in E^3 whose two principal curvatures at p are given respectively by a and $1/a$ with $a \geq 1$. Then the smaller eigenvalue of A_H at p is equal to $(a^2+1)/2a^2$ which approaches to $1/2$ when a approaches ∞.

For an n-dimensional submanifold M in E^m, let E^{n+1} be the linear subspace of dimension $n+1$ spanned by the tangent space at a point $p \in M$ and the mean curvature vector $H(p)$ at p. Geometrically, the shape operator A_{n+1} of M in E^m at p is the shape operator of the orthogonal projection of M^n into E^{n+1}. Moreover, it is known that if the shape operator of a hypersurface in E^{n+1} is definite at a point p, then it is strictly convex at p. For this reason a submanifold M in E^m is said to be *H-strictly convex* if the shape operator A_H is positive-definite at each point in M.

Theorem 15.1 implies immediately the following.

Corollary 15.3 *Let M be a submanifold of a Euclidean space. If there is an integer k, $2 \leq k \leq n$, such that k-Ricci curvatures of M are positive, then M is H-strictly convex, regardless of codimension.*

Theorem 15.1 and a classical result of W. Süss imply the following.

Corollary 15.4 *If M is a compact hypersurface of E^{n+1} with $\theta_k \geq 0$ (respectively, with $\theta_k > 0$) for a fixed k, $2 \leq k \leq n$, then M is embedded as a convex (respectively, strictly convex) hypersurface in E^{n+1}. In particular, if M has constant scalar curvature, then M is a hypersphere of E^{n+1}.*

Corollary 15.5 *If M is a compact hypersurface of E^{n+1} with nonnegative Ricci curvature (respectively, with positive Ricci curvature), then M is embedded as a convex (respectively, strictly convex) hypersurface in E^{n+1}. In particular, if M has constant scalar curvature, then M is a hypersphere of E^{n+1}.*

The following result from [11] establishes a sharp relationship between the Ricci curvature and the squared mean curvature.

Theorem 15.6 *Let $x : M \to R^m(\epsilon)$ be an isometric immersion of a Riemannian n-manifold M into a Riemannian space form $R^m(\epsilon)$. Then*

(1) *For each unit tangent vector $X \in T_pM$, we have*

$$H^2(p) \geq \frac{4}{n^2}\left\{Ric(X) - (n-1)\epsilon\right\}, \tag{15.10}$$

where $Ric(X)$ is the Ricci curvature of M at X.

(2) *If $H(p) = 0$, then a unit tangent vector X at p satisfies the equality case of* (15.10) *if and only if it lies in the relative null space N_p at p.*

(3) *The equality case of* (15.10) *holds identically for all unit tangent vectors at p if and only if either p is a totally geodesic point or $n = 2$ and p is a totally umbilical point.*

Theorem 15.6 can be applied to obtain sharp estimates of the squared mean curvature for some submanifolds with arbitrary codimension. For instance, Theorem 15.6 implies immediately the following.

Corollary 15.7 *Let $x : M \to E^m$ be an isometric immersion of a Riemannian n-manifold M in a Euclidean m-space with arbitrary codimension. Then*

$$H^2(p) \geq \left(\frac{4}{n^2}\right) \max_X Ric(X) \tag{15.11}$$

where X runs over all unit tangent vectors at p.

Remark 15.3 There exist many examples of submanifolds in a Euclidean m-space which satisfy the equality case of (15.11) identically. Two simple examples are spherical hypercylinder $S^2(r) \times E^1$ and round hypercone in E^4.

Remark 15.4 In general, given an integer k, $2 \leq k \leq n-1$, there *does not* exist a positive constant, say $C(n,k)$, such that

$$H^2(p) \geq C(n,k) \max_{L^k, X} Ric_{L^k}(X) \tag{15.12}$$

where L^k runs over all k-plane sections in T_pM and X runs over all unit tangent vectors in L^k. This fact can be seen from the following example:

Let $x : M^3 \to E^4$ be a minimal hypersurface whose shape operator is non-singular at some point $p \in M^3$. Then there exist two principal directions at p, say e_1, e_2, such that their corresponding principal curvatures κ_1, κ_2 are of the same sign. Thus, the sectional curvature K_{12} at p is positive. Now, consider the minimal hypersurface in E^{n+1} given by the product of $x : M^3 \to E^4$ and the identity map $\iota : E^{n-3} \to E^{n-3}$. Then, for any integer k, $2 \leq k \leq n-1$, the maximum value of the k-th Ricci curvatures of $M^n := M^3 \times E^{n-3}$ at a point (p,q), $q \in E^{n-3}$ is $K_{12} = \kappa_1\kappa_2$ which is positive. Since $H = 0$, this shows that there does not exist any positive constant $C(n,k)$ which satisfies (15.12).

In contrast, by applying Theorems 15.1 and 15.6, we have the following.

Corollary 15.8 *Let $x : M \to R^m(\epsilon)$ be an isometric immersion of a Riemannian n-manifold M in a Riemannian space form $R^m(\epsilon)$ of constant sectional curvature ϵ. Then, for any integer k, $2 \leq k \leq n$, we have*

$$H^2(p) \geq \frac{4(n-1)}{n^2}\left(\frac{\theta_k(p)}{k-1} - \epsilon\right). \tag{15.13}$$

The equality case of (15.13) *holds identically for all unit tangent vectors at p if and only if either p is a totally geodesic point or $k = n = 2$ and p is a totally umbilical point.*

Remark 15.5 The inequalities stated in Theorems 15.1 and 15.6 also hold for totally real submanifolds in a Kaehlerian space form with constant holomorphic sectional curvature 4ϵ as well.

Acknowledgment

The final version of this paper was written while the author was a visiting professor at Katholieke Universiteit Leuven. The author would like to take this opportunity to express his many thanks to Professor L. Verstraelen, Dr. F. Dillen and Dr. L Vrancken for their hospitality and for their valuable suggestions which improved the presentation of this paper. The author also thanks to Professors I. Dimitric and K. Ogiue for providing useful informations during the preparation of this paper.

References

1. D. E. Blair, *On the geometric meaning of the Bochner tensor,* Geometriae Dedicata **4**, 35–38 (1975).
2. D. E. Blair, F. Dillen, L. Verstraelen and L. Vrancken, *Calabi curves as holomorphic Legendre curves and Chen's inequality,* Kyunpook Math. J. **35**, 407–416 (1995/6) (A special issue dedicated to 60th birthday of Prof. U-Hang Ki).
3. J. Bolton, C. Scharlach, L. Vrancken and L. M. Woodward, *Lagrangian submanifolds of CP^3 satisfying Chen's equality,* (in preparation).
4. B. Y. Chen, Geometry of Slant Submanifolds, *Katholieke Universiteit Leuven* 1990.
5. B. Y. Chen, *Some pinching and classification theorems for minimal submanifolds* Archiv der Math. **60**, 568–578 (1993).
6. B. Y. Chen, *A Riemannian invariant and its applications to submanifold theory,* Results in Math. **27**, 17–26 (1995)
7. B. Y. Chen, *Jacobi's elliptic functions and Lagrangian immersions,* Proc. Royal Soc. Edinburgh Ser. A **126**, 687–704 (1996).
8. B. Y. Chen, *A report on submanifolds of finite type,* Soochow J. Math. **22**, 117-337 (1996).
9. B. Y. Chen, *Some new obstructions to minimal and Lagrangian isometric immersions,* Japan. J. Math. (to appear).
10. B. Y. Chen, *Complex extensors and Lagrangian submanifolds in complex Euclidean spaces,* Tôhoku Math. J. **49**, 277–297 (1997).
11. B. Y. Chen, *Relations between Ricci curvature and shape operator for submanifolds with arbitrary codimension,* Glasgow Math. J. **40** (1998).

12. B. Y. Chen, *A vanishing theorem for Lagrangian immersions into Einstein-Kaehler manifolds and its applications.* Soochow J. Math. **24**, 157–164 (1998).
13. B. Y. Chen, F. Dillen, *Totally real bisectional curvatures, Bochner-Kaehler and Einstein-Kaehler manifolds,* (preprint).
14. B. Y. Chen, F. Dillen, L. Verstraelen and L. Vrancken, *Totally real submanifolds of* CP^n *satisfying a basic equality,* Archiv der Math. **63**, 553–564 (1994).
15. B. Y. Chen, F. Dillen, L. Verstraelen and L. Vrancken, *Two equivariant totally real immersions into the nearly Kähler 6-sphere and their characterization,* Japan. J. Math. **21**, 207-222 (1995).
16. B. Y. Chen, F. Dillen, L. Verstraelen and L. Vrancken, *Characterization a class of totally real submanifolds of* S^6 *by their sectional curvatures,* Tôhoku Math. J. **47**, 185–198 (1995).
17. B. Y. Chen, F. Dillen, L. Verstraelen and L. Vrancken, *An exotic totally real minimal immersion of* S^3 *in* CP^3 *and its characterization,* Proc. Royal Soc. Edinburgh Ser. A **126**, 153–165 (1996).
18. B. Y. Chen, F. Dillen, L. Verstraelen and L. Vrancken, *Characterizations of Riemannian space forms, Einstein spaces and conformally flat spaces,* (preprint).
19. B. Y. Chen and Y. Tazawa, *Slant submanifolds in complex Euclidean space,* Tokyo J. Math. **14**, 101–120 (1991).
20. B. Y. Chen and L. Vrancken, *Lagrangian submanifolds satisfying a basic equality,* Math. Proc. Cambridge Phil. Soc. **120**, 291–307 (1996).
21. B. Y. Chen and L. Vrancken, *CR-submanifolds of complex hyperbolic spaces satisfying a basic equality,* Israel J. Math. (to appear).
22. B. Y. Chen and J. Yang, *Elliptic functions, Theta function and hypersurfaces satisfying a basic equality,* Math. Proc. Cambridge Phil. Soc. (to appear).
23. B. Y. Chen and K. Yano, *Manifolds with vanishing Weyl or Bochner curvature tensor,* J. Math. Soc. Japan **27**, 106–112 (1975).
24. S. S. Chern, Minimal Submanifolds in a Riemannian Manifold *Univ. of Kansas, Lawrence, Kansas,* 1968.
25. M. Dajczer and D. Gromoll, *Gauss parametrizations and rigidity aspects of submanifolds,* J. Differential Geometry **22**, 1–12 (1995).
26. M. Dajczer and D. Gromoll, *Real Kaehler submanifolds and uniqueness of the Gauss map* J. Differential Geometry **22**, 13–28 (1985).
27. M. Dajczer and L. A. Florit, *A class of austere submanifolds,* (preprint).
28. M. Dajczer and L. A. Florit, *On Chen's basic equality,* Illinois J. Math. **42**, 97–106 (1998).

29. F. Defever, I. Mihai and L. Verstraelen, *B. Y. Chen's inequality for C-totally real submanifolds in Sasakian space forms,* Boll. Un. Mat. Ital. (B) **11**, 365–374 (1997).
30. F. Defever, I. Mihai and L. Verstraelen, *A class of C-totally real submanifolds of Sasakian space forms,* J. Austral. Math. Soc. (to appear).
31. F. Dillen, M. Petrovic, L. Verstraelen, *The conharmonic curvature tensor and 4-dimensional catenoid,* Studia Univ. Babes-Boylai **23**, 16–23 (1988).
32. F. Dillen, M. Petrovic, L. Verstraelen, *Einstein, conformally flat and semi-symmetric submanifolds satisfying Chen's equality,* Israel J. Math. **100**, 163–169 (1997).
33. F. Dillen and L. Vrancken, *Totally real submanifolds in $S^6(1)$ satisfying Chen's equality,* Trans. Amer. Math. Soc. **348**, 1633–1646 (1996).
34. M. do Carmo and M. Dajczer, *Rotation hypersurfaces in spaces of constant curvature,* Trans. Amer. Math. Soc. **277**, 685-709 (1983).
35. N. Ejiri, *Minimal immersions of Riemannian products into real space forms,* Tokyo J. Math. **2**, 63–70 (1979).
36. M. Gromov, *A topological technique for the construction of solutions of differential equations and inequalities,* Intern. Congr. Math. (Nice 1970) **2**, 221–225 (1971).
37. R. S. Kulkarni, *Curvature structures and conformal transformations,* Bull. Amer. Math. Soc. **75**, 91–94 (1969).
38. K. Matsumoto and I. Mihai, *An obstruction to the existence of minimal totally real immersions in a locally conformal Kähler space form,* Bull. Yamagata Univ. Nat. Sci. **14**, 83–87 (1997).
39. T. Nagano, *On the minimum eigenvalues of the Laplacians in Riemannian manifolds,* Sci. Papers College Gen. Edu. Univ. Tokyo **11**, 177–182 (1961).
40. M. Obata, *Certain conditions for a Riemannian manifold to be isometric with a sphere,* J. Math. Soc. Japan **14**, 333-340 (1962).
41. M. Obata, *Riemannian manifolds admitting a solution of a certain system of differential equations,* Proc. U.S.-Japan Sem. Diff. Geom. Kyoto, Japan 101–114 (1965).
42. K. Ogiue, *Differential geometry of Kaehler submanifolds,* Advances in Math. **13**, 73–114 (1974).
43. R. Osserman, *Curvature in the eighties,* Amer. Math. Monthly **97**, 731–754 (1990).
44. R. C. Reilly, *On the first eigenvalue of the Laplacian for compact submanifolds of Euclidean space,* Comm. Math. Helv. **52**, 525–533 (1977).
45. I. M. Singer and J. A. Thorpe, *The curvature of 4-dimensional Einstein*

spaces, Global Analysis, Princeton University Press, 355–365 (1969).
46. Y. Tazawa, *Construction of slant submanifolds,* Bull. Inst. Math. Acad. Sinica **22**, 153–166 (1994).
47. Y. Tazawa, *Construction of slant submanifolds, II,* Bull. Soc. Math. Belg. (New Series) **1**, 569–576 (1994).

On some eigenvalue problems

Roger Chen [1]
Department of Mathematics
National Cheng Kung University
Tainan, Taiwan
rchen@mail.ncku.edu.tw

1 Introduction

Let M^m, $m \geq 3$, be a compact Riemannian manifold with $\partial M \neq \emptyset$. In local coordinates $(x^1, x^2, \ldots, x^m)$, the Riemannian metric is given by

$$ds^2 = \sum_{i,j=1}^{m} g_{ij}\, dx^i dx^j.$$

One defines on M a Schrödinger operator by

$$\Delta - q(x),$$

where

$$\Delta = \frac{1}{\sqrt{g}} \sum_{i,j=1}^{m} \frac{\partial}{\partial x^i}(\sqrt{g} g^{ij} \frac{\partial}{\partial x^j}),$$

is the the Laplace operator on M, $(g^{ij}) = (g_{ij})^{-1}$, $g = det(g_{ij})$ and $q(x)$ is a nonnegative C^2 function defined on M.
We consider the following Neumann eigenvalue problem:

$$\Delta u - qu = -\eta u \text{ in } M \tag{1}$$

$$\frac{\partial u}{\partial \nu} = 0 \text{ on } \partial M. \tag{2}$$

The spectrum of $\Delta - q$ is discrete and may be ordered as

$$\eta_1 \leq \eta_2 \leq \cdots \leq \eta_k \leq \cdots.$$

Also, it is well known that $\{\eta_k\}$ can be characterized by

[1]This research was partially supported by NSC 86-2115-M-006-030-T and its detailed proofs will be published elsewhere.

$$\eta_k = \inf_{u\not\equiv 0, u\perp u_j, j=1,\ldots,k-1} \frac{\int_M (|\nabla u|^2 + qu^2)}{\int_M u^2}, \tag{3}$$

where u_j is an eigenfunction with eigenvalue η_j. In particular, it is well known that Poincaré constant plays an important role in analysis and from the variational principle, one sees that, when $q \equiv 0$ in M, a lower estimate for the first nonzero η_1 will give an upper estimate for the Poincaré constant, C_p, in

$$\int_M |\nabla u|^2 \leq C_p \int_M (u - \bar{u})^2. \tag{4}$$

Therefore, it is interesting and useful if one may obtain computable lower estimates for η_1. Analogously, it is also interesting to find lower estimates for η_k, and for the gaps Γ_k.

It is the purpose of this paper to find a computable lower estimate for

$$\eta_k \text{ for } k = 1, 2, \ldots \text{ when} q \equiv 0, \tag{5}$$

or

$$\Gamma_k = \eta_k - \eta_1 \text{ for } k = 2, 3, \ldots \text{ otherwise}, \tag{6}$$

in terms of geometrical quantities of M and ∂M, the dimension of M, and the C^2 norm of q.

Acknowledgments

We would like to express our gratitude to the Organizing Committee of the third Pacific Rim Geometry Conference for the kind invitation and their hospitality.

2 Main Theorems

The estimates of η_1 or Γ_k have been studied extensively by many authors. Here, we only recall some of those most important works and we refer to [B] [1], [C] [2], [CnL] [4], [L] [10], [LT] [11], [LY1] [12], [LY2] [13], [M] [14], [SWYY] [15], [W] [16], [Y] [17], [ZY] [18], and [ZYu] [19] for further references. The following was proved by Li-Yau in [LY1] [12] and [LY2] [13].

Theorem 2.1 *When ∂M is convex, i.e. the second fundamental form elements is nonnegative,* $\mathrm{Ric}_M \geq 0$, *the first nonzero eigenvalue satisfies*

$$\eta_1 \geq \frac{\pi^2}{4d^2}, \tag{7}$$

and the $k-th$ nonzero eigenvalue satisfies

$$\eta_k \geq \frac{C(k+1)^{2/m}}{d^2}, \tag{8}$$

where d is the diameter of M and C is a constant depending only on m, the dimension of M.

It was conjectured by Li-Yau that the sharp lower estimate for η_1 should be $\frac{\pi^2}{d^2}$. Analogously, for the nontrivial potential function case, one is interested in estimating the gaps Γ_k. The following Dirichlet boundary condition case is proved in [SWYY] [15].

Theorem 2.2 *When $M \subset R^n$ is a strictly convex bounded domain and q is a nonnegative convex smooth function, then the first two Dirichlet eigenvalues will satisfy the gap estimate*

$$\Gamma_2 \geq \frac{\pi^2}{4d^2}. \tag{9}$$

The above conjecture of Li-Yau was finally proved by Zhong and Yang in [ZY] [18] by applying maximum principle to a judicious choice of test function. Analogously, using a similar type of argument, Zhong-Yu [ZYu] [19] were able to improve the lower estimate (9) to the following.

Theorem 2.3 *When $M \subset R^n$ is a strictly convex bounded domain and q is a nonnegative convex smooth function, then the first two Dirichlet eigenvalues satisfy the gap estimate*

$$\Gamma_2 \geq \frac{\pi^2}{d^2}. \tag{10}$$

Remark 2.1 *In the Dirichlet case, a very important feature as indicated in [SWYY][15] is that the ratio, $r = \frac{u_2}{u_1}$, of the first two eigenfunctions satisfies the Neumann condition on the boundary.*

For a nonconvex domain, the maximum principle does not apply. Therefore, some further information like symmetries or control on the size of "*interior rolling ball*" of the domain will be helpful. P. Buser [B] [1] first proved the Poincaré inequality for star-shaped domains in a Riemannian manifold by estimating the isoperimetric constant using a geometric method.

It would be interesting if one may give an analytic proof. Recently, we used a lower estimate for heat kernel and Cheeger-Yau's [CY] [5] heat kernel comparison theorem to get an analytic proof of the following theorem of Buser in [CnL] [4]. We prove in [CnL] [4] that

Theorem 2.4 *Let M^m be a manifold with $\partial M \neq \emptyset$. Assume that M is geodesically star-shaped with respect to $p \in M$, i.e. the interior of any of the distance minimizing geodesics from p to $q \in M$ lies in the interior of M, and that* $\mathrm{Ric_M} \geq -(\mathrm{m}-1)\mathrm{K}$ *for some constant $K \geq 0$. Let $B_p(R)$ be the largest geodesic ball centered at p contained in M, and $B_p(R_0)$ be the smallest geodesic ball centered at p containing M. Then there exists a constant $C_1 > 0$ depending only on m, such that the first nonzero Neumann eigenvalue η_1 has a lower bound given by*

$$\eta_1 \geq \frac{R^m}{R_0^{m+2}} \exp(-C_1(1 + R_0\sqrt{K})). \tag{11}$$

Remark 2.2 *This generalizes a lower estimate in [LY1] [12], [LY2] [13], and [ZY] [18]. It should also be noted that an important feature is that both Buser's and our results do not assume any bound on the second fundamental form of the boundary. Analogously, for the Shrödinger operator case, we also generalize a lower estimate in [ZYu] [19] to a nonconvex domain. In particular, we obtain the following theorem in [CnL] [4].*

Theorem 2.5 *Let $M^m, \partial M$ be as above. Let $\omega(q) = \sup q - \inf q$ be the oscillation of q over M. Then there exists a constant $\alpha_2 > 0$, depending only on m, $R_0\sqrt{K}$ and $\omega(q)\, R_0^2 + 1$, such that the gap of the first two Neumann eigenvalues has a lower bound given by*

$$\Gamma_2 \geq \frac{R^m}{R_0^{m+2}} \exp(-\alpha_2). \tag{12}$$

For a more general domain, Cheeger's dumbbell example [C] [2] shows that one needs some control on the size of "*interior rolling ball*" of the domain to obtain an positive lower bound of the first eigenvalue. The following condition is given in [M] [14] and [Cn2] [7].

Definition 2.1 *∂M is said to satisfy the "interior rolling R-ball" condition if for each point $p \in \partial M$ there is a geodesic ball $B_q(\frac{R}{2})$, centered at $q \in M$ with radius $\frac{R}{2}$, such that $p = B_q(\frac{R}{2}) \cap \partial M$ and $B_q(\frac{R}{2}) \subset M$.*

For a more general domain, [Cn2] [7], [M] [14] proved the following independently.

Theorem 2.6 *Let ∂M satisfy the "interior rolling R-ball" condition. Assume that* $\mathrm{Ric_M} \geq -(\mathrm{m}-1)\mathrm{K}$, *and the second fundamental form elements* $\mathrm{II} \geq -\mathrm{H}$. *Then the first nonzero Neumann eigenvalue satisfies*

$$\eta_1 \geq C(H, K, m, R, d), \tag{13}$$

where d is the diameter of M.

By using heat kernel estimates, and a method developed by S.Y. Cheng and P. Li, J. Wang [W][16] recently proved the following.

Theorem 2.7 *The k-th Neumann eigenvalue η_k of M satisfies*

$$\eta_k \geq C(K,H,m,R,d)k^{\frac{2}{m}} \tag{14}$$

for all $k = 0, 1, 2, \ldots$.

Analogously, we may generalize the above theorem to Schrödinger operator case. The following theorem is proved in [Cn3][8], [Cn4][9].

Theorem 2.8 *The gap of k-th Neumann eigenvalues η_k and η_1 satisfies*

$$\Gamma_k \geq C(K,H,R,m,d,q)\, k^{\frac{2}{m}} \tag{15}$$

for all $k = 2, 3, \ldots$.

3 Proof

In this section, we shall give an outline of the proof of the Theorem 2.8. Let $u(x,t)$ be a positive Neumann solution and $G(x,y,t)$ be the Neumann fundamental solution of

$$(\Delta - q - \frac{\partial}{\partial t})u(x,t) = 0 \tag{16}$$

on $M \times (0,\infty)$. To prove the theorem, we shall first prove the following gradient estimate for a positive solution of equation (16).

Lemma 3.1 *Let $(M^n, \partial M)$ be as in Theorem 2.6. Assume that $q \in C^2(M)$ satisfies*

1. $|\nabla q| \leq \gamma$
2. $\Delta q \leq \theta$

for some constants γ and θ. Then for a constant $\alpha > (1 + H + \frac{R\gamma}{2})^2$

$$\frac{|\nabla u|^2}{u^2} - \alpha\frac{u_t}{u} \leq C_1 + \frac{C_2}{t} \tag{17}$$

where C_1, C_2 are constants depending on $m, K, H, R, d, \gamma, \theta$.

Outline of Proof. We define a function on M by $\phi(x) = \psi(\frac{r(x)}{R})$, where $r(x)$ denotes the distance from $x \in M$ to ∂M and $\psi(r)$ is a nonnegative C^2-function defined on $[0,\infty)$ such that

$$\psi(r)\begin{cases}\le H+\frac{R\gamma}{2} \text{ if } r\in[0,1/2)\\[2ex] = H+\frac{R\gamma}{2} \text{ if } r\in[1,\infty)\end{cases}$$

with $\psi(0)=0,\quad 0\le\psi'(r)\le 2(H+\frac{R\gamma}{2}),\quad \psi'(0)=H+\frac{R\gamma}{2}$ and $\psi''(r)\ge -(H+\frac{R\gamma}{2})$.

Let $f=\log u$ and $\bar{q}=q-\sup q-1$. Then $(\Delta-\frac{\partial}{\partial t})f(x,t)=q-|\nabla f|^2(x,t)$. We consider the function

$$F(x,t)=t\{(1+\phi(x))^2(|\nabla f|^2(x,t)-\bar{q})-\alpha f_t(x,t)\}$$

For any fixed $T<\infty$, since $F(x,t)$ is continuous on $\bar{M}\times[0,T]$, there exists $(p,t_0)\in\bar{M}\times[0,T]$ at which F achieves its maximum. We may assume that $F(p,t_0)>0$ as otherwise our theorem follows trivially. **Claim:** $p\in\bar{M}\backslash\partial M$. In fact, if $p\in\partial M$, then $\dfrac{\partial F}{\partial\nu}(p,t_0)\ge 0$. Let $e_1,\ e_2,\ldots,e_m$ be an orthonormal frame at p with $e_m=\nu$. Notice that $f_m=f_\nu=\dfrac{\partial u}{\partial\nu}\dfrac{1}{u}=0$ on ∂M and $t_0>0$, therefore denoting $\varphi(x)=(1+\phi(x))^2$ and choosing $R<1$ we have

$$\begin{aligned}0 &\le \frac{\partial F}{\partial\nu}(p,t_0)=t_0\Big[\frac{\partial\varphi}{\partial\nu}(|\nabla f|^2-\bar{q})+(2\varphi\sum_{i=1}^{m}f_if_{i\nu}-q_\nu)-\alpha f_{\nu t}\Big](p,t_0)\\ &< t_0\left\{-\frac{2H+R\gamma}{R}\Big(|\nabla f|^2-\bar{q}\Big)+2H|\nabla f|^2+|\nabla q|\right\}\\ &<0.\end{aligned}$$

This is a contradiction and the claim follows. Thus, $F(x,t)$ achieves its maximum at $(p,t_0)\in(\bar{M}\backslash\partial M)\times(0,T]$.

At (p,t_0), we have

$$\nabla F=0,\ \tfrac{\partial F}{\partial t}\ge 0,\ \text{and}\Delta F\le 0.$$

Using a standard computation, we may prove the inequality (17).

By utilizing the gradient estimate (17), it is standard to apply those tools developed in [LY2] [13] to obtain the following Harnack inequality.

Theorem 3.1 *Let $(M^n,\partial M)$ be as in Theorem 2.6, q be as in Lemma 3.1, and choose R to be sufficiently small. Then $u(x,t)$ satisfies the inequality*

$$u(x_1,t_1)\le u(x_2,t_2)(\frac{t_2}{t_1})^{C_1}\exp\left(\frac{\alpha r^2(x_1,x_2)}{4(t_2-t_1)}+C_2(t_2-t_1)\right)\tag{18}$$

for any $\alpha > (1 + H + \frac{R\gamma}{2})^2$ *and* $0 < \beta < \frac{1}{2}$, $x_1, x_2 \in M$ *and* $0 < t_1 < t_2 < \infty$.
Outline of Proof. Let $\gamma : [0,1] \to M$ be a minimal geodesic joining from $\gamma(0) = x_2$, to $\gamma(1) = x_1$. Define a curve $\eta : [0,1] \to M \times [0,\infty)$ by

$$\eta(s) = (\gamma(s), (1-s)t_2 + st_1).$$

Then, we get $\eta(0) = (x_2, t_2)$ and $\eta(1) = (x_1, t_1)$. Let $f = \log(u(x,t)$, then

$$\begin{aligned} f(x_1,t_1) - f(x_2,t_2) &= \int_0^1 \frac{df(\eta(s))}{ds} ds \\ &= \int_0^1 \{\langle \dot{\gamma}, \nabla f \rangle - (t_2 - t_1) f_t\} \, ds \\ &\leq \int_0^1 \{|\dot{\gamma}||\nabla f| - (t_2 - t_1) f_t\} \, ds. \end{aligned}$$

If we denote $\rho = d(x_1, x_2)$, then $|\dot{\gamma}| = \rho$. We may apply the inequality (16) to estimate f_t, and prove the Harnack inequality (18).

Following an argument of Li-Yau's in [LY2] [13], we may get an upper estimate for the fundamental solution.

Theorem 3.2 *Let* $(M^n, \partial M)$ *be as in Theorem 2.6 and* q *be as in Lemma 3.1. Then for* $t > 0$ *and* $x, y \in M$,

$$G(x,y,t) \leq (1+\delta)^{2C_1} V_x^{-\frac{1}{2}}(\sqrt{t}) V_y^{-\frac{1}{2}}(\sqrt{t}) \exp\left(-\frac{\alpha r^2(x,y)}{(4+\delta)t} + C_2 \delta t + \frac{1+\delta}{\delta}\right) \tag{19}$$

Outline of Proof. By setting $t_2(1+\delta)t$, $t_1 = t$, $x_1 = y$, and x_2 to be any point in $B_x(\sqrt{t})$, and applying the inequality (18), we get inequality (19).

An immediate application of the Harnack inequality (18) is the following global Harnack inequality for the first eigenfunction.

Lemma 3.2 *Let* $(M^n, \partial M)$ *be as in Theorem 2.6,* q *as in Lemma 3.1, and* $u_1(x)$ *be the first Neumann eigenfunction with eigenvalue* η_1 *of the Schrödinger operator on* M. *Assume that* $\int_M u_1^2 = 1$. *Then for any* x, $y \in M$, *we have*

$$u_1(x) \leq u_1(y)\,(1+\delta)^{C_1} \exp\left(\sqrt{\alpha(C_2 - \eta_1)} r(x,y)\right) \tag{20}$$

for any $\delta > 0$, $\alpha > (1 + H + \frac{R\gamma}{2})^2$ *and* $0 < \beta < \frac{1}{2}$, x, $y \in M$. *Hence,*

$$(1+\delta)^{-C_1} \exp\left(-\sqrt{\alpha(C_2 - \eta_1)} d\right) V(M)^{-1}$$

$$\leq u_1(y) \qquad (21)$$
$$\leq (1+\delta)^{C_1} \exp\left(\sqrt{\alpha(C_2-\eta_1)}d\right) V(M)$$

where d denotes the diameter of M.

In order to improve the upper estimate (19) for the fundamental solution when time t is large, we may follow an argument developed by S. Y. Cheng and P. Li in [CL] [3] and we shall need the following lemma.

Lemma 3.3 *Let η_1 be the first Neumann eigenvalue. Then for any fixed t_0 and $t \geq t_0$,*

$$G(x,x,t) \leq G(x,x,t_0)\,[\exp(-\eta_1(t-t_0))]. \qquad (22)$$

Combining inequalities (17)- (22), we get a global upper estimate for the fundamental solution of the equation (16).

Theorem 3.3 *Let $(M^n, \partial M)$ be as in Theorem 2.6, and q be as in Lemma 3.1. Then for $t>0$ and any $x, y \in M$,*

$$G(x,y,t) \leq C_3 V_x^{-\frac{1}{2}}(\sqrt{t}) V_y^{-\frac{1}{2}}(\sqrt{t})\, u_1(x)\, u_1(y) \exp\left(-\alpha\frac{r^2(x,\,y)}{5t} - \eta_1 t\right), \qquad (23)$$

where C_3 is a positive constant which only depend on d, K, H, m, R, $\sup q$, γ, θ and it can be explicitly computed.

We shall also need the following volume comparison lemma proved by J. Wang in [W] [16].

Lemma 3.4 ([W]) *Let M be as in Theorem 2.6. Then $V_x(t) \geq C(d,K,H,m,R)\,t^m$ for some constant $C(d,K,H,m,R) > 0$ and all $x \in M$ and $t \leq R$.*

Combining this with (13), we obtain a desired upper estimate for the fundamental solution of the equation (16).

Theorem 3.4 *Let $(M^n, \partial M)$ be as in Theorem 2.6, and q be as in Lemma 3.1. Then for $t>0$ and $x \in M$,*

$$G(x,y,t) \leq C_4\, e^{-\eta_1 t}\, u_1(x)\, u_1(y)\, t^{-\frac{m}{2}} \qquad (24)$$

where C_4 is a positive constant depending on d, K, H, m, R, $\sup q$, γ, θ and it can be computed explicitly.

Proof of Theorem 2.8. Using the estimate (23) and employing an argument in [CnL] [4], we obtain the following upper estimate.

$$\sum_{k=1}^{\infty} e^{-\Gamma_k t} r_k(x) r_k(y) = \frac{G(x,y,t)}{e^{-\eta_1 t} u_1(x) u_1(y)} \leq C_4 t^{-\frac{m}{2}},$$

where $r_k(x) = \frac{u_k(x)}{u_1(x)}$. The proof is completed by setting $t = \frac{1}{\Gamma_k}$ for $k \geq 2$ and integrating with respect to the volume element $u_1^2(x)\,dx$ to get

$$C_4 \Gamma_k^{\frac{m}{2}} \geq \sum_{j=1}^{\infty} e^{-\frac{\Gamma_j}{\Gamma_k}} \geq \sum_{j=1}^{k} e^{-\frac{\Gamma_j}{\Gamma_k}} \geq k e^{-1}.$$

References

1. P. Buser, A note on the isoperimetric constant *Ann. Sci. École Norm. Sup.* **15** *(1982), 213-230.*
2. J. Cheeger, A lower bound for the smallest eigenvalue of the Laplacian, *In "Problems in analysis, a symposium in honor of S. Bochner", Princeton University Press, 1970.*
3. S. Y. Cheng and P. Li, Heat kernel estimates and lower bounds of eigenvalues, *Comment. Math. Helv.* **56** *(1981), 327-338.*
4. R. Chen and P. Li, On Poincaré type inequalities, *Trans. A.M.S.* **349** *(1997), 1561-1585.*
5. J. Cheeger and S. T. Yau, A lower bound for the heat kernel, *Comm. Pure Appl. Math.* **34** *(1981), 465-480.*
6. R. Chen, Eigenvalue estimate on a compact Riemannian manifold, *Amer. J. Math.* **111** *(1989), 769-781.*
7. R. Chen, Neumann eigenvalue estimate on a compact Riemannian manifold, *Proc. Amer. Math. Soc.* **108** *(1990),961-970.*
8. R. Chen, Gap estimates of Schrödinger operator, *Pacific J. Math.* **178** *(1997), 225-240.*
9. R. Chen, On global Schrödinger kernel estimate and eigenvalue problem, *Math. Z.* **227** *(1998), 69-81.*
10. P. Li, On the Sobolev constant and the p-spectrum of a compact Riemannian manifold, *Ann. Sci. École Norm. Sup.* **13** *(1980), 451-469.*
11. P. Li, P. and A. Treibergs, Applications of eigenvalue techniques to geometry, *Contemporary Geometry, J.-Q. Zhong Memorial Volume, H. H. Wu, ed. Plenum Press, 1991.*
12. P. Li and S. T. Yau, Estimates of eigenvalues of a compact Riemannian manifold, *AMS Proc. Symp. Pure Math.* **36** *(1980), 205-239.*
13. P. Li, P. and S. T. Yau, On the parabolic kernel of the Schrödinger operator, *Acta Math.* **156** *(1986), 153-201.*

14. D. Meyer, Minoration de la première valeur propre non nulle de problème de Neumann sur les variétés Riemanniennes à bord, *Ann. Inst. Fourier, Grenoble* **36** *(1986), 113-125.*
15. I. M. Singer, B. Wong, S. S. T. Yau and S. T. Yau, An estimate of the gap of the first two eigenvalues of the Schrödinger operator, *Ann. Scuo. Norm. Pisa, Cl. Sci. IV* **12** *(1985), 319-333.*
16. J. Wang, Global heat kernel estimates, *Pacific J. Math. 178 (1997), 377-398.*
17. S. T. Yau, Isoperimetric constants and the first eigenvalue of a compact Riemannian manifold, *Ann. Sci. École Norm. Sup.* **8** *(1975), 487-507.*
18. J. Q. Zhong and H. C. Yang, On the estimate of first eigenvalue of a compact Riemannian manifold, *Sci. Sinica Ser. A* **27** *(1984) , 1265-1273.*
19. J. Q. Zhong and Q. H. Yu, Lower bounds on the gap between the first and second eigenvalues of Schrödinger operator, *Proc. A.M.S.* **294** *(1986), 341-349.*

Grassmann geometry and J-holomorphic curves of a 6-dimensional sphere

Hideya Hashimoto
Nippon Institute of Technology

Mathematical Subject Classifications(1991): 53B25, 53C15

1 Introduction

This report is a short survey about submanifolds of 6-dimensional sphere. It is well known that a 6-dimensional sphere S^6 admits a homogeneous almost Hermitian structure. We recall some known results of submanifolds theory of 6-dimensional sphere.

In 1969, A. Gray ([Gr1]) proved that S^6 has no 4-dimensional almost complex submanifold.

In 1981, N.Ejiri ([E1]) proved that there exist a 3-dimensional totally real sphere with constant sectional curvature $\frac{1}{16}$ in S^6, and any totally real 3-dimenional submanifold of S^6 is automorphically minimal. In 1985, K.Mashimo ([M]) classified equivariant 3-dimenional totally real submanifolds from $SU(2)$ to S^6 by using transformation group theory and construct many examples of totally real submanifolds. In1990, F.Dillen, L.Verstralen and L.Vranken ([Dvv]) classified these submanifolds by using sectional curvature. In 1986, N.Ejiri ([E2]) constructed totally real submanifolds as tube of J-holomorphic curves. In 1996, F.Dillen and L.Vranken ([Dv]) studied geometrical properties of such submanifolds.

In 1982, R. Bryant ([Br1]) constructed superminimal J-holomorphic curves of any compact Riemann surface to S^6 by using twistor methods with respect to the G_2-moving frame. In 1983, K.Sekigawa ([Se1]) classified J-holomorphic curves by using Gauss curvature. In 1994 J.Bolton,L.Vranken and L.Woodward ([Bvw]) constructed another type, non-superminimal J-holomorphic curves of 2-dimensional tori to S^6 by using the soliton theory.

In 1984, K.Sekigawa ([Se2]) proved that there does not exist any CR-product submanifold in S^6 and construct 3-dimensional CR-submanifolds each distribution is integrable. Up to now, this is an only known example.

In 1995, J.Berndt, J. Bolton and L.Woodward ([Bbw]) costruct Hopf hypersurfaces in S^6 as a tube of J-holomorphic curves.

However there does not exist the unified theory of subamnaifolds of S^6. First we give the Grassmann geometry of a 6-dimensional sphere from this point of view.

2 Almost Hermitian structure of a 6-dimensional sphere

Let S^6 be the 6-dimensional unit sphere centered at the origin in 7-dimensional Euclidean space. We identified 7-dimensional Euclidean space with purely imaginary octonions ImO (or Cayley algebra). Taking account of algebraic properties of octonions we can define an almost Hermitian structure as follows.

Octonions is a non-commutative, non-associative, alternative normed (division) algebra (in the sense of Harvey-Lawson) . Thus, there is a vecctor cross product $\times$ in ImO such that $x \times y = \frac{1}{2}(\bar{y}x - \bar{x}y)$, where $\bar{x}$ denotes the conjugation of x in the octonions. In particular, if x is perpendicular to y, and $x, y \in ImO$, we have $x \times y = xy$. We define the section of endmorphism bundle as follows.

$$JX = X \times x,$$

where $X \in T_x S^6$ and x is a position vector. Then J satisfies the following properties $J^2 = -Id$, $< JX, JY > = < X, Y >$ for any $X, Y \in T_x S^6$, where $< \ , \ >$ is the standard metric on S^6. We remark that this almost complex structure is not integrable even locally. Also, the 2nd Betti number of S^6 is zero, so there does not exist symplectic structure, in particular, S^6 does not admit any Kähler structure. But the almost complex structure satisfy the nearly Kähler condition $((\nabla_X J)X = 0)$ where ∇ is the Levi-Civita connection of S^6, and X is any vector field of S^6. This condition means that the almost complex structure is parallel (with respect to the Levi-Civita connection) along a geodesic. We remark that Borel and Serre proved that, if even dimensional sphere S^{2n} admits an almost complex structure, then $n = 1$ *or* 3.

3 Grassmann geometry of a 6-dimensional sphere

We study submanifold theory of S^6 from the point of Grassmann Geometry. We recall the definition of Grassmann geometry. Let (M^m, g) be an m-dimensional Riemannian manifold. Fix a point p of M^m and $T_p M^m$ is a tangent space at p. Since the tangent space at p can be identified with the m-dimensional Euclidean space, we can define the Grassmann bundle over M^m as follows. We denote by $G^k(T_p M^m)$ the Grassmannian manifold of oriented k-planes in $T_p M^m$. Let

$$G^k(TM^m) = \bigcup_{p \in M^m} G^k(T_p M^m) \longrightarrow M^m$$

be the Grassmann bundle over M^m. Let V be a subbundle (or subset) of $G^k(TM^m)$.

Definition. Let S be a k-dimensional connected submanifold of M^m S is said to be a V-submanifold if $T_pS \in V$ for any $p \in S$.

The family of V-submanifolds of M^m constitutes the V-geometry of M^m. Such V-geometries are generally called the Grassmann geometries. The above definition, we see that the V-geometry coincides with the family of submanifolds with some restriction of 1st order differential, that is, the tangent space of V-submanifold is specified, or equivalently, for each S, V-submanifold admit the following restricted Gauss map;

$$S \ni p \longrightarrow T_pS \in V \subset G^k(TM^m).$$

To study the Grassmann Geometry, it is important to know the automorphism group of the $S^6 = (S^6, J, < \ , \ >)$. The following fact is proved by T.Fukami and S.Ishihara ([FI]) in 1955;

Proposition 3.1 The automorphism group of the almost Hermitian manifold S^6 coincide with the Lie group G_2 of automorphisms of the octonions.

From this Proposition, we can get the classification of Grassmann geometry.

Proposition 3.2 For typical submanifolds of S^6, the correspoding Grassmann geometries are given by the following table. We put $m(k) =$ dimension of $G^k(TS^6)$.

k	Submanifolds	V	$m(k)$	$m(k)$ − dim of V
2	J-holomorphic curves	$G_2/U(2)$	14	4
2	totally real surfaces	G_2/T^2	14	2
3	totally real submanifolds	$G_2/SO(3)$	15	4
3	CR submanifolds	G_2/T^1	15	2
4	almost complex submanifolds = $\{\emptyset\}$			
4	CR submanifolds	G_2/T^2	14	2
5	hypersurfaces	$G_2/SU(2)$	11	0

where T^k is a k-dimensional torus.

The fundamental problems about Grassmann geometries are the existence of submanifold to corresponding Grassmann geometry.

There exist many problems about Grassmann geometry.

1. How to construct examples, if possible, describe its moduli space.
2. Describe the obstruction of the existence of the typical submanifolds.
3. Describe the curvature properties of the examples.
4. Determine the geometrical invariants up to the action of G_2.
5. Do there exist relations among Grassmann geometries ?

4 J-holomorphic curves of a 6-dimensional sphere

We would like to forcus our attention to the J-holomorphic curves of S^6. Let $x : M^2 \to S^6$ be an isometric immersion. We call the immersion x a J-holomorphic curve if $x_* \circ J_0 = J \circ x_*$, where J_0 is the complex structure with respect to the induced metric. First, we remark that a J-holomorphic curve of S^6 is a minimal submanifold.

Next we shall give a construction of J-holomorphic curves which are obtained by R.L.Bryant ([Br1]).

It is known that the homogeneous space $G_2/U(2)$ coincides with the Grassmannian manifold $G_2(\mathrm{Im}O)$ of oriented 2-planes in $\mathrm{Im}O$. Then we have the CP^2-bundle over S^6 such that $\pi : G_2/U(2)\ (= G_2(\mathrm{Im}O)) \to G_2/SU(3)\ (= S^6)$. Let $H(q)$ be the horizontal subspace of $G_2(\mathrm{Im}O)$ at $q \in G_2(\mathrm{Im}O)$ with respect to the $SU(3)$ connection of S^6. Let $L_+(q)$ be the subspace (of $H(q)$) of the vectors on which π_* is complex linear with respect to the canonical complex structure of $G_2(\mathrm{Im}O)$ and the nearly Kähler structure on S^6. Then it can be shown that the dimension of the space $L_+(q)$is four and $\cup_{q \in G_2(\mathrm{Im}O)} L_+(q)$ is the subbundle of rank 4 of the tangent bundle of $G_2(\mathrm{Im}O)$. Let $\psi : M^2 \to G_2(\mathrm{Im}O)$ be a holomorphic map from a Riemann surface M^2 to the Grassmaniann manifold $G_2(\mathrm{Im}O)$ whose differential image $d\psi(T_pM^2)$ is included in $L_+(p)$ for any point $p \in M$. Then we have the superminimal J-holomorphic curve $\pi \circ \psi : M^2 \to S^6$, locally.

We write down the first order differential equation corresponding to the superminimal J-holomorphic curve of S^6.

Theorem 4.1. Let $\xi : M^2 \to C^7 \setminus \{0\} \simeq C \otimes \mathrm{Im}O \setminus \{0\}$ is a holomorphic map satisfying the following four conditions;

(1) $\quad < \xi, \xi >= 0,$

(2) $\quad \dfrac{d\xi}{dz} \neq 0,$

(3) $\quad \dfrac{d\xi}{dz} \times \bar{\xi} \neq 0,$

(4) $\quad < \dfrac{d\xi}{dz}, \xi \times \bar{\xi} >= 0,$

then the map

$$\frac{\sqrt{-1}\xi \times \bar{\xi}}{|\xi|^2} : M^2 \to S^6$$

is a super minimal J-holomorphic curve of S^6.

By using this theorem, we can get the J-holomorphic curves of S^6 as follows:

We set a basis of $\mathbf{C} \otimes_R \mathbf{ImO}$ by $\varepsilon = (0,1) \in \mathbf{H} \oplus \mathbf{H}$, $E_1 = iN$, $E_2 = jN$, $E_3 = -kN$, $\overline{E_1} = i\overline{N}$, $\overline{E_2} = j\overline{N}$ and $\overline{E_3} = -k\overline{N}$, where $N = (1 - \sqrt{-1}\varepsilon)/2$, $\overline{N} = (1 + \sqrt{-1}\varepsilon)/2 \in \mathbf{C} \otimes_R \mathbf{O}$ and $\{1, i, j, k\}$ is the canonical basis of $\mathbf{H}$. Then we have $< E_i, E_j > = < \overline{E_i}, \overline{E_j} > = < N, N > = < \overline{N}, \overline{N} > = 0$, $< E_i, \overline{E_j} > = \frac{1}{2}\delta_{ij}$, and $< N, \overline{N} > = \frac{1}{2}$.

Let $\xi : \mathrm{C} \to \mathrm{Q}^5$ be the holomorphic map defined as follows;

$$\begin{aligned} \xi &= E_1 2 + \overline{E_2}(2\sqrt{6}z) + \overline{E_3}(2\sqrt{-1}\sqrt{15}z^2) + \varepsilon(2\sqrt{10}z^3) \\ &+ E_3(2\sqrt{-1}\sqrt{15}z^4) + E_2(2\sqrt{6}z^5) + \overline{E_1}(-2z^6), \end{aligned}$$

where $\{\varepsilon, E_1, E_2, E_3, \overline{E_1}, \overline{E_2}, \overline{E_3}\}$ is the basis of $\mathrm{C}\otimes\mathrm{Im}O$ (defined above). Then the map

$$\frac{\sqrt{-1}\xi \times \bar{\xi}}{|\xi|^2} : P^1(\mathrm{C}) \to S^6$$

coincides with the Boruvka sphere (superminimal J-holomorphic curve of S^6) $\psi_0 : S^2(\frac{1}{6}) \to S^6(1)$ up to the action of G_2.

5 Local rigidity of J-holomorphic curves

Let $\varphi_1, \varphi_2 : M^2 \to S^6$ be two J-holomorphic curves of S^6. If there exists $g \in G_2$ such that $g \circ \varphi_1 = \varphi_2$, then the induced metrics and induced complex structures coincide. We want to know its inverse (the following theorem). To do this, we prepare some notations. Since M^2 is a 2-dimensional oriented manifold, then there exists a isothermal coordinate (U, z), and, we consider only local nature of J-holomorphic curves, we may treat J-holomorphic curves $\varphi_i : (U, z) \to S^6$. Let τ be the holomorphic 6-differential on (U, z) which is defined by

$$\tau = < \sigma_2(\frac{\partial}{\partial z}, \frac{\partial}{\partial z}, \frac{\partial}{\partial z}), \frac{\partial}{\partial z} \times \sigma(\frac{\partial}{\partial z}, \frac{\partial}{\partial z}) > (dz)^6 = F_i(z)(dz)^6$$

where σ (resp. σ_2) is the second (resp. third) fundamental form of the J-holomorphic curve $\varphi_i : (U, z) \to S^6$. We remark that the τ can be considered as a globally holomorphic 6-differential on M^2. We obtain the following

Theorem 5.1. Let $\varphi_1, \varphi_2 : (U, z) \to S^6$ be two J-holomorphic curves of S^6 where U is connected surface. If the induced metrics coincide (the Gauss curvature K satisfy $K < 1$) and the functions $F_1(z), F_2(z)$ satisfy $F_1(z) = F_2(z)$ on U. Then there exist an element $g \in G_2$ such that

$$g \circ \varphi_1 = \varphi_2.$$

6 Curvature properties of J-holomorphic curves

First, we shall give a theorem which is obtained by K.Sekigawa([Se]) and F.Dillen et al ([DVV]) concerned with curvature of M.

Theorem 6.1. Let M be a J-holomorphic curve of S^6.

(1). If M is complete and $\frac{1}{6} \leq K \leq 1$ then $K \equiv \frac{1}{6}$ or $K \equiv 1$,

(2). If M is compact and $0 \leq K \leq \frac{1}{6}$ then $K \equiv 0$ or $K \equiv \frac{1}{6}$.

By using the moving frame method of Bryant, we obtain the following

Theorem 6.2. Let $x : M \to S^6$ be a J-holomorphic curve in S^6. If its induced metric is complete and the Gauss curvature K satisfies the following conditions

(1) K is bounded from below,

(2) $K \leq \frac{1}{6}$,

(3) $\int_M (K^-)\Omega$ is finite,

(4) $|\,\mathbf{III}\,|^2 < \frac{1}{4}$.

Then the immersion x is congruent to the Veronese immersion of $S^2_{\frac{1}{6}}$ to S^6 up to G_2 action.

Theorem 6.3. Let $x : M \to S^6$ be a J-holomorphic curve in S^6. If its induced metric is complete and the Gauss curvature K satisfies the following conditions

(1) K is bounded from below,

(2) $K \leq 0$

(3) $\int_M (K^-)\Omega$ is finite,

(4) $|\,\mathbf{III}\,|^2 \leq \frac{1}{4}$.

Then $K \equiv 0$, $|\,\mathbf{III}\,|^2 \equiv \frac{1}{4}$ and the immersion x is congruent to the Kenmotsu surface $T^2 \to S^5$ up to G_2 action.

Lastly, we shall give the relation between the genus of M and superminimal points. A point $p \in M$ is called a superminimal point if the holomorphic 6-differential τ vanishes at p.

Theorem 6.4. Let $x : M \to S^6$ be a compact J-holomorphic curve in S^6 with genus g. Then we have

(1) The immersion x is superminimal if $g = 0$. ([Br1]).

(2) The immersion x is superminimal or otherwise the immersion x is nowhere superminimal on M if $g = 1$.

(3) If $K < 1$, $g \geq 2$ and there exists a point at which the immersion is not superminimal, then the multiplicity k_i of each zero of the third fundamental form is divisible by 6 and furthermore the equality

$$\sum_{i=1}^{l} k_i = 12(g-1)$$

holds, where p_i $(1 \leq i \leq g)$ are superminimal points of x with multiplicity k_i.

As a Corollary, we see that superminimal points are an obstruction to construct J-holomorphic curves of S^6

Corollary 6.5. Let $x : M^2 \to S^6$ be a compact J-holomorphic curve in S^6 with genus g. If $K < \frac{1}{6}$ and $g \geq 2$, there exist at most $2(g-1)$ superminimal points.

[Br1] R.L.Bryant. Conformal and minimal immersions of comppact surfaces into the 4-sphere. J. Diff. Geom., 17 (1982) 455-473

[Br2] R.L.Bryant. Submanifolds and special structures on the octonians. J. Diff. Geom., 17 (1982) 185-232.

[Bbw] J.Berndt, J.Boltonand L.Woodward . Almost complex curves and Hopf hypersurfaces in the nearly Kähler 6-sphere Geometriae Dedicata., 56 (1995) 237-247.

[Bvw] J.Bolton, L.Vrancken and L.Woodward . On almost complex curves in the nearly Kähler 6-sphere
Quart.J.Math., 45 (1994) 407-427.

[Dvv] F. Dillen, L.Verstralen and L.Vranken Classification of totally real 3-dimensional submanifolds of $S^6(1)$ with $K \geq \frac{1}{16}$. J.Math.Soc.Japan., 42(1990) 565-584.

[Dv] F. Dillen and L.Vranken Totally real submanifolds in $S^6(1)$ satisfying Chen's equality. Trans.A.M.S., 348(1996) 1633-1646.

[E1] N.Ejiri. Totally real submanifolds in a 6-sphere. Proc.A.M.S., 83 (1981) 759-763.

[E2] N.Ejiri. Equivariant minimal immersions of S^2 into $S^{2m}(1)$. Trans.A.M.S., 297 (1986) 105-124.

[FI] T.Fukami and S.Ishihara. Almost Hermitian structure on S^6. Tohoku Math J., 7 (1955) 151-156.

[Gr1] A.Gray. Almost complex submanifolds of the six sphere. Proc.A.M.S., 20(1969) 277-279.

[Gr2] A.Gray. Einstein-like manifolds which are not Einstein. Geometrae Dedicata., 7 (1978) 259-280.

[Hl] R.Harvey and H.B.Lawson. Calibrated geometries. Acta Math., 148 (1982) 47-157.

[H1] H.Hashimoto. J-holomorphic curves in a 6-dimensional sphere. (to appear)

[H2] H.Hashimoto. Explicit representation of super-minimal J-holomorphic curves in a 6-dimensional sphere. (to appear in Kodai Math. J.)

[Hs-L] W-Y.Hsiang and H.B.Lawson. Minimal submanifolds of low cohomogenity. J. Diff. Geom., 5(1971) 1-38.

[M] K.Mashimo. Homogeneous totally real submanifolds of S^6. Tsukuba J.Math (1985) 185-202.

[N1] H.Naitoh. Compact simple Lie algebras with two involutions ans submanifolds of compact symmmetric spaces I. To appear in Osaka Math. J.

[N2] H.Naitoh. Compact simple Lie algebras with two involutions ans submanifolds of compact symmmetric spaces II. To appear in Osaka Math. J..

[N3] H.Naitoh. Grassmann geometries on compact symmmetric spaces of classical type. To appear.

[Se1] K.Sekigawa Almost complex submanifoldds of a 6-dimensioanl sphere. Kodai Math.J., 6(1983) 174-185.

[Se2] K.Sekigawa Some CR-submanifolds in a 6-dimensioanl sphere. Tensor.N.S., 6(1984) 13-20.

[Se3] K.Sekigawa Notes on homogeneous almost Hermitian manifolds. Hokkaido Math.J, 7(1978) 206-213.

[Sp] M.Spivak A comprehensive introduction to differential geometry **IV.** Publish or Perish., 1975.

Nippon Institute of Technology
4-1, Gakuendai, Miyashiro,
Minami-Saitama Gun,
Saitama, 345, Japan.
E-mail address hideya@nit.ac.jp

Self-duality of neutral metrics on four-dimensional manifolds

H. KAMADA
Numazu College of Technology, 3600 Ooka, Numazu-shi, Shizuoka, 410-8501, Japan
E-mail: kamada@la.numazu-ct.ac.jp

1 Introduction

We discuss in this paper self-duality of oriented pseudo-Riemannian four-manifolds with metrics of signature (2,2), called *neutral* metrics. A neutral metric g on an oriented four-manifold M is said to be *self-dual* (resp. *anti-self-dual*) if its Weyl conformal curvature tensor, regarded as an endomorphism on the space of 2-forms, is self-dual (resp. anti-self-dual). It is significant to recognize a fact that self-duality of neutral metrics is conformally invariant (cf. Akivis-Goldberg [1], Kamada-Machida [18]).

For the existence of neutral metrics on four-manifolds, it is well-known that a four-manifold M admits a neutral metric if and only if M admits a two-plane field (see Steenrod [26]). Matsushita [22] studied the existence problem of orientable two-plane fields on compact oriented four-manifolds in connection with that of two kinds of almost complex structures, one being compatible with the given orientation and the other with opposite orientation. In particular, he showed that a compact complex surface admits an orientable two-plane field if and only if its Euler characteristic is even (cf. Bonome et al.[5]). Therefore we see that there exist many examples of oriented four-manifolds which admit neutral metrics. For instance, $K3$ surfaces and the total space of a fibre bundle over a Riemann surface with two-dimensional fibres admit neutral metrics. In fact, the product of two-dimensional spheres $S^2 \times S^2$, the product of Riemann surfaces $\Sigma_- \times \Sigma_+$ with genera ≥ 2, and the product of a circle and a three-dimensional sphere $S^1 \times S^3$ are the examples.

We now consider self-dual neutral metrics on four-manifolds. Conformally-flat neutral metrics are the typical examples of self-dual neutral metrics (cf. §6). We also know several examples of (non-compact) four-manifolds whose neutral metrics are not conformally-flat but self-dual (see Kamada-Machida[18]). The

---

1991 *Mathematics Subject Classification.* Primary 53C50; Secondary 53C15, 53C25, 53C55.
The author was supported by Grant-in-Aid for Encouragement of Young Scientists (No. 09740075), The Ministry of Education, Science, Sports and Culture, Japan.

existence of a neutral metric on a four-manifold depends, of course, on the topology of the manifold. It is natural to consider that the self-duality restricts, in addition, to the topology of the manifold with a neutral metric. We are then interested in the following question: Under what conditions can a neutral metric on a four-manifold be self-dual? Probably, this question still remains open.

In the present paper, we shall discuss on four-manifolds with self-dual neutral metrics, together with some interesting additional geometric properties. We are hoping that examples in §6 will be helpful for finding the conditions for a neutral metric be self-dual.

The paper is organized as follows. We first recall basic facts for self-duality of neutral metrics on four-manifolds (§2). We next introduce some geometric structures relevant to neutral metrics, e.g., a neutral Kähler structure, a split-quaternion structure, a neutral hyperhermitian structure and a neutral hyperkähler structure, and study relations with self-duality of neutral metrics (§§3–5). Some typical examples of (compact or non-compact) four-manifolds with self-dual neutral metrics are given in §6. We finally give some remarks in §7.

2 Preliminaries

We recall basic facts for neutral metrics on oriented four-manifolds (cf. Besse [4], Akivis-Goldberg [1], Kamada-Machida [18]). Let M be an oriented pseudo-Riemannian four-manifold with neutral metric g. The Hodge star operator $*$ of (M,g) satisfies $*^2 = \mathrm{Id}$ on the space of 2-forms $\bigwedge^2 = \bigwedge^2 T^*M$. Then $\bigwedge^2$ decomposes as $\bigwedge^2 = \bigwedge^+ \oplus \bigwedge^-$, where $\bigwedge^\pm = \{\alpha \in \bigwedge^2 \mid * \alpha = \pm\alpha\}$.

Let ∇ be the Levi-Civita connection of (M,g) and R the curvature tensor of (M,g) defined by $R(X,Y)Z = [\nabla_X, \nabla_Y]Z - \nabla_{[X,Y]}Z$ for arbitrary vector fields X, Y, Z on M. We denote respectively by Ric_0, s and W the traceless Ricci tensor, the scalar curvature and the Weyl conformal curvature tensor of (M,g). Namely, the Weyl conformal curvature tensor W is defined as follows:

$$\begin{aligned} g(W(X_1,X_2)X_3,X_4) = g(R(X_1,X_2)X_3,X_4) \\ -(1/2)\{ g(X_1,X_4)\mathrm{Ric}_0(X_2,X_3) - g(X_2,X_4)\mathrm{Ric}_0(X_1,X_3) \\ + g(X_2,X_3)\mathrm{Ric}_0(X_1,X_4) - g(X_1,X_3)\mathrm{Ric}_0(X_2,X_4)\} \\ -(s/12)\{ g(X_1,X_4)g(X_2,X_3) - g(X_2,X_4)g(X_1,X_3)\} \end{aligned}$$

for arbitrary vector fields X_1, X_2, X_3, X_4. Regarding R as an endomorphism on $\bigwedge^2 = \bigwedge^+ \oplus \bigwedge^-$, we have

$$R = W^+ \oplus W^- \oplus \mathrm{Ric}_0 \oplus (s/12)\mathrm{Id},$$

since W and Ric satisfy $*W = W*$ and $*\mathrm{Ric}_0 = -\mathrm{Ric}_0*$, respectively. We now define the following:

DEFINITION. A neutral metric g is said to be *self-dual* (resp. *anti-self-dual*) if $W^- \equiv 0$ (resp. $W^+ \equiv 0$). In particular, if $W \equiv 0$, then g is said to be *conformally-flat.*

Note that the notion of self-duality of neutral metrics is conformally invariant.

We now recall the significance of the self-duality of neutral metrics in relation to totally null planes (cf. Akivis-Goldberg [1]). Let (M,g) be an oriented pseudo-Riemannian four-manifold with neutral metric g. For a subspace σ of the tangent space T_xM at a point x of M, we denote by $\sigma^\perp$ the orthogonal complement of σ with respect to g. A subspace σ of T_xM is said to be ***totally null*** (or ***isotropic***) if $\sigma \subset \sigma^\perp$. It is known that $\dim \sigma \leq 2$ for any totally null subspace σ and that the equality is satisfied if and only if $\sigma = \sigma^\perp$. We call a two-dimensional totally null subspace a ***totally null*** (or ***maximal isotropic***) ***plane.*** Note that the condition for two-planes to be totally null is clearly conformally invariant. Let $\mathrm{Gr}_2(T_xM)$ be the Grassmannian manifold of two-dimensional subspaces in T_xM, and $\mathbf{P}(\bigwedge^2 T_x^*M)$ the real projective space of one-dimensional subspaces in $\bigwedge^2 T_x^*M$. We now define a map $\Phi : \mathrm{Gr}_2(T_xM) \longrightarrow \mathbf{P}(\bigwedge^2 T_x^*M)$ by $\Phi : \mathrm{span}\{u,v\} \mapsto [u^\flat \wedge v^\flat]$, where $\flat$ is the musical operator from T_xM to T_x^*M with respect to g and $[\cdot]$ denotes the equivalence class in $\mathbf{P}(\bigwedge^2 T_x^*M)$. Then the following fact is known (cf. Akivis-Goldberg[1]):

Proposition 2.1. *For any totally null plane σ, the image $\Phi(\sigma)$ belongs to* $\mathbf{P}(\bigwedge^\pm)$.

Hence we have the following:

DEFINITION. A totally null plane σ is said to be *self-dual* (resp. *anti-self-dual*) if $\Phi(\sigma)$ belongs to $\mathbf{P}(\bigwedge^+)$ (resp. $\mathbf{P}(\bigwedge^-)$).

Note that self-dual (resp. anti-self-dual) totally null planes are called ***α-planes*** (resp. ***β-planes***) in Akivis-Goldberg[1]. For any plane $\sigma = \mathrm{span}\{X,Y\}$, the signature of $g(R(X,Y)Y,X)$ is well-defined, and denoted by $\kappa(\sigma)$. By definition, it is clear that $g(R(\xi,\eta)\eta,\xi) = g(W(\xi,\eta)\eta,\xi)$ for any totally null plane $\sigma = \mathrm{span}\{\xi,\eta\}$. We then deduce that the signature $\kappa(\sigma)$ of a totally null plane σ on (M,g) is closely related to the Weyl part W of the curvature of g. In fact, the following is also known:

Proposition 2.2. *A neutral metric g on an oriented four-manifold M is self-dual (resp. anti-self-dual) if and only if $\kappa(\sigma)$ equals zero for any anti-self-dual* *(resp. self-dual) totally null plane σ at each point of M. In particular, g is conformally-flat if and only if $\kappa(\sigma)$ equals zero for any totally null plane σ.*

Dajczer-Nomizu [8] showed that if $\kappa(\sigma) = 0$ for any degenerate plane σ on (M, g), then all nondegenerate planes have the same sectional curvatures, i.e., (M, g) has constant curvature. Proposition 2.2 can be regarded as a conformal version of the result of Dajczer-Nomizu [8].

A subbundle of TM consisting of totally null planes is called a *totally null plane field.* For later convenience, we recall the following:

Lemma 2.3. *Let (M, g) be an oriented pseudo-Riemannian four-manifold with neutral metric g and σ a totally null plane field. If σ is integrable, then $\kappa(\sigma)$ equals zero.*

PROOF. We may assume that $\sigma = \mathrm{span}\{\xi, \eta\}$ (at least locally), where ξ, η are linearly independent vector fields on M such that $g(\xi, \xi) = g(\eta, \eta) = g(\xi, \eta) = 0$. If σ is integrable, then $[\xi, \eta] = u\xi + v\eta$ for some smooth functions u, v. It then follows that

$$\begin{aligned} 2g(\nabla_\xi \xi, \eta) &= 2\xi g(\xi, \eta) - \eta g(\xi, \xi) - 2g([\xi, \eta], \xi) = 0, \\ 2g(\nabla_\xi \xi, \xi) &= \xi g(\xi, \xi) = 0, \ 2g(\nabla_\xi \eta, \eta) = \xi g(\eta, \eta) = 0, \\ g(\nabla_\xi \eta, \xi) &= \xi g(\eta, \xi) - g(\eta, \nabla_\xi \xi) = 0, \end{aligned}$$

and hence that $\nabla_\xi \xi, \nabla_\xi \eta \in \Gamma(\sigma^\perp) = \Gamma(\sigma)$. Similarly, $\nabla_\eta \xi, \nabla_\eta \eta \in \Gamma(\sigma^\perp) = \Gamma(\sigma)$. Therefore we obtain

$$\begin{aligned} g(R(\xi, \eta)\eta, \xi) &= g(\nabla_\xi \nabla_\eta \eta, \xi) - g(\nabla_\eta \nabla_\xi \eta, \xi) - g(\nabla_{[\xi, \eta]} \eta, \xi) \\ &= \xi g(\nabla_\eta \eta, \xi) - g(\nabla_\eta \eta, \nabla_\xi \xi) - \eta g(\nabla_\xi \eta, \xi) + g(\nabla_\xi \eta, \nabla_\eta \xi) \\ &\quad - u g(\nabla_\xi \eta, \xi) - v g(\nabla_\eta \eta, \xi) \\ &= 0, \end{aligned}$$

that is, $\kappa(\sigma) = 0$. □

3 Neutral Kähler surfaces

In this section, we consider a complex structure on a four-manifold M with a neutral metric g. Let I be an almost complex structure on (M, g). We define the following:

DEFINITION. (M, g, I) is called a *neutral almost Hermitian* manifold if g is invariant by I (i.e., $g(IX, IY) = g(X, Y)$ for arbitrary vector fields X, Y on M). If I is also integrable, then (M, g, I) is called a *neutral Hermitian surface.* A neutral almost Hermitian manifold (M, g, I) is called a *neutral Kähler surface* if I is parallel with respect to the Levi-Civita connection of (M, g).

For a neutral Kähler surface (M, g, I), the anti-self-dual part W^- of the Weyl conformal curvature with respect to a suitable basis for $\bigwedge^-$ is expressed as follows:

$$W^- = \frac{s}{12}\begin{pmatrix} 2 & 0 & 0 \\ 0 & -1 & 0 \\ 0 & 0 & -1 \end{pmatrix}. \tag{1}$$

Thus we have the following (cf. Kamada-Machida[18], Derdziński[9], Itoh[16]):

Proposition 3.1. *Let (M, g, I) be a neutral Kähler surface with the natural orientation. Then g is self-dual if and only if g is scalar-flat.*

We consider compact neutral Kähler surfaces. Recalling the Gauss-Bonnet formula for a compact oriented pseudo-Riemannian four-manifold (M, g) with neutral metric g (see Avez[2], Chern[7]), we can write the Euler characteristic $\chi(M)$ in terms of the components of the curvature tensor as

$$\chi(M) = -\frac{1}{8\pi^2}\int_M \left\{ \|W_+\|^2 + \|W_-\|^2 - \frac{1}{2}\|\mathrm{Ric}_0\|^2 + \frac{s^2}{24} \right\} v_g, \tag{2}$$

$\|\cdot\|^2$ being the "squared" norm. We can also write the Hirzebruch signature $\tau(M)$ as

$$\tau(M) = \frac{1}{12\pi^2}\int_M \left\{ \|W_+\|^2 - \|W_-\|^2 \right\} v_g. \tag{3}$$

(see Matsushita[23]). For both formulae (2) and (3), see Law[20]. It should be noted that the norms $\|\cdot\|^2$ in these formulae are indefinite in general.

From (1), (2) and (3), we obtain the following (cf. Kamada-Machida[18]):

Proposition 3.2. *Let (M, g, I) be a compact neutral Kähler surface. If g is anti-self-dual, then $\tau(M)$ is nonpositive. In particular, $\tau(M) = 0$ if and only if g is scalar-flat and conformally-flat.*

We consider compact non-conformally-flat, anti-self-dual neutral Kähler surfaces. We first note that the fundamental form Ω_I of a neutral Kähler surface (M, g, I), defined by $\Omega_I = g(I\cdot, \cdot)$, gives a symplectic structure on M compatible with the opposite orientation of (M, I). For the existence of symplectic structures on compact complex surfaces compatible with the opposite orientation, we recall Draghici[10]'s recent result:

Theorem. *Let (M, I) be a compact complex surface that admits a positive-definite Kähler Einstein metric. Suppose also that $\tau(M)$ is negative and that $\bar{M}$, the manifold M with opposite orientation, admits an almost complex structure. Then the canonical bundle of any almost complex structure on $\bar{M}$ has* *vanishing Seiberg-Witten invariant. In particular, $\bar{M}$ admits no symplectic structures.*

From the theorem above, Enriques surfaces and $K3$ surfaces admit no symplectic structures compatible with the opposite orientation, and therefore they admit no neutral Kähler metrics. By Proposition 3.2, we see that any non-conformally-flat, anti-self-dual neutral Kähler surface (M, g, I) has the following properties: the Hirzebruch signature $\tau(M)$ is negative; $\bar{M}$ admits an almost complex structure; $\bar{M}$ admits a symplectic structure Ω_I. Then we see the following:

Proposition 3.3. *If a compact complex surface admits a non-conformally-flat, anti-self-dual neutral Kähler metric, then it admits no positive-definite Kähler Einstein metrics.*

From Proposition 3.3 and the Enriques-Kodaira classification of compact complex surfaces (cf. Barth et al.[3]), we also see the following:

Proposition 3.4. *If a compact minimal complex surface (M, I) admits a non-conformally-flat, anti-self-dual neutral Kähler metric, then (M, I) is biholomorphic to one of the following possibilities:*
(i) a surface of class VII_0 *with positive Euler characteristic,*
(ii) a minimal properly elliptic surface with positive Euler characteristic.

For these candidates, see §7.2.

We next consider a neutral Kähler Einstein metric on a compact complex surface. From (1), (2) and (3), we have the following (cf. Kamada-Machida[18], see §7.3):

Proposition 3.5. *Let (M, g, I) be a compact neutral Kähler surface. If g is Einstein, then $c_1^2(M, I)$, the square of the first Chern class, is nonpositive (i.e., $-2\chi(M) \geq 3\tau(M)$). If g is also anti-self-dual, then $\chi(M)$ is nonpositive. In particular, $\chi(M) = 0$ if and only if g is flat.*

From Proposition 3.5, we see that if a compact complex surface (M, I) admits a neutral Kähler Einstein metric with nonzero scalar curvature, then $c_1^2(M, I)$ must be negative. In particular, such a surface admits no positive-definite Einstein metrics. From the Enriques-Kodaira classification of compact complex surfaces, a minimal compact complex surface with negative Euler characteristic is a ruled surface of genus ≥ 2. However, there exist no non-flat, anti-self-dual Einstein neutral Kähler metrics on any ruled surface of genus ≥ 2, since its signature is zero. Hence we see that any compact complex surface with a non-flat, anti-self-dual Einstein neutral Kähler metric must contain an exceptional curve.

4 Split-quaternion structures

In this section, we first recall *split-quaternion* (or *paraquaternionic*) structures on four-manifolds (cf. Kamada[17], García-Río et al.[13]).

DEFINITION. A triple $(I, 'J, 'K)$ of endomorphisms on the tangent bundle TM of a four-manifold M is called a *split-quaternion* (or *paraquaternionic*) structure if $I, 'J, 'K$ satisfy the following:

$$I^2 = -\mathrm{Id}, \; 'J^2 = 'K^2 = \mathrm{Id} \quad \text{and} \quad I'J = -'JI = 'K.$$

A neutral metric g on a four-manifold M with a split-quaternion structure $(I, 'J, 'K)$ is said to be *compatible* if the quadruple $(g, I, 'J, 'K)$ satisfies the following I-invariance and $'J, 'K$-skew-invariance:

$$g(X,Y) = g(IX, IY) = -g('JX, 'JY) = -g('KX, 'KY)$$

for arbitrary vector fields X, Y on M. A four-manifold M endowed with a split-quaternion structure $(I, 'J, 'K)$ admits a compatible neutral metric g if and only if the structure group of TM can be reduced to $\mathrm{GL}_0(1, '\mathbf{H}) := \{p + qi + r'j + s'k \mid p^2 + q^2 > r^2 + s^2, \; p, q, r, s \in \mathbf{R}\}$, where i, $'j$, $'k$ satisfy the split-quaternion relations: $i^2 = -1, 'j^2 = 'k^2 = 1, i'j = -'ji = 'k$. Note that if we regard $\mathrm{GL}_0(1, '\mathbf{H})$ as a subgroup of $\mathrm{GL}(2, \mathbf{C}) (\subset \mathrm{GL}(4, \mathbf{R}))$ by

$$\iota : p + qi + r'j + s'k \mapsto \sqrt{p^2 + q^2 - r^2 - s^2} \cdot \frac{1}{\sqrt{p^2 + q^2 - r^2 - s^2}} \begin{pmatrix} p + q\sqrt{-1} & r + s\sqrt{-1} \\ r - s\sqrt{-1} & p - q\sqrt{-1} \end{pmatrix},$$

then the image of ι ($\cong \mathbf{R}_+ \times \mathrm{SU}(1,1)$) is contained in the conformal group $\mathrm{CO}(2,2)$. We therefore see that for any neutral metric g compatible with $(I, 'J, 'K)$, its conformal class is uniquely determined by the triple $(I, 'J, 'K)$ (cf. §7.1).

Such a quadruple $(g, I, 'J, 'K)$ is called a *neutral almost hyperhermitian* (abbreviated as *n.a.h.h.*) structure. For an n.a.h.h. manifold $(M, g, I, 'J, 'K)$, we define three nondegenerate 2-forms $\Omega_I, \Omega_{'J}, \Omega_{'K}$, called the *fundamental forms*, as follows:

$$\Omega_I(X,Y) := g(IX, Y), \; \Omega_{'J}(X,Y) := g('JX, Y), \; \Omega_{'K}(X,Y) := g('KX, Y)$$

for arbitrary vector fields X, Y. Note that the skew-invariance of g by $'J, 'K$ is essential for the definition of $\Omega_{'J}, \Omega_{'K}$.

Recall that the fundamental forms $\Omega_1 := \Omega_I, \Omega_2 := \Omega_{'J}, \Omega_3 := \Omega_{'K}$ of an n.a.h.h. manifold $(M, g, I, 'J, 'K)$ satisfy the following relation:

$$-\Omega_1^2 = \Omega_2^2 = \Omega_3^2, \quad \Omega_l \wedge \Omega_m \equiv 0 \; (l \neq m; l, m = 1, 2, 3). \tag{4}$$

Conversely, we can show the following (cf. Geiges [14], Geiges-Gonzalo [15]):

Proposition 4.1. For any triple $(\Omega_1, \Omega_2, \Omega_3)$ satisfying (4), there exists a unique n.a.h.h. structure $(g, I, \mathcal{J}, \mathcal{K})$ such that $\Omega_I = \Omega_1, \Omega_{\mathcal{J}} = \Omega_2, \Omega_{\mathcal{K}} = \Omega_3$.

PROOF. We define $(g, I, \mathcal{J}, \mathcal{K})$ as follows:

$$\Omega_3(I\cdot,\cdot) = \Omega_2,\ \Omega_1(\mathcal{J}\cdot,\cdot) = \Omega_3,\ \Omega_2(\mathcal{K}\cdot,\cdot) = -\Omega_1 \tag{5}$$

and $g := \Omega_1(\cdot, I\cdot)$. It follows from $\Omega_2^2 = \Omega_3^2$ and $\Omega_2 \wedge \Omega_3 \equiv 0$ that $I^2 = -\mathrm{Id}$. Similarly, it also follows from $-\Omega_1{}^2 = \Omega_3{}^2$ and $\Omega_1 \wedge \Omega_3 \equiv 0$ that $\mathcal{J}^2 = \mathrm{Id}$. By definition, we see that $(I, \mathcal{J}, \mathcal{K})$ is a split-quaternion structure. It can also be verified that Ω_1 is invariant by I, and hence that g is compatible with $(I, \mathcal{J}, \mathcal{K})$. Thus we obtain the desired n.a.h.h. structure $(g, I, \mathcal{J}, \mathcal{K})$. □

For n.a.h.h. structures on four-manifolds, we show the following:

Proposition 4.2. *For an n.a.h.h. four-manifold $(M, g, I, \mathcal{J}, \mathcal{K})$, there exists a positive-definite almost Hermitian metric h on (M, I) such that*

$$-\Omega_1^2 = \Omega_2^2 = \Omega_3^2 = \omega_1^2, \quad \omega_1 \wedge \Omega_l \equiv 0 \ (l = 1, 2, 3),$$

where $\Omega_1 = \Omega_I, \Omega_2 = \Omega_{\mathcal{J}}, \Omega_3 = \Omega_{\mathcal{K}}$, and where ω_1 denotes the fundamental form of (M, h, I). It then follows that Ω_1 is anti-self-dual and Ω_2, Ω_3 are self-dual with respect to the metric h.

PROOF. The proof is very simple. We first note that $\Omega_2 + \sqrt{-1}\Omega_3$ belongs to $\bigwedge^{2,0}$, the space of (2,0)-forms on (M, I), and hence that Ω_2, Ω_3 are self-dual with respect to any (positive-definite) almost Hermitian metric h on (M, I). Take any almost Hermitian metric h_0 with respect to I such that the fundamental form ω_0 of h_0 satisfies $\omega_0^2 = -\Omega_1^2 = \Omega_2^2 = \Omega_3^2 =: 2v_g$. Then we see that $\omega_0 \wedge \Omega_l \equiv 0$ $(l = 2, 3)$. Let φ denote the function on M defined by $\omega_0 \wedge \Omega_1 = \varphi\Omega_1^2$. Then $(\omega_0 - \varphi\Omega_1) \wedge \Omega_l \equiv 0$ $(l = 1, 2, 3)$ and $(\omega_0 - \varphi\Omega_1)^2 = 2(1 + \varphi^2)v_g$. Then the 2-form $\omega_1 := (\omega_0 - \varphi\Omega_1)/\sqrt{1 + \varphi^2}$ is invariant by I (i.e., $\omega_1(IX, IY) = \omega_1(X, Y)$), since ω_0 and Ω_1 are invariant by I. Define a metric h by $h(X, Y) := \omega_1(X, IY)$. The metric h must be definite, since $\omega_1^2 = 2v_g$ clearly holds. If necessary, by replacing h with $-h$, we may assume that h is positive-definite. Then h is a required almost Hermitian metric. It follows that Ω_1 is orthogonal to $\bigwedge_h^+$, the space of self-dual 2-forms on (M, h), since $\bigwedge_h^+$ is spanned by $\omega_1, \Omega_2, \Omega_3$. Namely, Ω_1 is an anti-self-dual 2-form on (M, h). □

For an n.a.h.h. four-manifold $(M, g, I, \mathcal{J}, \mathcal{K})$, its fundamental forms $\Omega_1 := \Omega_I, \Omega_2 := \Omega_{\mathcal{J}}, \Omega_3 := \Omega_{\mathcal{K}}$ define three isomorphisms $\wedge\Omega_l : \bigwedge^1 \longrightarrow \bigwedge^3$ $(l = 1, 2, 3)$. Then we define 1-forms $\beta_1, \beta_2, \beta_3$, by $d\Omega_l = \beta_l \wedge \Omega_l$ $(l = 1, 2, 3)$.

These 1-forms $\beta_1, \beta_2, \beta_3$, called the *Lee forms*, are related to the integrability of $I, \mathcal{J}, \mathcal{K}$. An involution S on TM (i.e., $S^2 = \mathrm{Id}, S \neq \mathrm{Id}$) is said to be *integrable* if the bundles $\mathcal{F}_S^{\pm} := \{v \in TM |\ Sv = \pm v\}$ are integrable. The integrability of S is equivalent to $N_S \equiv 0$, where N_S is the *Nijenhuis tensor* of S defined by

$$N_S(X,Y) := [SX, SY] + S^2[X,Y] - S[SX,Y] - S[X,SY],$$

for arbitrary vector fields X, Y on M. This is also true for the integrability of an almost complex structure I. Namely, I is integrable if and only if $N_I \equiv 0$, where N_I is the Nijenhuis tensor of I defined by replacing S with I in the definition of N_S. An n.a.h.h. manifold $(M, g, I, \mathcal{J}, \mathcal{K})$ is called a *neutral hyperhermitian* (abbreviated as *n.h.h.*) surface if $I, \mathcal{J}, \mathcal{K}$ are integrable. We show the following needed later (cf. Boyer [6]):

Proposition 4.3. *Let $(g, I, \mathcal{J}, \mathcal{K})$ be an n.a.h.h. structure on a four-manifold. Then $I, \mathcal{J}, \mathcal{K}$ are integrable if and only if the Lee forms satisfy $\beta_1 = \beta_2 = \beta_3$.*

We set $\beta := \beta_1 = \beta_2 = \beta_3$, and call it the *Lee form*. To show Proposition 4.3, we need the following:

Lemma 4.4. *Set $\Omega_1 := \Omega_I, \Omega_2 := \Omega_{\mathcal{J}}, \Omega_3 := \Omega_{\mathcal{K}}$ and $N_1 := N_I, N_2 := N_{\mathcal{J}}, N_3 := N_{\mathcal{K}}$. Then they satisfy* (5) *and the following equations:*

$$\begin{aligned}
\Omega_1(X,Y) &= \Omega_1(IX,IY) = \Omega_1(\mathcal{J}X,\mathcal{J}Y) = \Omega_1(\mathcal{K}X,\mathcal{K}Y),\\
\Omega_2(X,Y) &= -\Omega_2(IX,IY) = -\Omega_2(\mathcal{J}X,\mathcal{J}Y) = \Omega_2(\mathcal{K}X,\mathcal{K}Y),\\
\Omega_3(X,Y) &= -\Omega_3(IX,IY) = \Omega_3(\mathcal{J}X,\mathcal{J}Y) = -\Omega_3(\mathcal{K}X,\mathcal{K}Y).
\end{aligned} \tag{6}$$

$$\begin{aligned}
d\Omega_2(X,Y,Z) + d\Omega_2(IX,IY,Z) &= d\Omega_3(X,Y,IZ) + d\Omega_3(IX,IY,IZ)\\
&\quad + \Omega_3(N_1(Y,Z),IX) + \Omega_3(N_1(Z,X),IY),
\end{aligned} \tag{7}$$

$$\begin{aligned}
d\Omega_3(X,Y,Z) - d\Omega_3(\mathcal{J}X,\mathcal{J}Y,Z) &= d\Omega_1(X,Y,\mathcal{J}Z) - d\Omega_1(\mathcal{J}X,\mathcal{J}Y,\mathcal{J}Z)\\
&\quad - \Omega_1(N_2(Y,Z),\mathcal{J}X) - \Omega_1(N_2(Z,X),\mathcal{J}Y),
\end{aligned} \tag{8}$$

$$\begin{aligned}
-d\Omega_1(X,Y,Z) + d\Omega_1(\mathcal{K}X,\mathcal{K}Y,Z) &= d\Omega_2(X,Y,\mathcal{K}Z) - d\Omega_2(\mathcal{K}X,\mathcal{K}Y,\mathcal{K}Z)\\
&\quad - \Omega_2(N_3(Y,Z),\mathcal{K}X) - \Omega_2(N_3(Z,X),\mathcal{K}Y),
\end{aligned} \tag{9}$$

where X, Y, Z are arbitrary vector fields on M. □

PROOF OF Proposition 4.3. We only show that I is integrable if and only if $\beta_2 = \beta_3$. Set $\beta_{23} := \beta_2 - \beta_3$ and define $B_{23}(X,Y) := \beta_{23}(X)Y - \beta_{23}(Y)X$. By (7), we see that

$$\Omega_2(B_{23}(X,Y) + B_{23}(IX,IY), Z) = \Omega_2(N_1(Y,Z),X) + \Omega_2(N_1(Z,X),Y),$$

where X, Y, Z are arbitrary vector fields on M.

If $\beta_2 = \beta_3$, then we have $\Omega_2(N_1(Y,Z),X) + \Omega_2(N_1(Z,X),Y) \equiv 0$. By changing X, Y, Z cyclically, we also have the following:

$$\Omega_2(N_1(Z,X),Y) + \Omega_2(N_1(X,Y),Z) \equiv 0,$$
$$\Omega_2(N_1(X,Y),Z) + \Omega_2(N_1(Y,Z),X) \equiv 0,$$

so we see that $\Omega_2(N_1(X,Y),Z) \equiv 0$ for arbitrary vector fields X, Y, Z. Therefore $N_1 \equiv 0$, i.e., I is integrable.

Conversely, if I is integrable, then we have $B_{23}(X,Y) + B_{23}(IX,IY) \equiv 0$ If we set $Y = IX$, then $B_{23}(X,IX) = \beta_{23}(X)IX - \beta_{23}(IX)X \equiv 0$. Note that if X is nonzero at a point $x \in M$, then $\{X, IX\}$ is linearly independent in T_xM and hence $\beta_{23}(X) = \beta_{23}(IX) = 0$ at x. Since X is arbitrary, we see that $\beta_{23} \equiv 0$, i.e., $\beta_2 = \beta_3$. This completes the proof. □

Concerning self-duality of n.a.h.h. metrics, we see the following (for the Riemannian analogue, see e.g., Pedersen-Swann [25]):

Proposition 4.5. *Let $(M, g, I, \mathcal{J}, \mathcal{K})$ be an n.a.h.h. four-manifold. If $I, \mathcal{J}, \mathcal{K}$ are integrable, then g is self-dual.*

PROOF. For any constant θ, set $\mathcal{J}_\theta := \cos\theta\mathcal{J} + \sin\theta\mathcal{K}$ and $\mathcal{K}_\theta := -\sin\theta\mathcal{J} + \cos\theta\mathcal{K}$. Then $(g, I, \mathcal{J}_\theta, \mathcal{K}_\theta)$ is also an n.a.h.h. structure on M, and, moreover, its fundamental forms $\Omega_I, \Omega_{\mathcal{J}_\theta}, \Omega_{\mathcal{K}_\theta}$ satisfy

$$d\Omega_I = \beta \wedge \Omega_I, \quad d\Omega_{\mathcal{J}_\theta} = \beta \wedge \Omega_{\mathcal{J}_\theta}, \quad d\Omega_{\mathcal{K}_\theta} = \beta \wedge \Omega_{\mathcal{K}_\theta}.$$

It then follows from Proposition 4.3 that these $I, \mathcal{J}_\theta, \mathcal{K}_\theta$ are integrable. Since $\mathcal{K}_\theta = \mathcal{J}_{(\theta+\pi/2)}$ for any θ, we may consider only $\mathcal{J}_\theta$. Setting, for simplicity, $\mathcal{F}_\theta^\pm := \mathcal{F}_{\mathcal{J}_\theta}^\pm$, we easily see that each $\mathcal{F}_\theta^\pm$ is an integrable totally null plane field on (M, g). ¿From Lemma 2.3, the signature $\kappa(\mathcal{F}_\theta^\pm)$ vanishes for each θ. Moreover we can verify that each $\mathcal{F}_\theta^\pm$ is an anti-self-dual totally null plane field, since $\Phi(\mathcal{F}_{\mathcal{J}}^\pm) = [-\Omega_I \pm \Omega_{\mathcal{K}}] \in \mathbf{P}(\bigwedge^-)$ and $\Phi(\mathcal{F}_\theta^\pm)$ depends continuously on θ. Note that for any anti-self-dual totally null plane σ_x ($x \in M$), there exists a constant θ such that $(\mathcal{F}_\theta^\pm)_x = \sigma_x$. Thus we see that $\kappa(\sigma) = 0$ for any anti-self-dual totally null plane σ, and hence that g is self-dual. □

We remark that if I is integrable and if $\mathcal{F}_{\mathcal{J}}^+$ and $\mathcal{F}_{\mathcal{K}}^+$ are integrable, then $\mathcal{F}_{\mathcal{J}}^-$ and $\mathcal{F}_{\mathcal{K}}^-$ are also integrable. To show this, we suppose that $\mathcal{F}_{\mathcal{J}}^+$ and $\mathcal{F}_{\mathcal{K}}^+$ are integrable. Note that for any vector $V \in \mathcal{F}_{\mathcal{J}}^+$, the vector IV belongs to $\mathcal{F}_{\mathcal{J}}^-$, and vice versa. For arbitrary vector fields X, Y tangent to $\mathcal{F}_{\mathcal{J}}^-$, we see that $\mathcal{K}(X - IX) = X - IX$, $\mathcal{K}(Y - IY) = Y - IY$. ¿From the integrability of $\mathcal{F}_{\mathcal{J}}^+$ and $\mathcal{F}_{\mathcal{K}}^+$, we have

$$\mathcal{J}[IX, IY] = [IX, IY], \tag{10}$$
$$\mathcal{K}[X - IX, Y - IY] = [X - IX, Y - IY]. \tag{11}$$

By (10) and the integrability of I, we can rewrite (11) as

$$\begin{aligned} \mathcal{K}([X,Y]-I[X,Y]) &= [X,Y]-I[X,Y] \\ \text{(or equivalently} \quad -\mathcal{K}(I[X,Y]+[X,Y]) &= I[X,Y]+[X,Y]). \end{aligned} \tag{12}$$

From this, we obtain that $\mathcal{J}[X,Y]=-[X,Y]$, i.e., $\mathcal{F}_{\mathcal{J}}^{-}$ is integrable. Similarly, we can show that $\mathcal{F}_{\mathcal{K}}^{-}$ is also integrable.

5 Neutral hyperkähler structures

In this section, we consider neutral hyperkähler structures on four-manifolds. We begin with the definition as follows:

DEFINITION. An n.a.h.h. four-manifold $(M,g,I,\mathcal{J},\mathcal{K})$ is called a *neutral hyperkähler* surface if $I,\mathcal{J},\mathcal{K}$ are parallel with respect to the Levi-Civita connection of (M,g).

Then we see the following:

Proposition 5.1. *An n.a.h.h. structure $(g,I,\mathcal{J},\mathcal{K})$ on a four-manifold is neutral hyperkähler if and only if $I,\mathcal{J},\mathcal{K}$ are integrable and $\Omega_I,\Omega_{\mathcal{J}},\Omega_{\mathcal{K}}$ are closed.*

PROOF. We only show that $\nabla\mathcal{J}\equiv 0$ if and only if $\Omega_{\mathcal{J}}=N_{\mathcal{J}}\equiv 0$. Note that for arbitrary vector fields X,Y,Z on M, the following hold:

$$\nabla\Omega_{\mathcal{J}}(X,Y)=g((\nabla\mathcal{J})X,Y), \tag{13}$$

$$2g((\nabla_X\mathcal{J})Y,Z)=d\Omega_{\mathcal{J}}(X,Y,Z)+d\Omega_{\mathcal{J}}(X,\mathcal{J}Y,\mathcal{J}Z)-g(\mathcal{J}X,N_{\mathcal{J}}(Y,Z)). \tag{14}$$

If $\nabla\mathcal{J}\equiv 0$, then, from (13), we see that $\Omega_{\mathcal{J}}$ is closed. ¿From (14), we also see that $N_{\mathcal{J}}\equiv 0$. Conversely, if $\Omega_{\mathcal{J}}=N_{\mathcal{J}}\equiv 0$, then, from (14), we see that $\nabla\mathcal{J}\equiv 0$. □

Here we remark that an n.h.h. surface $(M,g,I,\mathcal{J},\mathcal{K})$ is neutral hyperkähler if and only if its Lee form β identically vanishes. Furthermore, if β is closed (resp. exact), then $(M,g,I,\mathcal{J},\mathcal{K})$ is locally (resp. globally) conformally neutral hyperkähler. Noting the remark above, we obtain the following:

Proposition 5.2. *Let $(\Omega_1,\Omega_2,\Omega_3)$ be a triple of three symplectic structures on a four-manifold M satisfying the relation (4). Then there exists a unique neutral hyperkähler structure $(g,I,\mathcal{J},\mathcal{K})$ on M such that $\Omega_I=\Omega_1$, $\Omega_{\mathcal{J}}=\Omega_2$, $\Omega_{\mathcal{K}}=\Omega_3$.*

PROOF. We can define the n.a.h.h. structure $(g,I,\mathcal{J},\mathcal{K})$ by (5). It follows from the closedness of Ω_l that $\beta_l\equiv 0$ $(l=1,2,3)$. By Proposition 4.3, we see that $I,\mathcal{J},\mathcal{K}$ are integrable and hence that $(g,I,\mathcal{J},\mathcal{K})$ is the required neutral hyperkähler structure on M. □

Concerning the curvature of neutral hyperkähler surfaces, we see the following:

Proposition 5.3. *Any neutral hyperkähler surface* $(M, g, I, \mathbb{J}, \mathbb{K})$ *is Ricci-flat and self-dual. Moreover its first Chern class* $c_1(M, I)$ *equals zero.*

PROOF. The self-duality of a neutral hyperkähler metric can be verified from Proposition 4.5. Thus we first show the Ricci-flatness of a neutral hyperkähler metric. Put $\lambda_1^- := -\Omega_I$, $\lambda_2^- := \Omega_{\mathbb{J}}$, $\lambda_3^- := \Omega_{\mathbb{K}}$. Then $\lambda_1^-, \lambda_2^-, \lambda_3^-$ form a basis for $\bigwedge^-$. In general, for a neutral Kähler surface (M, I), the curvature form R_∇ of the connection on $\bigwedge^-$ induced by the Levi-Civita connection ∇ of (M, g) determines the Ricci form of (M, g, I). It follows from Proposition 5.1 that $\lambda_1^-, \lambda_2^-, \lambda_3^-$ are parallel with respect to ∇. Thus we see that the Ricci curvature of (M, g) vanishes; in particular, the real first Chern class $c_1(M, I)_{\mathbf{R}}$ equals zero.

The complex-valued 2-form $\Omega_{\mathbb{J}} + \sqrt{-1}\Omega_{\mathbb{K}}$ is a nonvanishing closed $(2, 0)$-form (i.e., holomorphic 2-form) on (M, I), and hence gives a trivialization of the canonical bundle $K_{(M,I)}$, as a holomorphic vector bundle. Thus we see that $c_1(M, I) = 0$. □

We now consider compact neutral hyperkähler surface. From the arguments above and the Enriques-Kodaira classification, we see that if a compact complex surface admits a neutral hyperkähler structure, then it is biholomorphic to one of the following possibilities:

(a) a complex torus, (b) a $K3$ surface, (c) a primary Kodaira surface.

From Draghici's result, any $K3$ surface admits no neutral Kähler structures, so the candidate (b) cannot occur. For the candidates (a) and (c), we see that there exist neutral hyperkähler structures on any complex torus and a primary Kodaira surface of certain type (see §§6.1, 6.7 and §7.4).

6 Examples

We now give some typical examples of four-manifolds with self-dual (rather conformally-flat) neutral metrics. For other examples, see Kamada-Machida[18] and Kamada[17] (cf. §7.4).

We will use the following notation and convention (cf. Wolf[28]). Let $\mathbf{R}^n_s$ denote the vector space of real n-tuples $x = (x_0, \ldots, x_{n-1})$ with the inner product

$$(g_0)^n_s(x, y) = -(x_0y_0 + \cdots + x_{s-1}y_{s-1}) + (x_sy_s + \cdots + x_{n-1}y_{n-1}).$$

Then $\mathbf{R}^n_s$ can be regarded as a pseudo-Riemannian manifold of signature $(n - s, s)$, called the *pseudo-Euclidean space* of signature $(n - s, s)$. Moreover we define

$$S^n_s = \{x \in \mathbf{R}^{n+1}_s | (g_0)^{n+1}_s(x,x) = 1\}, \quad H^n_s = \{x \in \mathbf{R}^{n+1}_{s+1} | (g_0)^{n+1}_{s+1}(x,x) = -1\}.$$

It is well-known that S^n_s and H^n_s are complete pseudo-Riemannian manifolds of signature $(n-s, s)$ and of constant curvatures $+1$ and -1, respectively. S^n_s is called the *pseudo-Riemannian sphere* and H^n_s the *pseudo-Riemannian hyperbolic space.*

6.1 Pseudo-Euclidean space and flat tori

$$M := \mathbf{R}^4 \cong \mathbf{C}^2 \text{ or } T^4 = \mathbf{R}^4/\Gamma \cong \mathbf{C}^2/\Gamma,$$
$$g := (g_0)^4_2 = -dx_0^2 - dx_1^2 + dx_2^2 + dx_3^2 = -|dz_0|^2 + |dz_1|^2$$

(Γ denotes a lattice in $\mathbf{R}^4 \cong \mathbf{C}^2$). Clearly, g is flat, and can also be obtained as a neutral hyperkähler metric on $\mathbf{C}^2$. Indeed, the following 2-forms $\Omega_1, \Omega_2, \Omega_3$ defines a neutral hyperkähler structure $(g, I, \mathcal{J}, \mathcal{K})$:

$$\Omega_1 := -dx_0 \wedge dx_1 + dx_2 \wedge dx_3,$$
$$\Omega_2 := dx_0 \wedge dx_2 - dx_1 \wedge dx_3, \quad \Omega_3 := dx_0 \wedge dx_3 - dx_2 \wedge dx_1,$$

where $z_0 := x_0 + x_1\sqrt{-1}$, $z_1 := x_2 + x_3\sqrt{-1}$ are the standard complex coordinates of $\mathbf{C}^2 \cong \mathbf{R}^4$. Note that action of the complex structure I on $\mathbf{C}^2$ coincides with that of the multiplication of $\sqrt{-1}$.

Taking a lattice Γ in $\mathbf{C}^2$, we see that the quotient space $\mathbf{C}^2/\Gamma$ is a complex torus with the flat neutral hyperkähler structure induced by $(g, I, \mathcal{J}, \mathcal{K})$ on $\mathbf{C}^2$.

6.2 Pseudo-Riemannian sphere

$$M := S^4_2 = \{(x_0, \ldots, x_4) \in \mathbf{R}^5_3 | -x_0^2 - x_1^2 + x_2^2 + x_3^2 + x_4^2 = 1\},$$
$$g := g_{S^4_2} = (-dx_0^2 - dx_1^2 + dx_2^2 + dx_3^2 + dx_4^2)|_{S^4_2}.$$

It has constant curvature $+1$. It is clearly conformally-flat and Einstein. We also note that $(S^4_2, -g_{S^4_2})$ coincides with the pseudo-Riemannian hyperbolic space H^4_2 (cf. Kamada-Machida[18]).

6.3 Product space of two-dimensional spheres

$$M := S^2 \times S^2,$$
$$g := -h_{S^2} \oplus h_{S^2},$$

where h_{S^2} is the standard Riemannian metric of the unit round sphere S^2. This example is related to §6.2. Indeed, set $(x_0, x_1) = \sinh\tau(\cos\varphi, \sin\varphi)$, and $(x_2, x_3, x_4) = \cosh\tau(u_2, u_3, u_4)$, where $\tau \geq 0$, $0 \leq \varphi \leq 2\pi$ and $u_2^2 + u_3^2 + u_4^2 = 1$. Then $g_{S_2^4}$ can be expressed as

$$g_{S_2^4} = -(d\tau^2 + \sinh^2\tau d\varphi^2) + \cosh^2\tau h_{S^2}.$$

Setting $\tan\theta = \sinh\tau$, we obtain the following:

$$(\cos^2\theta) g_{S_2^4} = -(d\theta^2 + \sin^2\theta d\varphi^2) + h_{S^2}.$$

The right hand side can be naturally extended to the religion $0 \leq \theta \leq \pi$ and $0 \leq \varphi \leq 2\pi$, and hence can be regarded as the product neutral metric $-h_{S^2} \oplus h_{S^2}$ on $S^2 \times S^2$. Clearly, g is conformally-flat. The standard complex structure on S^2 induces a complex structure I on $S^2 \times S^2$. Then $(S^2 \times S^2, g, I)$ is a neutral Kähler surface. In particular, g is scalar-flat (but not Einstein).

6.4 Product space of two-dimensional hyperbolic spaces

$$M := H^2 \times H^2 \text{ or } \Sigma_- \times \Sigma_+,$$
$$g := -h_{H^2} \oplus h_{H^2},$$

where $\Sigma_\pm$ are Riemann surfaces of genera ≥ 2, and h_{H^2} is the hyperbolic metric of curvature -1. The metric g can be obtained as follows (cf. Kulkarni [19]): We first consider the time-like cone of the three-dimensional Minkowski space:

$$\mathbf{R}^3_{1+} = \{(x_1, x_2, x_3) \in \mathbf{R}^3 | \, x_1^2 - (x_2^2 + x_3^2) > 0, \ x_1 > 0\}$$

with the standard Lorentzian metric $(g_0)^3_1 = (-dx_1^2 + dx_2^2 + dx_3^2)|_{\mathbf{R}^3_{1+}}$. Note that $(g_0)^3_1$ can be expressed as $(g_0)^3_1 = -d\rho^2 + \rho^2 h_{H^2}$, where $\rho := (x_1^2 - (x_2^2 + x_3^2))^{1/2}$ on $\mathbf{R}^3_{1+}$. Denote by $g_{\mathbf{R}\times\mathbf{R}^3_{1+}}$ the product metric on $\mathbf{R}\times\mathbf{R}^3_{1+}$. Noting $\mathbf{R}\times\mathbf{R}^3_{1+} \cong \mathbf{R} \times \mathbf{R}_+ \times H^2$, we set

$$(1/\rho^2) g_{\mathbf{R}\times\mathbf{R}^3_{1+}} = -(dx_0^2 + d\rho^2)/\rho^2 + h_{H^2},$$

where x_0 is the standard coordinate of $\mathbf{R}$. Then the space $(\mathbf{R} \times \mathbf{R}^3_{1+}, g)$ is isometric to $H^2 \times H^2$ with the product neutral metric $g = -h_{H^2} \oplus h_{H^2}$ and hence is conformally-flat.

Take Fuchsian groups $\Gamma_\pm$ and set $\Sigma_\pm = H^2/\Gamma_\pm$. Then $g_{H^2 \times H^2}$ can be regarded as a neutral metric on $\Sigma_- \times \Sigma_+$. In a similar way to that of §6.3, we can see that the metric is scalar-flat and neutral Kähler.

6.5 Product space of a circle and a three-dimensional Lorentzian space-form

$$M := S^1 \times N_1^3,$$
$$g := -dt^2 \oplus g_{N_1^3},$$

where dt is the nonsingular 1-form on S^1 and $(N_1^3, g_{N_1^3})$ is a three-dimensional Lorentzian space-form (e.g., $N_1^3 = S_1^3, \mathbf{R}_1^3$ or H_1^3). Then the product metric g is conformally-flat.

In the case where $N_1^3 = H_1^3$, we can define a neutral hyperhermitian structure $(g, I, \mathcal{J}, \mathcal{K})$ on $M = S^1 \times H_1^3$. To see this, we first note that H_1^3 can be identified with the Lie group $SU(1,1)$ as follows:

$$H_1^3 \ni (x_0, x_1, x_2, x_3) \mapsto \begin{pmatrix} x_0 + x_1\sqrt{-1} & x_2 + x_3\sqrt{-1} \\ x_2 - x_3\sqrt{-1} & x_0 - x_1\sqrt{-1} \end{pmatrix} \in SU(1,1).$$

Let $\alpha^1, \alpha^2, \alpha^3$ be left-invariant 1-forms on $SU(1,1)$, which form a basis for the dual of the Lie algebra of $SU(1,1)$, satisfying the following structure equation:

$$d\alpha^1 = -2\alpha^2 \wedge \alpha^3, \quad d\alpha^2 = 2\alpha^3 \wedge \alpha^1, \quad d\alpha^3 = 2\alpha^1 \wedge \alpha^2.$$

(Note that $\alpha^1, \alpha^2, \alpha^3$ were denoted by τ_1, τ_2, τ_3 in Kamada-Machida[18], respectively.) Define three nondegenerate 2-forms $\Omega_1, \Omega_2, \Omega_3$ on $S^1 \times H_1^3$ as follows:

$$\Omega_1 := -dt \wedge \alpha^1 + \alpha^2 \wedge \alpha^3, \ \Omega_2 := dt \wedge \alpha^2 - \alpha^1 \wedge \alpha^3, \ \Omega_3 := dt \wedge \alpha^3 - \alpha^2 \wedge \alpha^1.$$

From these 2-forms, we obtain a neutral hyperhermitian structure $(g, I, \mathcal{J}, \mathcal{K})$, whose neutral metric g is conformally-flat.

6.6 Punctured complex plane and Hopf surface

$$M := \mathbf{C}^2 \setminus \{0\},$$
$$g := 4(-|dz_0|^2 + |dz_1|^2)/(|z_0|^2 + |z_1|^2),$$

where (z_0, z_1) is the standard complex coordinate system on $\mathbf{C}^2$ (cf. Vaisman [27]). Then g is a conformally-flat neutral metric on $\mathbf{C}^2 \setminus \{0\}$. Let λ, μ be complex numbers satisfying $|\lambda| = |\mu| > 1$. Define an action of $G = \mathbf{Z}$ on $\mathbf{C}^2 \setminus \{0\}$ as follows:

$$k \cdot (z_0, z_1) := (\lambda^k z_0, \mu^k z_1) \ (k \in \mathbf{Z}).$$

Then $(\mathbf{C}^2 \setminus \{0\})/G$ is a Hopf surface. Moreover the metric g on $\mathbf{C}^2 \setminus \{0\}$ can be regarded as a neutral metric on $(\mathbf{C}^2 \setminus \{0\})/G$, since g is invariant under the action of G. Thus g is conformally-flat. Note that the scalar curvature s_g of (M, g) is non-constant and given by $s_g = 60(|z_0|^2 - |z_1|^2)/(|z_0|^2 + |z_1|^2)$.

We can also define a neutral hyperhermitian structure on a Hopf surface of some particular type. For example, if $\mu = \bar{\lambda}$ ($|\lambda| = |\mu| > 1$), then the neutral hyperhermitian structure $(g, I, \mathcal{J}, \mathcal{K})$ on $\mathbf{C}^2 \setminus \{0\}$ induces a neutral hyperhermitian structure on $(\mathbf{C}^2 \setminus \{0\})/G$.

6.7 Product space of a circle and a three-dimensional unimodular Lie group

$$M := S^1 \times G^3,$$
$$g := -(dt \otimes \alpha^1 + \alpha^1 \otimes dt) + (\alpha^2 \otimes \alpha^3 + \alpha^3 \otimes \alpha^2),$$

where dt is the nonsingular 1-form on S^1 and G^3 is a three-dimensional unimodular Lie group with left-invariant 1-forms $\alpha^1, \alpha^2, \alpha^3$, which form a basis for the dual of the Lie algebra of G^3, satisfying the following structure equation:

$$d\alpha^1 = \epsilon_1 \alpha^2 \wedge \alpha^3, \quad d\alpha^2 = \epsilon_2 \alpha^3 \wedge \alpha^1, \quad d\alpha^3 = \epsilon_3 \alpha^1 \wedge \alpha^2$$

($\epsilon_1, \epsilon_2, \epsilon_3$ are the structure constants). Direct calculation shows that g is conformally-flat and scalar-flat.

Furthermore, let I be an almost complex structure on $S^1 \times G^3$ defined by

$$I\alpha^1 = -\alpha^2, \quad I\alpha^2 = \alpha^1, \quad I\alpha^3 = -dt, \quad Idt = \alpha^3.$$

Then we see that I is integrable if and only if $\epsilon_1 = \epsilon_2$. On the other hand, we also see that the fundamental form Ω_I is closed (i.e., (g, I) is almost neutral Kähler) if and only if $\epsilon_2 = 0$. In particular, if $\epsilon_1 = \epsilon_2 = 0$, then (g, I) is a neutral Kähler structure on $S^1 \times G^3$. Note that this construction for a neutral Kähler structure (g, I) on $S^1 \times G^3$ appeared in Fernández-Gotay-Gray [12]. If further $\epsilon_3 \neq 0$, then G is a nilpotent Lie group, and hence a suitable compact quotient of $S^1 \times G^3$ can be regarded as a primary Kodaira surface of particular type (see Fernández et al.[12]).

As mentioned in Kamada [17], for the neutral Kähler structure (g, I) on $S^1 \times G^3$ ($\epsilon_1 = \epsilon_2 = 0$) above, we can also define involutions $\mathcal{J}, \mathcal{K}$ such that $(g, I, \mathcal{J}, \mathcal{K})$ is a neutral hyperkähler structure on $S^1 \times G^3$. In fact, the following closed 2-forms $\Omega_1, \Omega_2, \Omega_3$ defines $(g, I, \mathcal{J}, \mathcal{K})$:

$$\Omega_1 := \alpha^1 \wedge \alpha^3 + \alpha^2 \wedge dt,$$
$$\Omega_2 := \alpha^1 \wedge \alpha^3 - \alpha^2 \wedge dt, \quad \Omega_3 := \alpha^1 \wedge dt + \alpha^2 \wedge \alpha^3.$$

7 Remarks

7.1 Neutral structures

We first observe that if g is a neutral metric, then so is $-g$. This leads us to the notion of *neutral structures*. Roughly speaking, a neutral structure is to be defined as an equivalence class of a collection of local neutral metrics which coincide, up to sign, on common domains. It is clear that a neutral metric on M determines a neutral structure. Conversely, if $H^1(M; \mathbf{Z}_2) = 0$, then for any neutral structure on M, there exists a neutral metric which determines the given neutral structure. For details, see Law [21].

For a geometric quantity Υ defined for (M, g), we write Υ^g for Υ. Then we see that the curvature tensors, the Ricci tensors, the Weyl conformal curvature tensors and the scalar curvatures of (M, g) and $(M, -g)$ satisfy the following: $R^{-g} = R^g, \mathrm{Ric}^{-g} = \mathrm{Ric}^g, W^{-g} = W^g$ and $s^{-g} = -s^g$, and moreover that the Hodge star operators satisfy $*^{-g} = *^g$ on $\bigwedge^2$. Thus we can define the notion of self-duality for neutral structures on oriented four-manifolds, in a similar way to that for neutral metrics.

We also note that a split-quaternion structure on a four-manifold uniquely determines a neutral structure, except for conformal factor (cf. §4).

7.2 Anti-self-dual neutral Kähler surfaces

According to Proposition 3.4, a minimal compact complex surface which admits an anti-self-dual but not conformally-flat neutral Kähler metric must be biholomorphic to a surface of class VII_0 or a minimal properly elliptic surface, and its Euler characteristic is positive. The author does not know whether there exist non-conformally-flat, anti-self-dual Kähler structures on these complex surfaces at present.

We note that if an anti-self-dual neutral Kähler surface (M, g, I) has constant scalar curvature, then (M, g) satisfies that $\nabla R \equiv 0$. This can be seen by an argument similar to that in the Riemannian case (cf. Derdziński [9]).

7.3 The Hitchin-Thorpe type inequality

In Proposition 4.5, we obtain the inequality $c_1^2(M, I) = 2\chi(M) + 3\tau(M) \leq 0$ for any compact Einstein neutral Kähler surface (M, g, I). Note that for compact neutral Einstein four-manifolds of certain types, the Hitchin-Thorpe type inequality $-2\chi(M) \geq 3|\tau(M)|$ is obtained in Law [20], Matsushita-Law [24]. Proposition 4.5 is a corollary of their results. So we are led to the following question: *Which compact neutral Einstein four-manifolds do attain the equality in the Hitchin-Thorpe type inequality?*

7.4 Examples of self-dual neutral metrics

All of the examples in §6 are not only self-dual but also conformally-flat. Several examples of non-compact four-manifolds with non-conformally-flat, self-dual neutral metrics are known (see Kamada-Machida [18]). For instance, the indefinite complex hyperbolic space $H_1^2(\mathbf{C})$ and the cotangent bundle T^*H^2 of H^2 have non-conformally-flat, self-dual neutral metrics, like the Fubini-Study metric on the complex projective plane $P^2(\mathbf{C})$ and the Eguchi-Hanson metric on the cotangent bundle T^*S^2 of S^2 in the Riemannian case (cf. Eguchi-Hanson [11]).

For compact examples, the author [17] has recently constructed non-flat neutral hyperkähler structures on some particular types of complex tori and primary Kodaira surfaces. Clearly, these non-flat neutral hyperkähler surfaces give examples of compact four-manifolds with non-conformally-flat, self-dual neutral metrics. Note that these examples satisfy the equality $-2\chi = 3|\tau|(= 0)$.

7.5 Neutral hyperhermitian structures

In the Riemannian case, it is known that any compact hyperhermitian surface is locally conformally hyperkähler, and moreover that it is globally conformal to one of the following: a complex torus with the flat metric, a $K3$ surface with the Ricci-flat Calabi-Yau metric, a Hopf surface of certain type with a conformally-flat metric (see Boyer [6]).

In the neutral case, we obtain several examples of neutral hyperhermitian surfaces in §6. By definition, these examples are also locally conformal to neutral hyperkähler surfaces. So we have questions: *Is it true that any compact neutral hyperhermitian surface is locally conformal to a neutral hyperkähler surface*? Furthermore, *which compact complex surfaces do admit neutral hyperhermitian structures*? We shall discuss these problems elsewhere.

Final notes

After this article was submitted, the author knew J. Petean's recent work (*Indefinite Kähler-Einstein metrics on compact complex surfaces*, Commun. Math. Phys. **189** (1997), 227–235). By making use of his results, we can refine some of our statements in §3. In particular, we should note that no compact complex surfaces admit non-flat, anti-self-dual Einstein neutral Kähler metrics.

E. García-Río pointed out that the total space of the tangent bundle of any closed Riemann surface admits a non-conformally-flat, scalar-flat, neutral Kähler metric. For details, see J. Cendan-Verdes, E. García-Río and

M.E. Vázquez-Abal, *On the semi-Riemannian structures of the tangent bundle of a two-point homogeneous space*, Riv. Mat. Univ. Parma (5) **3** (1994), 253–270.

Acknowledgments

The author would like to thank Professor Yasuo Matsushita for many helpful suggestions and encouragement. The author also would like to thank Professor Eduardo García-Río for valuable comments.

References

1. M.A. Akivis and V.V. Goldberg, Conformal Differential Geometry and its Generalizations, Wiley, New York, Chichester, Brisbane, Toronto, Singapore, 1996.
2. A. Avez, *Formule de Gauss-Bonnet-Chern en métrique de signature quelconque*, De Revista de la Unión Mat. Argentina **21** (1963), 191–197.
3. W. Barth, C. Peters and A. Van de Ven, Compact complex surfaces, Springer-Verlag, Berlin, Heidelberg, New York, Tokyo, 1984.
4. A.L. Besse, Einstein manifolds, Springer-Verlag, Berlin, Heidelberg, New York, 1987.
5. A. Bonome, R. Castro, E. García-Río, L. Hervella and Y. Matsushita, *The Kähler-Einstein metrics on a $K3$ surface cannot be almost Kähler with respect to an opposite almost complex structure*, Kodai Math. J. **18** (1995), 506–514.
6. C.P. Boyer, *A note on hyperhermitian four-manifolds*, Proc. Amer. Math. Soc. **102** (1988), 157–164.
7. S.S. Chern, *Pseudo-Riemannian geometry and the Gauss-Bonnet formula*, An. da Acad. Brasileira de Ciências, **35** (1963), 17–26.
8. M. Dajczer and K. Nomizu *On sectional curvature of indefinite metrics. II*, Math. Ann. **247** (1980), 279–282.
9. A. Derdziński, *Self-dual Kähler manifolds and Einstein manifolds of dimension* 4, Compositio Math. **49** (1983), 405–433.
10. T. Draghici, *New examples of compact* 4*-manifolds which do not admit symplectic structures*, preprint (1995).
11. T. Eguchi and A.J. Hanson, *Self-dual solutions to Euclidean gravity*, Ann. of Phys. **120** (1979), 82–106.
12. M. Fernández, M.J. Gotay and A. Gray, *Compact parallelizable four-dimensional symplectic and complex manifolds*, Proc. Amer. Math. Soc. **103** (1988), 1209–1212.

13. E. García-Río, Y. Matsushita and R. Vázquez-Lorenzo, *Paraquaternionic Kähler manifolds*, preprint (1998).
14. H. Geiges, *Symplectic couples on 4-manifolds*, Duke Math. J. **85** (1996), 701–711.
15. H. Geiges and J. Gonzalo, *Contact geometry and complex surfaces*, Invent. Math. **121** (1995), 147–209.
16. M. Itoh, *Self-duality of Kähler surfaces*, Compositio Math. **51** (1984), 265–273.
17. H. Kamada, *Examples of neutral hyperkähler structures on compact four-dimensional manifolds*, preprint (1997).
18. H. Kamada and Y. Machida, *Self-duality of metrics of type* $(2, 2)$ *on four-dimensional manifolds*, Tôhoku Math. J. **49** (1997), 259–275.
19. R.S. Kulkarni, *Conformal structures and Möbius structures*, in Conformal geometry (R. S. Kulkarni and U. Pinkall, eds.), Aspects of Math. **E12**, Vieweg, Braunschweig, 1988, pp. 1–39.
20. P. Law, *Neutral Einstein metrics in four dimensions,* J. Math. Phys. **32** (1991), 3039–3042.
21. P. Law, *Neutral structures on even-dimensional manifolds*, Rocky Mountain J. Math. **23** (1993), 979–998.
22. Y. Matsushita, *Fields of* 2*-planes and two kinds of almost complex structures on compact* 4*-dimensional manifolds*, Math. Z. **207** (1991), 281–291.
23. Y. Matsushita, *Thorpe-Hitchin inequality for compact Einstein* 4*-manifolds of metric signature* $(++--)$ *and the generalized Hirzebruch index formula*, J. Math. Phys. **24** (1983), 36–40.
24. Y. Matsushita and P. Law, *Hitchin-Thorpe type inequalities for pseudo-Riemannian* 4*-manifolds of metric signature* $(++--)$, preprint (1997).
25. H. Pedersen and A. Swann, *Riemannian submersions, four-manifolds and Einstein-Weyl geometry,* Proc. London Math. Soc. **66** (1993), 381–399.
26. N. Steenrod, Topology of fiber bundles, Princeton Univ. Press, 1958.
27. I. Vaisman, *Generalized Hopf manifolds*, Geom. Dedicata **13** (1982), 231–255.
28. J.A. Wolf, Spaces of constant curvature, Publish or Perish, Wilmington, Delaware, 1984.

Marked length spectrum on the finite set of elements determines the irreducible representation in the isometry group of rank one symmetric space of noncompact type

Inkang Kim
Department of Mathematics
Korea Advanced Institute of Science and Technology
373-1 Kusong-dong Yusong-ku
Taejon 305-701, Korea
E-mail:inkang@mathx.kaist. ac.kr

In this paper we show that the space of irreducible representations from a finitely presented group into the group of isometries of rank one symmetric space of non-compact type (other than Cayley hyperbolic plane), embeds into R^n for some n, where the coordinates are the translation lengths of isometries in the representation. The ingredients of the proof consist of two facts that the representation is determined by its marked length spectrum and that the nested sequence of algebraic subvarieties is stabilized at a finite step by the Noetherian property of the polynomial ring.

1 Introduction

It is known that the lengths of $9g-9$ closed geodesics of the Riemann surface S of genus g, determine the hyperbolic structure of the surface in the Teichmuller space [FLP]. So there is a smooth map from the Teichmuller space of the genus-g surface into R^{9g-9} where a point ρ corresponds to the point $(l_\rho(g_1), \cdots, l_\rho(g_{9g-9}))$. Here $l_\rho(g)$ is the length in ρ-metric of the closed geodesic in the homotopy class of $g \in \pi_1(S)$.

It was not known that such a result holds for higher dimensional cases. In this paper we answer the question affirmatively for all dimensions of rank one symmetric spaces of noncompact type. The theorem is much more general than the question, namely such a parameterization is true even for the space $\Re^0$ of non-elementary irreducible representations.

The translation length of an isometry is a minimum distance it moves. The marked length spectrum of a representation is a function associating the translation length of the isometry $\rho(g)$, to each g in the group. See section 3 for definitions.

[a]1991 *Mathematics Subject Classification.* 51M10,57S.
[b]*Key words and phrases.* marked length spectrum, Noetherian ring, hyperbolic metric.

The idea of the proof is very simple. By [Kim1], it is known that an irreducible representation into the group of isometries of hyperbolic n-space is determined up to conjugacy by its marked length spectrum. So the task is to choose a finite set of elements of the group such that the marked length spectrum on this set determines the representation up to conjugacy. This can be done using the Noetherian property of the polynomial ring. More precisely we find some polynomial equations defining the subvariety of $\Re \times \Re$, which consists of pairs of representations having the same marked length spectrum on the finite set of elements in the group, using Lorentzian model of hyperbolic space. As we increase the elements of the group, subvarieties defined by these elements are decreasing. But this should be stable after a finite step by the Noetherian property of the polynomial ring. We can state the main theorem in the algebraic point of view as follows.

Theorem *Let $R^0(G)$ be the space of irreducible representations from a finitely presented group G into $Iso(X)$ where X is a real, complex or quaternionic hyperbolic space. Then there is a finite set of elements, $g_1, \cdots, g_N$ such that if $l(\rho(g_i)) = l(\phi(g_i))$ for $i = 1, \cdots, N$, then they are conjugate, and so $\chi_\rho = \chi_\phi$ for any $\rho, \phi \in R^0(G)$ where χ_ϕ is the character of ϕ when $Iso(X)$ is identified with some subgroup in $GL_n(F)$ where $F = R$ or C.*

2 The space of representations: Algebraic variety

Let $G = \langle g_i, i = 1, \cdots, n; r_1 = \cdots = r_k = 1 \rangle$. The space of representations $R(G)$ from G to $GL_m(F)$ is a set of homomorphisms from G into $GL_m(F)$. It is a set of points $(\rho(g_1), \cdots, \rho(g_n))$ in $GL_m(F)^n \subset F^{m^2 \times n}$ which satisfy $\rho(r_i) = 1$. Then it is easy to see that $R(G)$ is an algebraic variety.

In this paper we are particularly interested when representations are in $Iso(H^n_F)$. Lie group $Iso(H^n_F)$ acts on $R(G)$ by conjugation. A representation is irreducible if it does not leave invariant any totally geodesic subspace except H^n_F. Let $R^0(G)$ be the set of irreducible representations. Then the action of $Iso(H^n_F)$ is free on $R^0(G)$ since $\alpha\rho\alpha^{-1} = \rho$ for some $\alpha \neq 1 \in Iso(H^n_F)$ only if ρ is reducible and α is an identity on the totally geodesic subspace which ρ leaves invariant. Furthermore $R^0(G)$ is open since a small perturbation of an irreducible representation remains as an irreducible representation. So $\Re^0(G) = R^0(G)/Iso(H^n_F)$ is a smooth open set in $\Re(G) = R(G)/Iso(H^n_F)$. See [Gun], [Fal] for references. In this paper we want to find a smooth parametrization of $\Re(G)^0$ by the translation lengths of the finite set of isometries.

3 Isometries of hyperbolic space and the marked length spectrum

The translation length of an isometry α acting on a hyperbolic space H equipped with a hyperbolic metric d, is defined as:

$$l(\alpha) = \inf_{x \in H}\{d(x, \alpha(x))\}.$$

There are three kinds of isometries in hyperbolic space. When $l(\alpha) = 0$ and it is realized in the space, α is called *elliptic*. When $l(\alpha) = 0$ but it is not realized in the space, α is called *parabolic*. Finally when $l(\alpha) > 0$, α is called *hyperbolic*.

When we identify $Iso(H^n_F)$ with a component of $O_F(n, 1)$ when $F = C$ or R, the trace of a parabolic element is $n + 1$ and the trace of an elliptic or a hyperbolic element is $2\sum \cos\theta_i \pm \epsilon + 2\cosh r$, $\epsilon = 0, 1$ for real hyperbolic, $\sum e^{i\theta_j} + 2\cosh r$ for complex hyperbolic, where r is a translation length of the isometry.

Definition 1 *Two representations ρ, $\phi : G \to Iso(H)$ have the same marked length spectrum iff $l(\rho(g)) = l(\phi(g))$ for all $g \in G$.*

A representation is non-elementary if the image under the representation is non-elementary, i.e, its limit set is infinite. Here is the theorem that we need in this paper.

Theorem 1 *Let ρ, ϕ be non-elementary, irreducible representations from a group G into $Iso(X)$ where X is a rank one symmetric space of non-compact type. If they have the same marked length spectrum, then they are conjugate.*

The proof is quite long and needs understanding of subriemannianic nature of the ideal boundary of rank one symmetric spaces. Since it is a separate issue, we just refer the readers to [Kim1].

4 Ring of symmetric functions

The subring of symmetric functions in polynomial ring is finitely generated. A well-known basis of the ring is the set of elementary symmetric functions:

$$s_1 = \sum_{i=1}^{n} x_i;$$

$$s_2 = \sum_{1 \le i < j \le n} x_i x_j;$$

$$\cdots$$

$$s_n = x_1 x_2 \cdots x_n.$$

To introduce another well-known basis, we need the following concepts. A partition λ of n is any finite sequence of non-negative integers in decreasing order: $\lambda = (\lambda_1, \cdots, \lambda_k)$ such that $\sum \lambda_i = n$. Let

$$P_k(x_1, \cdots, x_n) = \sum_{i=1}^{n} x_i^k.$$

Define $P_\lambda = \Pi P_{\lambda_i}$ for a partition $\lambda = (\lambda_1, \cdots, \lambda_k)$. It is known that the set of symmetric functions of degree n is generated by P_λ as a vector space, where λ is a partition of n. For example when $n = 3$, a basis of degree-3 symmetric functions is P_1^3, P_2P_1, P_3. So the ring of symmetric functions of degree up to n is generated by $P_1, \cdots, P_n$.

The following lemma is elementary, yet important to define the algebraic subvarieties in later discussion.

Lemma 1 *For any positive integer n, $P_{n+1}(x_1, \cdots, x_n)$ can be expressed as some polynomial in $P_i(x_1, \cdots, x_n)$ for $i = 1, \cdots, n$.*

Proof: Consider the ring of symmetric functions in variables $x_1, \cdots, x_n$. Then the elementary symmetric function s_{n+1}, is equal to zero since there are only n variables. Yet s_{n+1} can be expressed as the linear combination of P_λ where λ are partitions of $n+1$. This implies that P_{n+1} can be expressed as a polynomial in P_k where $k = 1, \cdots, n$. ■

5 Parametrization of the space of irreducible representations by a finite set of translation lengths

Lemma 2 *Let A be a hyperbolic isometry in $Iso(H_F^n)$ where $F = R$ or C. Then there is a polynomial relation among trA^k and $\cosh r$ where r is the translation length of a hyperbolic isometry.*

Proof: If we identify $Iso(H_F^n)$ with a component of $O_F(n,1)$ in Lorentzian model, conjugating the isometry, we can put A into the form:

$$\begin{bmatrix} \cos\theta_1 & \sin\theta_1 & 0 & 0 & 0 & 0 \\ -\sin\theta_1 & \cos\theta_1 & \cdots & 0 & 0 & 0 \\ 0 & 0 & 0 & \pm 1 & 0 & 0 \\ 0 & 0 & 0 & \cdots & 0 & 0 \\ 0 & 0 & 0 & 0 & \cosh r & \sinh r \\ 0 & 0 & 0 & 0 & \sinh r & \cosh r \end{bmatrix}$$

in real hyperbolic case. In complex hyperbolic case, A can be conjugated into the form:

$$\begin{bmatrix} e^{i\theta_1} & 0 & 0 & 0 & 0 \\ 0 & \cdots & 0 & 0 & 0 \\ 0 & 0 & 0 & \cosh r & \sinh r \\ 0 & 0 & 0 & \sinh r & \cosh r \end{bmatrix}$$

Then

$$trA^k = \sum_i \cos k\theta_i + 2\cosh kr$$

for real hyperbolic, and

$$trA^k = \sum_i e^{ki\theta_i} + 2\cosh kr$$

for complex hyperbolic. Since $\cos k\theta$ can be expressed in terms of $\cos^m \theta$ for $m \leq k$, we get

$$\sum_i \cos^k \theta_i = P_k(trA^m, \cosh mr)$$

for some polynomial P_k and $m = 1, \cdots, k$ for real hyperbolic space. Then by lemma 1, there is a polynomial P such that $P(trA^m, \cosh mr) = 0,\ 1 \leq m \leq [\frac{n-1}{2}] + 1$. For complex hyperbolic case, lemma 1 can be used directly. ■

Now we are ready to prove the main theorem for real and complex case.

Theorem 2 *Let $\Re^0(G)$ be the space of irreducible representations, up to conjugacy, from a finitely presented group G into $Iso(X)$ where $X = H^n_R$ or H^n_C. Then there is a smooth embedding f from $\Re^0(G)$ into R^k and $g_1, \cdots, g_k \in G$ for some k, such that $f(\phi) = (l(\phi(g_1)), \cdots, l(\phi(g_k)))$ for $\phi \in \Re^0(G)$.*

Proof: From section 2, we know $R(G) \times R(G)$ is an algebraic variety. Then for any given finite set $S_k = \{g_1, \cdots, g_k\} \subset G$, the set V_k consisting of pairs of representations which have the same marked length spectrum on S_k, is an algebraic subvariety by lemma 2 since if ρ and ϕ have the same marked length spectrum on S_k, there are polynomials P_i and P_i' such that $P_i(tr\rho(g_i^m), \cosh mr_i) = 0 = P_i'(tr\phi(g_i^m), \cosh mr_i)$ for $1 \leq i \leq k$. As we increase the sets $S_k \subset S_{k+1}$ to exhaust G, the corresponding algebraic subvarieties V_k are decreasing but by the Noetherian property of the polynomial ring, $V_N = V_\infty$ for some positive integer N. Let $\mathcal{V}_k^0 = \{V_k \cap (R^0(G) \times R^0(G))\}/(Iso(H^n_F) \times Iso(H^n_F))$. By Theorem 1 saying that an irreducible representation is determined up to conjugacy by its marked length spectrum, $\mathcal{V}_\infty^0$ is the subset of a diagonal of $\Re^0(G) \times \Re^0(G)$. So we conclude that there is a finite set of elements $g_1, \cdots, g_N \in G$ such that an irreducible representation is determined up to conjugacy by its marked length spectrum on this set.

Since the function $l_g : \Re^0(G) \to R$, defined by $l_g(\phi) = l(\phi(g))$ is a smooth function, the map f is smooth. ■

It is worthwhile to mention the following corollary, in dimension 2 and 3, which can be viewed as a strong generalization even for hyperbolic surfaces. Note that even in 2-dimensional case, the surface might not be geometrically finite. But our main theorem guarantees that the finite set of lengths of closed geodesics determine the hyperbolic structure on the surface.

Corollary 1 *There is an embedding which is smooth except at the reducible representations, from the space of non-elementary representations from a finitely presented group into the group of isometries of real hyperbolic 2 or 3 space, to* R^k *where the mapping is given by the translation lengths of some* k *elements in the group.*

Proof: It follows from the fact that in dimension 2 or 3, we do not need irreducibility of a representation to conclude that the representation is determined by its marked length spectrum. [Kim] or [Kim1]. ∎

In 3-dimensional topology, it is important to determine when two homotopy equivalent 3-manifolds are homeomorphic. Corollary 1 in turn implies the following corollary.

Corollary 2 *Let* M, N *be two hyperbolic 3-manifolds which are homotopy equivalent.* M *and* N *are isometric, so homeomorphic, if the lengths of closed geodesics of some finite set of conjugacy classes in* $\pi_1(M) = \pi_1(N)$ *are the same.*

6 Quaternionic and Cayley hyperbolic cases

Since Quaternions and Cayley numbers are not commutative, most of the properties which hold in real or complex field break down. But we want to prove that Theorem 2 holds even in these two division rings.

6.1 Quaternionic case

There is a natural way to identify quaternions with a pair of complex numbers. Since $x_0 + x_1 i + x_2 j + x_3 k = x_0 + x_1 i + (x_2 + x_3 i)j$, $x_0 + x_1 i + x_2 j + x_3 k$ can be identified to $(x_0 + x_1 i, x_2 + x_3 i)$. So n-dimensional quaternions H^n can be identified with C^{2n}. By this identification there is a group monomorphism sending a matrix $A + Bj$ in $GL_n(H)$ acting on $(x_1 + y_1 j, \cdots, x_n + y_n j)$ on the right, to the matrix in $GL_{2n}(C)$

$$\begin{bmatrix} A & B \\ -\bar{B} & \bar{A} \end{bmatrix}$$

which acts on $(x_1, \cdots, x_n; y_1, \cdots, y_n)$ on the right. Then $Sp(n,1)$ can be identified with the subset of $U(n,1;n,1)$ which preserve $(n,1;n,1)$ hermitian form and the symplectic form, i.e. the form

$$\langle (x_1^1, \cdots, x_1^{n+1}; y_1^1, \cdots, y_1^{n+1}), (x_2^1, \cdots, x_2^{n+1}; y_2^1, \cdots, y_2^{n+1}) \rangle$$
$$= (\sum_{l=1}^{n} y_1^l x_2^l - x_1^l y_2^l) - (y_1^{n+1} x_2^{n+1} - x_1^{n+1} y_2^{n+1}).$$

Note the trace of $A+Bj$ is not invariant under conjugation, but $Re(tr(A+Bj))$ is invariant under conjugation since the trace of

$$\begin{bmatrix} A & B \\ -\bar{B} & \bar{A} \end{bmatrix}$$

is $2Re(trA) = 2Re(tr(A+Bj))$.

Theorem 3 *Let $\Re^0(G)$ be the space of irreducible representations, up to conjugacy, from a finitely presented group G into $Iso(H_H^n)$. Then there is a smooth embedding f from $\Re^0(G)$ into R^k and $g_1, \cdots, g_k \in G$ for some k, such that $f(\phi) = (l(\phi(g_1)), \cdots, l(\phi(g_k)))$ for $\phi \in \Re^0(G)$.*

Proof: A hyperbolic isometry in $Iso(H_H^n)$ which fixes 0 and ∞ is of the form:

$$\begin{bmatrix} M & 0 & 0 \\ 0 & \cosh r & \sinh r \\ 0 & \sinh r & \cosh r \end{bmatrix}$$

where $C + Dj = M \in Sp(n-1)$, which can be identified with

$$N = \begin{bmatrix} \begin{pmatrix} C & 0 & 0 \\ 0 & \cosh r & \sinh r \\ 0 & \sinh r & \cosh r \end{pmatrix} & \begin{pmatrix} D & 0 \\ 0 & 0 \end{pmatrix} \\ \begin{pmatrix} -\bar{D} & 0 \\ 0 & 0 \end{pmatrix} & \begin{pmatrix} \bar{C} & 0 & 0 \\ 0 & \cosh r & \sinh r \\ 0 & \sinh r & \cosh r \end{pmatrix} \end{bmatrix}$$

Since

$$\begin{bmatrix} C & D \\ -\bar{D} & \bar{C} \end{bmatrix}$$

is in $U(2n-2)$, there is a matrix

$$J = \begin{bmatrix} E & F \\ G & H \end{bmatrix}$$

so that

$$J \cdot \begin{bmatrix} C & D \\ -\bar{D} & \bar{C} \end{bmatrix} \cdot J^{-1}$$

is equal to a diagonal matrix $(e^{i\theta_1}, \cdots, e^{i\theta_{2n-2}})$. Then

$$\begin{bmatrix} \begin{pmatrix} E & 0 \\ 0 & I \end{pmatrix} & \begin{pmatrix} F & 0 \\ 0 & 0 \end{pmatrix} \\ \begin{pmatrix} G & 0 \\ 0 & 0 \end{pmatrix} & \begin{pmatrix} H & 0 \\ 0 & I \end{pmatrix} \end{bmatrix}$$

conjugates N into the form:

$$N' = \begin{bmatrix} \begin{pmatrix} A & 0 & 0 \\ 0 & \cosh r & \sinh r \\ 0 & \sinh r & \cosh r \end{pmatrix} & \begin{pmatrix} 0 & 0 \\ 0 & 0 \end{pmatrix} \\ \begin{pmatrix} 0 & 0 \\ 0 & 0 \end{pmatrix} & \begin{pmatrix} B & 0 & 0 \\ 0 & \cosh r & \sinh r \\ 0 & \sinh r & \cosh r \end{pmatrix} \end{bmatrix}$$

where A, B are diagonal matrices. Then $trN' = \sum \cos\theta_i + 4\cosh r$. Now we can conclude the same conclusion as in lemma 2, saying that there is a polynomial relation among trN and $\cosh r$. Viewing $\Re^0(G)$ as a subset of C^N for some N, we can proceed the proof as in Theorem 2. ■

6.2 Cayley hyperbolic case

In Cayley hyperbolic case, there is no good identification of $Iso(H^2_O)$ in $O_F(n,1)$ even though the adjoint representation Ad gives a linear embedding into $GL_{52}(R)$. But the calculation is far complicated. We will deal with this in a separate paper.

It is a well-known fact that the irreducible representation ϕ, from some group G into $GL_n(F)$, is determined up to conjugacy by its character χ_ρ, where $\chi_\rho(g) = tr(\rho(g))$. In general, the translation length of an isometry does not determine the trace of the isometry, but in the group of isometries of hyperbolic space, knowing large enough number of translation lengths of the group is sufficient to determine its character by our theorem. So we get the following.

Corollary 3 *Let $R^0(G)$ be the space of irreducible representations from a finitely presented group G into $Iso(X)$ where X is a rank one symmetric space of non-compact type other than Cayley hyperbolic plane. Identify $Iso(X)$ with some linear subgroup in $GL_n(F)$ where F is real or complex. Then there is a finite set of elements, $g_1, \cdots, g_N$ such that if $l(\rho(g_i)) = l(\phi(g_i))$ for $i = 1, \cdots, N$, then $\chi_\rho = \chi_\phi$ for any $\rho, \phi \in R^0(G)$.*

Proof: By our main theorems, there are such elements which determine the representations up to conjugacy. But the character is the conjugacy invariant for real or complex field. ■

In [Kim1], it is also shown that

Theorem 1′ *Let X be a rank one symmetric space of non-compact type. Let $\rho, \phi : G \to Iso(X)$ be two irreducible representations having the proportional marked length spectrum, i.e. $l(\rho(g)) = cl(\phi(g))$ for some constant c. Then $c = 1$ and so they are conjugate.*

This theorem is plausible in the hyperbolic surface, since if we pinch a closed geodesic, the closed geodesics crossing that closed geodesic transversely tend to get longer. This theorem asserts that it is also true for higher dimensional case. ¿From this theorem and the same proof as in theorem 2, we get

Corollary 3′ *Let $R^0(G)$ be the space of irreducible representations from a finitely presented group G into $Iso(X)$ where X is a rank one symmetric space of non-compact type other than Cayley hyperbolic plane. Identify $Iso(X)$ with some linear subgroup in $GL_n(F)$ where F is real or complex. Then there is a finite set of elements, $g_1, \cdots, g_N$ such that if $l(\rho(g_i)) = cl(\phi(g_i))$ for $i = 1, \cdots, N$ and a fixed constant c, then $\chi_\rho = \chi_\phi$ for any $\rho, \phi \in R^0(G)$.*

In the spectral theoretic point of view, it amounts to say that a finite set of the marked largest eigenvalues of the matrices in the image of a representation, determines the irreducible representation up to conjugacy.

References

[Fal] G. Faltings, Real projective structures on Riemann surfaces, Comp. Math. 48 (1983), 223-269.

[FLP] A. Fathi, F. Laudenbach,and V. Poenaru, Travaux de Thurston sur les surfaces, volume 66-67, Asterisque, 1979.

[Gun] R. Gunning, Lectures on Vector Bundles over Riemann Surfaces, Princeton University Press, 1967.

[Kim] I. Kim, Geometric Structures on Manifolds and the Marked length spectrum, Thesis, U.C. Berkeley, 1996.

[Kim1] I. Kim, Rigidity of rank one symmetric spaces and their product, preprint.

Compactification and Construction of asymptotically Euclidean scalar-flat kähler surfaces

Jongsu Kim
Department of Mathematics
Sogang University
Seoul, 121-742, Korea
jskim@ccs.sogang.ac.kr

1 Introduction

In recent years some progress has been made in the study of Kähler metrics with zero scalar curvature on compact complex surfaces, for instance [7, 8, 14]. In this paper we are interested in noncompact cases, especially those of asymptotically Euclidean scalar-flat Kähler metrics. Some examples were found by C. LeBrun [12] on $\mathbf{C}^2$ blown up at a finite number of distinct points. He also proved that any complex surface admitting such a metric is biholomorphic to $\mathbf{C}^2$ blown up at a finite number of points (which may overlap). P. Li and S.T. Yau's result [15] also implies this.

For the classification of asymptotically Euclidean scalar-flat Kähler metrics there is a conjecture [12] that a global Torelli type theorem holds, which states as follows. Let M be any blow up of $\mathbf{C}^2$. Then each choice of Kähler class on M is represented by a unique asymptotically Euclidean scalar-flat Kähler form.

It is hence necessary to have general constructions together with uniqueness theorems. The main purpose of this note is to explain a construction of which the method possibly constructs all of such metrics modulo Kähler classes. For completeness sake, we first make a brief review of known examples and properties, together with some characterizations. In particular we will outline conformal compactification of such metrics.

Then we deal with the problem of existence. We explain a construction of asymptotically Euclidean scalar-flat Kähler metrics in a simple but informative case. The body of arguments in construction contains most of key ideas necessary for general case.

The structure of this paper is organized as follows.
In section 2, we define the (locally) asymptotically Euclidean metrics and exhibit some examples of (locally) asymptotically Euclidean scalar-flat Kähler surfaces.
In section 3, we show that by some elliptic regularity argument an asymptoti-

cally Euclidean scalar-flat Kähler metric on a complex surface can be smoothly conformally compactified with the infinity point.
In section 4, we describe some twistor theory associated to these anti-self-dual metrics, i.e. metrics with the positive part of Weyl curvature being zero. Then we discuss applications of the compactification property of section 3 as above.
In section 5, we explain a construction of asymptotically Euclidean scalar-flat Kähler metrics. We modify the construction in [7] of compact scalar-flat Kähler metrics. For simplicity we restrict to the case of $\mathbf{C}^2$ blown up at one point.

2 Preliminaries

We define a Riemannian metric g on a noncompact manifold M to be aymptotically locally Euclidean if M has just one end and if there exists a compact subset K of M such that a finite covering $\tilde{V}$ of $M-K$ is diffeomorphic to the complement of the unit ball in $\mathbf{R}^4$; the pulled-back Riemannian metric g_{ij} on $\tilde{V}$ is required to approximate asymptotically the Euclidean metric δ_{ij} on $\mathbf{R}^4$, so that in the natural coordinates x_i, one has

$$g_{ij} = \delta_{ij} + a_{ij} \tag{1}$$

with $\partial^p a_{ij} = O(r^{-2-p})$, $p = 0, 1, 2$ where $r^2 = \sum x_i^2$ and ∂ denotes differentiation with respect to the coordinates x_i.
We call g *asymptotically Euclidean* if we can take $M-K$ itself as the covering $\tilde{V}$ in the above.

Now we describe some known examples of asymptotically Euclidean scalar flat Kähler surfaces. Let $r \in (1, \infty)$ and $\sigma_1, \sigma_2, \sigma_3$ be a left invariant coframe for S^3. Then the following metric on $(1, \infty) \times S^3$

$$g = \frac{dr^2}{1-\frac{1}{r^2}} + r^2(\sigma_1^2 + \sigma_2^2 + (1 - \frac{1}{r^2})\sigma_3^2). \tag{2}$$

can be completed with S^2 corresponding to $r = 1$. The normal bundle of this S^2 has Euler number -1. So the completed manifold is diffeomorphic to $\mathbf{C}^2$ blown up at one point. Indeed g is scalar-flat Kähler on $\mathbf{C}^2$ blown up at one point and called Burns meric. It is clear that g is aymptotically Euclidean.

Next, there are metrics found by LeBrun [13], which can be viewed as superpositions of (2) metrics, on a blow-up of $\mathbf{C}^2$ at a finite number of distinct points along a complex line in $\mathbf{C}^2$; it is of the form

$$g = f^2(Vh + V^{-1}\omega^2) \tag{3}$$

where h is a hyperbolic metric, $\Delta_h V = 0$ and $d\omega = \star_h dV$ with $\star_h$ being the Hodge-star operator with respect to h. In [12], LeBrun perturbed these metrics to ones on blow-ups of $\mathbf{C}^2$ at a finite number of any distinct points.
These so far are essentially all the known asymptotically Euclidean examples.

There are some more asymptotically locally Euclidean examples; there are ricci-flat multi-Eguchi Hanson metrics, see [5]. Generalizing these, Kronheimer [10] classified all asymptotically locally Euclidean ricci-flat Kähler surfaces.
There are locally asmptotically Euclidean scalar-flat Kähler metrics on the total spaces of some complex line bundles over S^2, generalizing (2) metric.
The classfication of asymptotically locally Euclidean scalar-flat Kähler surfaces is more general problem beyond the subject of this note.

3 Compactification

Now suppose that g is an asymptotically Euclidean scalar-flat Kähler metric on a noncompact complex surface M. We use the notation of (1). Let $h = \frac{1}{\phi^2} g$ where ϕ is positive, smooth and equal to r^2 outside some compact subset of M. A scalar-flat Kähler metric of real dimension 4 is anti-self-dual, see e.g. [14]. As anti-self-duality is a conformally invariant property, h is anti-self-dual. The following lemma is well known in relativity [17, p127], [9].

Lemma 3.1 *The anti-self-dual metric h satisfies the following system of partial differential equations, called the Bach equation.*

$$\delta^D d^D(ric - \frac{1}{6}sh) = 2\widetilde{W}ric. \tag{4}$$

Here ric, s and W are the ricci, scalar and Weyl curvature tensors respectively. D is the Levi-Civita connection of h and the differential operator d^D acts on symmetric 2-tensors ψ by

$$d^D\psi(x, y, z) = D_x\psi(y, z) - D_y\psi(x, z),$$

whereas δ^D is its dual operator. $\widetilde{W}ric$ is a product term of W and ric ; for a tensor of curvature type A (e.g. W and the Riemannian curvature tensor R), $\widetilde{A}$ is the linear map on symmetric 2-tensors such that $(\widetilde{A}\psi)(x, y) = \sum_{i=1}^n \psi(A(x, e_i)y, e_i)$ for an orthonormal basis $e_1, e_2, .., e_n$.

Then we have formulas [3]

$$\begin{aligned} \delta^D d^D(ric) &= 2D^*Dric - 2\widetilde{R}ric + 2ric \circ ric + Dds & (5) \\ \delta^D d^D(sh) &= (\Delta s)h & (6) \end{aligned}$$

where $(ric \circ ric)_{ij} = \sum ric_i^k ric_{kj}$ in local coordinates.

From (3.1) and (3.2) we get

$$(D^*D - \frac{6}{5}DD^*)(ric - \frac{1}{12}sh) = \widetilde{W}ric - ric \circ ric + \widetilde{R}ric. \quad (7)$$

Now substituting the variables by $y_i = x_i/r^2$, $i = 1, .., 4$, we can view the above equation as defined on an origin-deleted ball in $\mathbf{R}^4$ with y_i coordinates. One can then show from the asymptotic condition (1)

Lemma 3.2 *The metric h can be extended up to the origin as a $L^{2,\infty}$ weak solution of (7).*

We are going to apply some elliptic regularity argument to this elliptic system (7) as in [1]. Set $u = ric - \frac{1}{12}sh$. Now u is a weak L^∞ solution to the system (7) when it was viewed as a second order elliptic system in u. Consider also the ricci curvature equation in harmonic coordinates

$$h^{ij}\frac{\partial^2 h_{rs}}{\partial x_i \partial x_j} + Q_{rs}(h, \partial h) = ric_{rs} \quad (8)$$

where Q is a quadratic term in h and its first derivatives, of the general form $h^{-2}(\partial h)^2$,

Using elliptic regularity [16] to (7) and (8) one can show that h is in W_k^p for any $k, p > 2$, and so h is in C^∞. See [6] for details.

So we have:

Theorem 3.1 *Let g be an asymptotically Euclidean scalar-flat Kähler metric on a noncompact complex surface M. Then g can be smoothly conformally compactified.*

4 Application To Uniqueness

In this section we discuss some application of theorem 3.1, also referring to [11].
To prove this application one needs to bring in some complex 3-dimensional geometry, so called twistor theory. We make a brief sketch of the arguments.

The twistor space Z of $(M^4, h, orientation)$ is the S^2-bundle of pointwise almost complex structures on M which are compatible with both h and the orientation; Z admits a natural almost complex structure which makes it into a complex 3-manifold exactly when h is anti-self-dual [2].

Consider $t : Z \to M$ the bundle projection and then a hermitian structure J on M is exactly a section of t whose image we denote by D while another hermitian structure $-J$ is another section whose image is $\bar{D}$. Important point is that the pair (Z, D) or triple $(Z, D, \bar{D})$ completely determines the conformal class $[h]$ of the metric as well as the complex structure J.

Now we return to the compactified manifold $\bar{M} = M \cup \{\infty\}$ of section 3. We denote the conformally compactified metric on $\bar{M}$ by h as before. Also let the closures of D and $\bar{D}$ in $Z_{\bar{M}}$ be denoted by $cl(D)$ and $cl(\bar{D})$ respectively. Then we have

Proposition 4.1 *In the twistor space of $(\bar{M}, h)$, the divisor $cl(D)$ intersects $cl(\bar{D})$ along a complex projective line l.*

This l corresponds to the ∞ point of $\bar{M}$ and one can show $M \cup l$ is a compact smooth complex surface with the self-intersection number of l being $+1$. By surface classification $M \cup l$ is a complex projective plane with a finite number of points blown up.

Therefore we get

Theorem 4.1 (12,15) *If (M, g) is an asymptotically Euclidean scalar flat Kähler surface, then M is biholomorphic to $\mathbf{C}^2$ blown up at a finite number of points.*

Let M be $\mathbf{C}^2$ blown up at 1, 2 or 3 distinct points. Then the one point compactification $\bar{M}$ is oriented diffeomorphic to a connected sum of copies of 1, 2, or 3 complex projective planes with reversed orientation. The twistor space $Z_{\bar{M}}$ of $(\bar{M}, h)$ has the divisors $cl(D)$ and $cl(\bar{D})$. By Pontecorvo's study [18], the type of the conformal class $[h]$ is positive. Then by Poon's work [19], these cases are completely understood as well as the case when $\mathbf{C}^2$ is blown up at a finite number of points along a complex line. So we conclude

Theorem 4.2 *Any aymptotically Euclidean scalar-flat Kähler metric on* $\mathbf{C}^2$ *blown up at 1, 2 or 3 distinct points or on* $\mathbf{C}^2$ *blown up at a finite number of points along a complex line is unique up to isometry in each Kähler class. In particular, it is one of the metrics in (3).*

5 Construction

In this section we discuss construction of aymptotically Euclidean scalar-flat Kähler metrics. We modify the construction in [7] of compact scalar-flat Kähler metrics.

Our argument in noncompact case is not much different from above compact case. Here we deal only with one-point blowing up of $\mathbf{C}^2$ case. Then the resulting metric will be nothing but the Burns metric. But even in this case, the body of arguments in construction contains most of key ideas necessary for general case. Of course this construction is different from that of (2) or (3).

We let Z_1 be the twistor space of $\overline{\mathbf{CP}}_2$, the complex projective plane with reversed orientation, equipped with the Fubini-Study metric. This Z_1 is biholomorphic to the complete flag manifold $\mathbf{F}$ of $\mathbf{C}^3$. Z_1 has a degree 1 divisor D_1 with its conjugate $\bar{D}_1$ meeting along a real twistor line L_1. We let Z_2, D_2, $\bar{D}_2$, L_2 to be a copy of Z_1, D_1, $\bar{D}_1$, L_1 respectively. Also let $\tilde{Z}_1$ be the blow up of Z_1 along a real twistor line l_1 distinct from L_1, but in contrast $\tilde{Z}_2$ be the blow up of Z_2 along L_2 which we denote by l_2.

We denote by $\tilde{D}_1$ and $\tilde{D}_2$ the proper transform in $\tilde{Z}_1$ and $\tilde{Z}_2$, of D_1 and D_2 respectively. Similarly by $\tilde{\bar{D}}_1$ and $\tilde{\bar{D}}_2$ we mean the proper transform in $\tilde{Z}_1$ and $\tilde{Z}_2$, of $\bar{D}_1$ and $\bar{D}_2$ respectively. Then $\tilde{D}_1$ and $\tilde{\bar{D}}_1$ are biholomorphic to D_1 and $\bar{D}_1$ with one point blown up respectively while $\tilde{D}_2$ and $\tilde{\bar{D}}_2$ are biholomorphic to D_2 and $\bar{D}_2$ respectively.

As a twistor line, l_i for $i = 1, 2$ has the normal bundle $\nu_{l_i/Z_i} \cong \mathcal{O}(1) \oplus \mathcal{O}(1)$. The exceptional divisor Q_i of the blowing up l_i is then its projectivization which is a smooth quadric hypersurface with normal bundle $\nu_{Q_i/Z_i} \cong \mathcal{O}_{\mathbf{CP}_1 \times \mathbf{CP}_1}(1, -1)$, $i = 1, 2$.

We will glue $\tilde{Z}_1$ and $\tilde{Z}_2$ together with divisors $\tilde{D}_1$, $\tilde{\bar{D}}_1$ and $\tilde{D}_2$, $\tilde{\bar{D}}_2$. We identify Q_1 and Q_2 in such a way that the two $\mathbf{CP}_1$-factors, the base and the fiber, of each quadric are reversed. But in order to glue $\tilde{D}_1$, $\tilde{\bar{D}}_1$ and $\tilde{D}_2$, $\tilde{\bar{D}}_2$ simultaneously note the two pairs of rational curves to be identified: $C_1 := \tilde{D}_1 \cap Q_1$ and $\bar{C}_1 := \tilde{\bar{D}}_1 \cap Q_1$ are the bases with normal bundle $\mathcal{O}(-1)$ in $\tilde{D}_1$ and $\tilde{\bar{D}}_1$ respectively while $C_2 := \tilde{D}_2 \cap Q_2$ and $\bar{C}_2 := \tilde{\bar{D}}_2 \cap Q_2$ are fibers whose normal bundle $\mathcal{O}(1)$ in $\tilde{D}_2$ and $\bar{\tilde{D}}_2$. We can glue C_1, $\bar{C}_1$ with C_2, $\bar{C}_2$ respectively when we glue Q_i in the manner described above.

The end result is a complex 3-dimensional space $\mathcal{Z} = \tilde{Z}_1 \cup_Q \tilde{Z}_2$ with a singular hypersurface $\mathcal{D} = \tilde{D}_1 \cup_{\mathbf{CP}_1} \tilde{D}_2$ and $\bar{\mathcal{D}} = \bar{\tilde{D}}_1 \cup_{\mathbf{CP}_1} \bar{\tilde{D}}_2$ such that $\mathcal{Z}$ has an anti-holomorphic involution σ_0, that $\mathcal{D}$ and $\bar{\mathcal{D}}$ intersect along L_1 and that $\mathcal{Z}$, $\mathcal{D}$ and $\bar{\mathcal{D}}$ have only mild singularities. Further, $\mathcal{Z}$ satisfies the *d−semistable* condition $\nu_{Q/Z} \cong \mathcal{O}_{Q_1}(1,-1) \otimes \mathcal{O}_{Q_2}(-1,1) = \mathcal{O}_Q$. This *d − semistability* is a necessary condition for the complex space $\mathcal{Z}$ to admit "smoothing" such that the total space of the smoothing is smooth. $\mathcal{D}$ satisfies the *d − semistable* condition along $C_1 := C_2$ and $\bar{C}_1 := \bar{C}_2$; $\nu_{\mathbf{CP}_1/\mathcal{D}} = \mathcal{O}(1) \otimes \mathcal{O}(-1) = \mathcal{O}$ respectively, but along L_1, $\nu_{\mathbf{CP}_1/\mathcal{D}} = \mathcal{O}(1) \otimes \mathcal{O}(1) = \mathcal{O}(2)$ so the *d−semistable* condition does not hold along L_1. By our assumption on Z_1 and Z_2, $\mathcal{Z} = \tilde{\mathbf{F}} \cup_Q \tilde{\mathbf{F}}$ with divisors $\mathcal{D} = \widetilde{\mathbf{CP}}_2 \cup_{\mathbf{CP}_1} \widetilde{\mathbf{CP}}_2$ and $\bar{\mathcal{D}} = \overline{\widetilde{\mathbf{CP}}}_2 \cup_{\mathbf{CP}_1} \overline{\widetilde{\mathbf{CP}}}_2$.

Denote the divisor $\mathcal{D} + \bar{\mathcal{D}}$ by $\mathcal{D}\bar{\mathcal{D}}$. We now study the deformations of the pair $(\mathcal{Z}, \mathcal{D}\bar{\mathcal{D}})$ into a smooth twistor space Z with two degree 1 divisors D and $\bar{D}$ in it. As $\bar{\mathcal{D}}$ is the image of $\mathcal{D}$ under the real structure σ_0 of $\mathcal{Z}$, the pair $(\mathcal{Z}, \mathcal{D}\bar{\mathcal{D}})$ does have a real structure so that we can consider its real deformations.

We compute

$$H^2(\Theta_{\tilde{\mathbf{F}}} \otimes \mathcal{I}_{\tilde{D}_i \bar{\tilde{D}}_i}) = H^2(\Theta_{\tilde{\mathbf{F}}, \tilde{D}_i \bar{\tilde{D}}_i}) = H^2(\Theta_{\tilde{\mathbf{F}}}) = 0.$$

where $\mathcal{I}_{\tilde{D}_i \bar{\tilde{D}}_i}$ is the ideal sheaf of $\tilde{D}_i \bar{\tilde{D}}_i$, for $i = 1, 2$.

We denote the sheaf of derivations of X by τ_X^0 and the sheaf of relative derivations of X with respect to Y by $\tau_{X,Y}^0$.

We compute

Lemma 5.1 $H^2(\tau^0_{\mathcal{D}\bar{\mathcal{D}}}) = H^2(\tau^0_{\mathcal{Z}}) = H^2(\tau^0_{\mathcal{Z},\mathcal{D}\bar{\mathcal{D}}}) = H^2(\tau^0_{\mathcal{Z}|\mathcal{D}\bar{\mathcal{D}}}) = 0.$

Proof : Consider the "normalization" short exact sequence.

$$0 \to \tau^0_{\mathcal{D}\bar{\mathcal{D}}} \to q_*(\Theta_{\tilde{D}_1, C_1} \oplus \Theta_{\bar{\tilde{D}}_1, \bar{C}_1} \oplus \Theta_{\tilde{D}_2, C_2} \oplus \Theta_{\bar{\tilde{D}}_2, \bar{C}_2}) \to i_*(\Theta_{L_1} \oplus \Theta_{C_2} \oplus \Theta_{\bar{C}_2}) \to 0$$

where $q : \tilde{D}_1 \amalg \bar{\tilde{D}}_1 \amalg \tilde{D}_2 \amalg \bar{\tilde{D}}_2 \to \mathcal{D}\bar{\mathcal{D}}$ is the normalization map and $i : L_1 \amalg C_2 \amalg \bar{C}_2 \to \mathcal{D}\bar{\mathcal{D}}$ the inclusion. $\amalg$ means disjoint union. One can then prove $H^2(\tau^0_{\mathcal{D}\bar{\mathcal{D}}}) = 0$ from the long exact sequence of cohomology groups associated to above short exact sequence . One can prove vanishing of other cohomology groups by similar arguments. □

Due to Lemma 5.1 we can use the following diagram as a basic structure of arguments. For details see [7]. The middle row is the Ext sequence of deformations from [20]; Infinitesimal deformations of the map $f : \mathcal{D}\bar{\mathcal{D}} \hookrightarrow \mathcal{Z}$ are given by T^1_f with obstructions lying in T^2_f.

$$
\begin{array}{ccccccc}
0 & & 0 & & 0 & & \\
\uparrow & & \uparrow & & \uparrow & & \\
\Delta & \stackrel{d'\oplus d}{\longrightarrow} & H^0(\tau^1_{\mathcal{D}\bar{\mathcal{D}}}) \oplus H^0(\tau^1_{\mathcal{Z}}) & \stackrel{\hat{\psi}'-\hat{\psi}}{\longrightarrow} & H^0(\tau^1_{\mathcal{Z}|\mathcal{D}\bar{\mathcal{D}}}) & \to 0 & \\
\uparrow \beta & & \uparrow r\oplus\rho & & \uparrow \varrho & & \\
T^1_f & \stackrel{\phi'\oplus\phi}{\longrightarrow} & T^1_{\mathcal{D}\bar{\mathcal{D}}} \oplus T^1_{\mathcal{Z}} & \stackrel{\psi'-\psi}{\longrightarrow} & Ext^1(\Omega^1_{\mathcal{Z}|\mathcal{D}\bar{\mathcal{D}}}, \mathcal{O}_{\mathcal{D}\bar{\mathcal{D}}}) & \to T^2_f & \\
\uparrow \alpha & & \uparrow i\oplus j & & \uparrow \iota & & \\
H^1(\tau^0_{\mathcal{Z},\mathcal{D}\bar{\mathcal{D}}}) & \stackrel{\phi'_-\oplus\phi_-}{\longrightarrow} & H^1(\tau^0_{\mathcal{D}\bar{\mathcal{D}}}) \oplus H^1(\tau^0_{\mathcal{Z}}) & \stackrel{\psi'_- - \psi_-}{\longrightarrow} & H^1(\tau^0_{\mathcal{Z}|\mathcal{D}\bar{\mathcal{D}}}) & \to 0 & \\
\uparrow & & \uparrow & & \uparrow & & \\
0 & & 0 & & 0 & &
\end{array}
$$

This diagram is commutative with exact rows and columns. From the description of $\mathcal{D}$, $\bar{\mathcal{D}}$ and $\mathcal{Z}$, we see that $\tau^1_{\mathcal{D}\bar{\mathcal{D}}} \cong \mathcal{O}_{C_1} \oplus \mathcal{O}_{\bar{C}_1} \oplus \mathcal{O}_{L_1}(2)$ so that $H^0(\tau^1_{\mathcal{D}\bar{\mathcal{D}}}) \cong \mathbf{C} \oplus \mathbf{C} \oplus \mathbf{C}^3$. There is a canonical imbedding of $H^0(\tau^1_{\mathcal{Z}}) \cong \mathbf{C}$ into $H^0(\tau^1_{\mathcal{D}\bar{\mathcal{D}}})$ and of $H^0(\tau^1_{\mathcal{Z}|\mathcal{D}\bar{\mathcal{D}}})$ into $H^0(\tau^1_{\mathcal{D}\bar{\mathcal{D}}})$. Note that the map $\hat{\psi}'$ is the projection while $\hat{\psi}$ is the diagonal imbedding.

From the surjectivity of $\hat{\psi}' - \hat{\psi}$, we can see that $T^2_f = 0$. One can then compute the dimension of T^1_f to be $t^1_f = h^1(\tau^0_{\mathcal{Z},\mathcal{D}\bar{\mathcal{D}}}) + 4$.

Now observe that Δ is isomorphic to the kernel of $\hat{\psi}' - \hat{\psi}$ which consists of the elements (c, c, x, c) in $H^0(\tau^1_{\mathcal{D}\bar{\mathcal{D}}}) \oplus H^0(\tau^1_{\mathcal{Z}}) \cong H^0(\mathcal{O}_{C_1}) \oplus H^0(\mathcal{O}_{\bar{C}_1}) \oplus H^0(\mathcal{O}_{L_1}(2)) \oplus H^0(\tau^1_{\mathcal{Z}})$, so $d' \oplus d$ is a monomorphism.

By a similar proof to that of Lemma 5.1, we can prove

Lemma 5.2 *The map* $H^1(\tau^0_{\mathcal{Z},\mathcal{D}\bar{\mathcal{D}}}) \stackrel{\phi'_-}{\longrightarrow} H^1(\tau^0_{\mathcal{D}\bar{\mathcal{D}}})$ *is surjective.*

We also need the following

Lemma 5.3 *There exists a deformation* $\pi : \mathcal{M} \to \Delta \subset \mathbf{C}$ *of* $\mathcal{D}\bar{\mathcal{D}}$ *such that* $\pi^{-1}(t)$ *is biholomorphic to the blow up of* $D_1\bar{D}_1$ *at two points* $p \in D_1$ *and* $\sigma(p) \in \bar{D}_1$ *for nonzero* $t \in \Delta$.

Proof : We can explicitly construct such deformation. Let S be the blow up of $D_1\bar{D}_1$ at two points $p \in D_1$ and $\sigma(p) \in \bar{D}_1$ with corresponding exceptional curves denoted by l and $\bar{l}$. Let $\mathcal{M}$ be the blow up of $S \times \mathbf{C}$ along $l \times \{0\}$

and $\bar{l} \times \{0\}$. $\mathcal{M}$ is an effectively parametrized family with the central fiber biholomorphic to $\mathcal{D}\bar{\mathcal{D}}$. □

We are now ready for the following theorem.

Theorem 5.1 *We get a family Z_t of smooth twistor spaces with a pair of degree 1 divisors, D_t, $\bar{D}_t$ for small t such that D_t intersects $\bar{D}_t$ along a projective line and that D_t and $\bar{D}_t$ are biholomorphic to the blow-up of D_1 and $\bar{D}_1$ respectively.*

Proof : As ϕ'_- is onto, by diagram chasing we can find an element of T^1_f which yields the deformation of Lemma 5.3. By the compatibility of all maps with the real structure, we may assume that the element is real. It is easy to see from the above description that this element gives rise to smoothings of $\mathcal{Z}$ also. □

It is easy to observe that Z_t is the Poon's twistor space [19] and that it is conformal to an asymptotically Euclidean scalar-flat Kähler metric on $\mathbf{C}^2$ blown up at a point. This metric is the Burns métric.
Above construction arguments would provide a basic frame in constructing more asymptotically Euclidean scalar-flat Kähler metrics. Indeed, generalizing above arguments we may prove that there exist asymptotically Euclidean scalar-flat Kähler metrics on a blow-up of $\mathbf{C}^2$ at a finite number of any (not necessarily distinct) points.

Acknowledgments

The author wishes to acknowledge the financial support of the Korea Research Foundation made in the program year of 1997.

References

1. M. Anderson, *Degeneration of Metrics with Bounded Curvature and Applications to Critical Metrics of Riemannian Functionals,* Proc. Symp. Pure Math., A.M.S. **54**, 53–79 (1993)
2. Atiyah, M.F., Hitchin, N.J., Singer, I.M. *Self-duality in four dimensional riemannian geometry,* Proc. R. Soc. Lond., Ser. A **362**, 425–461 (1978)
3. Besse, A. *Einstein Manifolds, Springer, Berlin, Heidelberg, New York, 1987*
4. Donaldson, S.K., Friedman, R. *Connected sums of self-dual manifolds and deformations of singular spaces,* Nonlinearity **2**, (1989)
5. Gibbons, G.W., Hawking, S.W. *Gravitational Muti-Instantons,* Physics Letters **78B, No. 4**, 517 (1978)
6. J. Kim *Conformal Compactification of Asymptotically Euclidean Scalar Flat Kähler Surfaces,* J. Korean Math. Soc. **34, No. 4**, 1037–1048 (1997)
7. J. Kim and M. Pontecorvo *A New Method of Constructing Scalar-Flat Kähler Metrics,* Journal of Differential Geometry **42, No. 2**, 449–478 (1995)
8. J. Kim, C. LeBrun and M. Pontecorvo, *Scalar-Flat Kähler Surfaces of All Genera,* J. Reine Angew. Math. **486**, 69-95 (1997)
9. Kozameh, C.N., Newman, E.T. and Tod, K.P. *Conformal Einstein Spaces,* Gen. Rel. Grav. **17**, 343-52 (1985)
10. Kronheimer, P. *A Torelli Type Theorem for HyperKahler Metrics,* Journal of Differential Geometry **29, No. 3**, 685–698 (1989)
11. LeBrun, C., Poon, Y.S. *Self-Dual Manifolds with Symmetry,* Proceedings of Symposia in Pure Mathematics **54, Part 2**, 365–377 (1993)
12. LeBrun, C. *Twistors, Kähler manifolds and bimeromorphic geometry I,* J. Am. Math. Soc. **5**, 289–316 (1992)
13. LeBrun, C. *Explicit self-dual metrics on* $\mathbf{CP}_2\#\cdots\#\mathbf{CP}_2$*,* J. Differential Geom. **34**, 223–253 (1991)
14. LeBrun, C., Singer, M. *Existence and deformation theory for scalar-flat Kähler metrics on compact complex surfaces,* Inv. Math. **112**, 273-313 (1993)
15. Li, P., Yau, S.T. Lecture Notes in Pure and Appl. Math. 143, Dekker, New York , 131–144 (1993)
16. Morrey, C.B. *Multiple Integrals in the Calculus of Variations,* Springer-Verlag, Berlin, Heidelberg, New York , (1966)
17. Penrose, R. and Rindler, W. *Spinors and Space-time, volume 2, Cambridge University Press, Cambridge,* 1986

18. Pontecorvo, M *Algebraic Dimension of Twistor Spaces and Scalar Curvature of Anti-Self-Dual Metrics,* Mathematische Annalen **291**, 113-122 (1991)
19. Poon Y.S. *On the algebraic structure of twistor spaces,* J. Differential Geom. **36** , 451–491 (1992)
20. Ran, Z. *Deformation of maps,* Algebraic geometry proceedings (Trento, 1988), Ballico, E., Ciliberto, C., Lectures Notes in Math., Springer, Berlin **1389**, 247–253 (1989)

THE WU METRIC AND MINIMUM ELLIPSOIDS

KANG-TAE KIM
Department of Mathematics
Pohang University of Science and Technology
Pohang 790-784 The Republic of Korea
E-mail: kimkt@postech.ac.kr

The goal of this article is twofold: we provide first a comprehensive and intuitive introduction to the Wu metric on Kobayashi hyperbolic complex manifolds. Then we present with concise proofs several fundamental characteristic merits of the Wu metric, partly in conjunction with a conjecture of Kobayashi raised in 1970. In addition, we discuss a striking curvature behavior of the Wu metric on Thullen domains following the work of Cheung and Kim [8 9].

1 Introduction

By an *invariant metric*, we mean an *assignment* of a metric to each member of a broad collection of complex manifolds in such a way that the biholomorphic mappings between them become isometries. Directly from its definition, it is not surprising at all to see that invariant metrics play a crucial role in both complex differential geometry and several complex variables.

The standard list of extensively studied well-known examples up to the time of this writing includes the Bergman metric,[4] the Einstein-Kähler metric (= Calabi-Yau metric),[6] the Carathéodory metric and the Kobayashi metric.[13] As much as these metrics enjoy their own distinctive merits, they also possess some weaknesses. The first two are smooth Kählerian metrics, but they do not exhibit any good behavior under general holomorphic mappings. The latter two have the well-known *distance-decreasing property*; i.e. all holomorphic mappings are distance decreasing with respect to them. They are defined on all complex manifolds. However, they are far from being Hermitian. Regularity is also poor.

Thus enters the Wu metric. The Wu metric is defined on all complex manifolds. It enjoys the distance decreasing property up to a fixed constant. It is always Hermitian. It can be semi-definite, but it is positive definite whenever the Kobayashi metric is definite. It is always continuous. It coincides with the Kählerian invariant metrics on the homogeneous manifolds up to a constant multiple, for obvious reasons. In short, the Wu metric offers an elegant interpolation between the sub-Finslerian and the Kählerian invariant metrics.

Nevertheless, the Wu metric has not yet received its due attention partly owing to its rather complicated (although general) definition given in Wu's original article.[17] Subsequent papers by Cheung and Kim[8 9] did not put much effort into reconstructing the Wu metric and its basic properties in a more comprehensive way. Thus, we set an aim for this article to introduce the Wu metric as intuitively and comprehensively as possible in terms of minimum volume ellipsoids, and deduce from it several fundamental but rather striking properties. In order to achieve this, we restrict ourselves to the Kobayashi hyperbolic manifolds only.

Setting such an aim, we organize this article as follows.

First, we give concise proofs to the basic properties of the Wu metric from the viewpoints of the minimum volume ellipsoids in Sections 2 and 3. In Wu's paper,[17] the proofs are not given, although the key ideas are indicated. In these sections, we exploit known facts about the minimum volume ellipsoids and the Kobayashi metrics to give a concise and descriptive proofs to several basic facts.

The Wu metric raises several intriguing questions, some of which are closely related to well-known old problems such as a conjecture of Kobayashi posed in 1970. (See Kobayashi.[13]) We will discuss some of these problems in Section 4.

Then we present an analysis of the asymptotic behavior of the Wu metric in Section 5, which involves a refinement of Graham's analysis of the Kobayashi metric. We use the scaling technique in several complex variables, which is similar to the analysis of Kim and Yu.[12] We prove that the Wu metric as well as the Kobayashi metric converges to the 'Levi metric' at the boundary.

As the exposition progresses, there arise several intriguing questions which should be the subject of future research. We include some of them in this article.

2 The Wu metric on Kobayashi Hyperbolic Manifolds

Although the Wu metric is defined on all complex manifolds (semi-definite in general), we restrict ourselves to complex manifolds that are Kobayashi hyperbolic, in the sense that the Kobayashi metric is positive definite. We first start with the well-known Kobayashi metric.

2.1 *The Kobayashi-Royden metric*

Let M be a complex manifold of dimension n. Denote by B^n the open unit ball in the complex Euclidean space $\mathbf{C}^n$. For each $p \in M$ and $v \in T_pM$, consider the collection

$$\mathcal{F}(p; v) = \{f : B^1 \to M \mid f \text{ holomorphic}, \\ f(0) = p, f'(0) = \frac{v}{\lambda_{f,v}} \text{ for some } \lambda_{f,v} \in \mathbf{C} \setminus \{0\}\}.$$

Then the Kobayashi-Royden metric $k_M(p; v)$ of $v \in T_pM$ is defined by

$$k_M(p; v) = \inf_{f \in \mathcal{F}(p;v)} |\lambda_{f,v}|.$$

It is well-known that bounded domains in $\mathbf{C}^n$ and complex Kähler manifolds with negative holomorphic curvatures are Kobayashi hyperbolic.

Although the triangle inequality does not hold for $k_M(p; \cdot)$ on T_pM, it defines the arc length of a piecewise C^1 smooth curve, and hence a distance function. [15] Thus the concept of completeness is naturally defined for the positive definite Kobayashi metric, in the sense of Cauchy completeness for the Kobayashi integrated distance.

It is also known that this metric is upper semi-continuous on the tangent bundle, and possesses the distance decreasing property. For more details, we refer to the articles by Kobayashi, [13] and Royden. [15]

2.2 *Minimum Ellipsoid for the Kobayashi Indicatrix*

As mentioned earlier, we now introduce a construction of the Wu metric on Kobayashi hyperbolic manifolds, following Wu's paper. [17]

Let M be a complex manifold with its Kobayashi metric k_M positive definite on every tangent space. Then consider the set

$$\mathcal{I}(M, p) = \{v \in T_pM \mid k_M(p; v) \leq 1\}$$

which is often called the *Kobayashi indicatrix.*

Fix the point $p \in M$ momentarily. Introduce a basis $\{u_1, \ldots, u_n\}$ for the complex vector space T_pM and identify it with $\mathbf{C}^n$ via this basis. Then consider a complex ellipsoid in T_pM, which is a convex body in $T_pM = \mathbf{C}^n$ defined by

$$\sum_{\alpha,\beta=1}^{n} g_{\alpha\bar{\beta}} u_\alpha u_{\bar{\beta}} \leq 1$$

for a certain positive definite Hermitian matrix $(g_{\alpha\bar{\beta}})$. Notice that this matrix gives rise to a Hermitian form, say g, on $T_pM = \mathbf{C}^n$ given by

$$g_{\alpha\bar{\beta}} = g(u_\alpha, u_\beta).$$

Staying with such a basis-dependent setting, we consider the *ellipsoid with the minimum volume among the complex ellipsoids that contain the Kobayashi indicatrix* $\mathcal{I}(M,p)$.

Now we exploit Fritz John's analysis[11] on the minimum volume ellipsoids, which shows that there exists a unique complex ellipsoid containing $\mathcal{I}(M,p)$ that has the minimum volume with respect to the Euclidean volume element for T_pM identified with $\mathbf{R}^n$ by a choice of the basis. Notice that our argument so far depends upon the Euclidean volume which in turn depends upon the choice of the basis for T_pM through which its identification with $\mathbf{C}^n$ has been introduced. However, it is important to notice that the choice of minimum ellipsoid is in fact independent of the choice of the basis or the resulting identification of T_pM with $\mathbf{C}^n$. It can be seen in what follows:

For two complex ellipsoids E and F corresponding to the Hermitian forms g and h respectively in a complex linear space, suppose that the volume of E is greater than or equal to that of F. This is equivalent to the inequality

$$\det g(u_\alpha, u_\beta) \leq \det h(u_\alpha, u_\beta) \tag{1}$$

where the determinants above are those of the matrices representing the Hermitian forms g and h, respectively, with respect to the basis $\{u_1, \cdots, u_n\}$. Notice that if we take another basis, say $\{v_1, \cdots, v_n\}$ for T_pM, we have

$$(g(v_\alpha, v_\beta)) = Q\,(g(u_\alpha, u_\beta))\,Q^*$$

where Q is the matrix for change of basis and Q^* its conjugate transpose. This shows that (1) above is equivalent to the inequality

$$\det g(v_\alpha, v_\beta) \leq \det h(v_\alpha, v_\beta).$$

This shows that the minimum ellipsoid in the sense of Fritz John containing the Kobayashi indicatrix $\mathcal{I}(M,p)$ is uniquely determined *independently of the choices of basis.*

As in Wu's article[17], we may therefore summarize the above as: *the determinant of a Hermitian form is not defined coordinate-free, but the order relation between the determinants is well-defined and coordinate-free.*

2.3 The Wu metric

Now we are ready to define the Wu metric on M. For each $p \in M$, consider the Hermitian form h_p that defines the minimum complex ellipsoid containing the Kobayashi indicatrix $\mathcal{I}(M, p)$, following the discussion in the preceding section. Then the correspondence

$$p \mapsto h_p$$

defines an assignment of a unique Hermitian inner product on each T_pM. This defines the *Wu metric* on M.

At this stage, it is not even evident whether this metric should possess any smoothness. We will discuss various fundamental properties including the regularity, invariance, distance-decreasing property and several other properties in the subsequent sections.

3 Fundamental Properties of The Wu metric

We start with the most basic properties. Although the key ideas to their proofs were sketched in Wu's paper, [17] comprehensive proofs are worth presenting here. It seems that the proofs are better understood through the results of F. John [11] concerning the minimum volume ellipsoids of convex bodies in the Euclidean spaces. We demonstrate this assertion in what follows.

3.1 Continuity

The following result shows that the Wu metric possesses better regularity than the Kobayashi-Royden metric. Before we proceed further, we emphasize again that the contents of this section follow essentially from Wu's line of reasoning and arguments. [17]

Proposition 3.1 (Wu) *The Wu metric on a complete Kobayashi hyperbolic manifold is continuous.*

Keys to the proof have been indicated by Wu in p. 663 in his paper, [17] essentially using the ellipsoids. However, we would like to point out the following helpful fact which provides a foundation to an intuitive proof of the above proposition: although the Kobayashi-Royden infinitesimal metric is in general only upper semicontinuous, its integrated distance, when positive definite and complete, is proper, in the sense that the manifold topology is the same as the one generated by the Kobayashi distance. [3] In particular the Kobayashi distance function is continuous.

Then, there is a standard way of 'differentiating' the distance function to create an infinitesimal metric. When the distance function is continuous, so is its infinitesimal metric constructed in this way. For the Kobayashi distance function, the differentiated infinitesimal metric is the norm on each tangent space given by the Minkowski function associated with the convex hull of the Kobayashi-Royden indicatrix introduced in the preceding section.

Notice that the minimum ellipsoid for the Kobayashi-Royden indicatrix is the same as the minimum ellipsoid for its convex hull. Since this convex hull, being the indicatrix of the differentiated infinitesimal metric of the continuous complete Kobayashi distance, changes continuously as the base point changes on the manifold in consideration, it is now immediate to see that the minimum ellipsoid varies continuously with respect to the base points. This proves the proposition above.

3.2 Invariance and Distance-decreasing property

The Wu metric enjoys the merit of possessing distance-decreasing property. We demonstrate this in the following.

Proposition 3.2 (Wu) *Let $f : M \to N$ be a holomorphic mapping between Kobayashi hyperbolic manifolds. Let h_M and h_N, respectively, denote the Wu metrics of M and N. Say $m = \dim_{\mathbf{C}} M$. Then*

$$f^* h_N \leq \sqrt{m}\, h_M.$$

If f is a biholomorphic mapping, then we have

$$f^* h_N = h_M.$$

The proof of this result follows easily from a distinctive property of the minimum ellipsoid. In case f is a biholomorphism, it is a Kobayashi isometry. Therefore, the complex linear map $df_p : T_pM \to T_{f(p)}N$ preserves the Kobayashi indicatrices and their convex hulls. Namely,

$$df_p(\mathcal{I}(M,p)) = \mathcal{I}(N, f(p)).$$

It is a simple matter then to prove that the minimum ellipsoids are preserved by df_p. This shows that the Wu metric is an invariant metric, proving the second statement of the proposition.

To see the distance-decreasing property, let $f : M \to N$ denote a holomorphic mapping which is not necessarily biholomorphic. Then the distance decreasing property of the Kobayashi metric is equivalent to

$$df_p(\mathcal{I}(M,p)) \subset \mathcal{I}(N, f(p)).$$

Since df_p is linear, the same inclusion holds for the convex hulls, i.e.,

$$df_p(\hat{\mathcal{I}}(M,p)) \subset \hat{\mathcal{I}}(N, f(p))$$

where $\hat{\mathcal{I}}$ denotes the convex hull of the Kobayashi indicatrix $\mathcal{I}$. Notice that the minimum ellipsoid of a subset is not in general contained in the minimum ellipsoid of the superset. Thus, a direct comparison between the minimum ellipsoids cannot be obtained.

To get around this difficulty, we recall a standard fact from the minimum ellipsoid theory. First, call a set A in a complex vector space *circular* if $e^{i\theta}v \in A$ for all $\theta \in \mathbf{R}$ and $v \in A$. Then it is known [11] that the minimum complex ellipsoid, say E_A of a circular convex set A in $\mathbf{C}^m$, satisfies the relation

$$E_A \subset \sqrt{m}\, A.$$

Denote by $\mathcal{W}(M,p)$ the unit Wu metric ball in T_pM. Then, we have the following consequence:

$$\mathcal{W}(M,p) \subset \sqrt{m}\,\hat{\mathcal{I}}(M,p)$$

where $m = \dim_{\mathbf{C}} M$, since the convex hull $\hat{\mathcal{I}}(M,p)$ of the Kobayashi indicatrix is circular.

Therefore, we arrive at the formula

$$\begin{aligned} df_p(\mathcal{W}(M,p)) &\subset df_p(\sqrt{m}\,\hat{\mathcal{I}}(M,p)) = \sqrt{m}\, df_p(\hat{\mathcal{I}}(M,p)) \\ &\subset \sqrt{m}\,\hat{\mathcal{I}}(N,f(p)) \subset \sqrt{m}\,\mathcal{W}(N,f(p)) \end{aligned}$$

which proves the remaining assertion of the proposition.

Remark 3.1 The dilatation arguments of the minimum ellipsoids show that the (integrated) Wu distance is equivalent to the Kobayashi distance. In particular, on a Kobayashi hyperbolic manifold, *the Wu metric is complete if and only if the Kobayashi metric is complete.*

3.3 Covering Property

The Wu metric, with the Kobayashi metric in its background, also has the following quite useful property:

Proposition 3.3 *Let M and N be Kobayashi hyperbolic manifolds, and let $\pi : M \to N$ a holomorphic covering mapping. Denote by h_M, h_N the Wu metrics of M and N, respectively. Then, $\pi^* h_N = h_M$.*

We first recall the covering property of the Kobayashi distance. (See Proposition 1.6 in p. 48 of Kobayashi's book. [13]) Then the pull-back by π of differentiated infinitesimal metric of the Kobayashi distance of N coincides with the differentiated infinitesimal metric of the Kobayashi distance of M. Since $d\pi_p$ ($p \in M$) is a complex linear isomorphism preserving the convex hulls of the corresponding Kobayashi indicatrices, it preserves the Wu metric as well.

We remark that such covering property does not hold for the Bergman metric. For instance, examine e^z which maps the left half plane onto the punctured disk.

3.4 Further smoothness of the Wu metric

For further smoothness, several rather obvious facts are worthwhile to remark. On the open unit ball B^n in $\mathbf{C}^n$, the Wu metric coincides with the Kobayashi metric, since the Kobayashi metric itself is a smooth Kählerian metric in this case. On the bounded symmetric domains we arrive at the same conclusion. Moreover, the Wu metric in this case is obviously real analytic smooth, due to the transitive real analytic automorphism group action.

However, beyond such homogeneous manifolds that are Kobayashi hyperbolic, the explicit computation of the Wu metric appears as difficult as for the other invariant metrics. Differentiability of the Wu metric beyond the C^0 continuity is yet to be investigated in general. Nonetheless, a striking behavior has been observed in the case of Thullen domains by Cheung and Kim, [8 9] which we discuss in the following section.

4 The Case of Thullen Domains

4.1 Thullen Domains

By a *Thullen domain*, we mean the following set

$$E_\lambda = \{(z, w) \in \mathbf{C}^2 \mid |z|^2 + |w|^{2\lambda} < 1\}.$$

As λ traverses the set of positive real numbers, the entire collection of E_λ, $0 < \lambda < \infty$, constitute a one parameter family of bounded domain of holomorphy that are holomorphically *distinct.* (This was originally shown by Thullen. For a modern reference, we refer to Sunada's paper. [16])

It is known that the automorphism group of E_λ with $\lambda \neq 1$ does not act transitively on E_λ. In fact, it is a real 4 dimensional Lie group generated by the mappings

$$(z,w) \to \left(\frac{z-a}{1-\bar{a}z}, \left(\frac{\sqrt{1-|a|^2}}{1-\bar{a}z} \right)^{\frac{1}{\lambda}} \cdot w \right)$$

and

$$(z,w) \to (e^{i\theta}z, e^{i\psi}w)$$

where $a \in \mathbf{C}$, $|a| < 1$ and $\theta, \psi \in \mathbf{R}$. Moreover, the generic orbits are real 3 dimensional sets, represented by the equation $|z|^2 + c|w|^{2\lambda} < 1$ for an appropriate $c > 1$.

Thus, this domain is a natural candidate to study beyond the homogeneous domains. Earlier articles of Azukawa and Suzuki [2] and Bland [5] feature an analysis of the curvature of the Bergman metric and the Einstein-Kähler metric on the Thullen domains, respectively. The results indeed show certain differences of the Thullen domains with $\lambda > 1$ in contrast to the case of the ball. However, the analysis of the Wu metric shows genuinely distinctive information as explained below.

4.2 Results of Cheung and Kim

Let h_λ denote the Wu metric on the Thullen domain E_λ, for $\lambda > 0$. Then the following results have been obtained by Cheung and Kim.

Theorem 4.1 (Cheung-Kim [8]) *Let $\lambda > 1$. The Wu metric h_λ is real analytic at points of $E_\lambda \setminus (Z \cup S_\lambda)$ where $S_\lambda = \{f(0, 2^{-1/2\lambda}) \mid f \in \operatorname{Aut} E_\lambda\}$ and $Z = \{(z,0) \mid |z| < 1\}$. At points of Z, the Wu metric h_λ is of Hölder class $C^{[2\lambda], 2\lambda - [2\lambda]}$, where the brackets represent the Gauss symbol. Finally at points of S_λ, the Wu metric is of class C^1 but not of C^2.*

Before we state the next result, we set some notation. We denote by

$$S_\lambda = \{(z,w) \in \mathbf{C}^2 \mid |z|^2 + 2|w|^{2\lambda} = 1\}.$$

and

$$S_\lambda^+ = \{(z,w) \in E_\lambda \mid |z|^2 + 2|w|^{2\lambda} > 1\}$$
$$S_\lambda^- = \{(z,w) \in E_\lambda \mid |z|^2 + 2|w|^{2\lambda} < 1\}.$$

Then we state

Theorem 4.2 (Cheung-Kim [8]) *The Wu metric is Kähler on S_λ^+ with constant negative holomorphic curvature -2, and non-Kähler Hermitian on S_λ^- with bounded non-constant negative holomorphic curvature. Overall on E_λ, the Wu metric has bounded negative holomorphic curvature in the sense of currents.*

For the case of $0 < \lambda < 1$ we have the following results.

Theorem 4.3 (Cheung-Kim [8 9]) *For $0 < \lambda < 1$, the Wu metric h_λ for the Thullen domain E_λ is real analytic except at points of $Z = \{(z, 0) \in \mathbf{C}^2 \mid |z| < 1\}$. It is continuous everywhere. The Wu metric is nowhere Kähler. Its holomorphic curvature is bounded, negative and nonconstant.*

Referring to the articles by Cheung and Kim cited above for more detailed arguments and the proofs of the preceding theorems, we put several remarks in this article concerning the consequences of the above theorems.

Remark 4.1 First, notice that as $\lambda \searrow 1$, the domains E_λ converge to the unit ball in the sense of Hausdorff distance between convex sets. However, notice that the convergence is not C^2. For such a C^1 perturbation of boundaries, the set S_λ clearly shows that the intrinsic geometry of the Wu metric deforms precisely in C^1 sense. It is also striking to see that the Wu metric becomes non-Kähler Hermitian when the ball is deformed to a Thullen domain.

Remark 4.2 It also appeals to us that one should not ignore the fact that, when $\lambda > 1$, E_λ is weakly pseudoconvex at the boundary points of type $(e^{i\theta}, 0)$, $\theta \in \mathbf{R}$. Usually the other invariant distances show differences mainly in its growth rates at the boundary points. However, the Wu metric exhibits the phenomenon that the singularity set S_λ propagates from the weakly pseudoconvex boundary points into the interior of the domain. It should be worthwhile to know whether this phenomenon is indeed general for all the weakly pseudoconvex boundary points. Thus the weakly pseudoconvex boundary points are "visible" from inside of E_λ. There are certain cases other than the Thullen domains that show such a behavior, supporting such a guess.

Remark 4.3 In the above cited articles by Cheung and Kim, explicit formulae of the Wu metric for E_λ are given, for all the positive values of λ. It should be interesting to see whether one can perform a similar analysis for other types of circular domains such as the ones defined by $|z|^p + |w|^q < 1$ where neither p nor q equals 2. The case $p = q$ may be special, in the light of the work of Azukawa and Suzuki. [2]

Remark 4.4 Recall that Kobayashi, in his work [13], conjectured that *every complete Kobayashi hyperbolic manifold admits a complete Hermitian metric of negative holomorphic curvature.* Considering that the distance decreasing property is a consequence of bounded negative curvatures (See the versions of generalized Schwarz's Lemma by Ahlfors, [1] Chern, [7] Yau [18]) the aim of this conjecture appears to attempt to explain the distance-decreasing property of the sub-Finslerian Kobayashi metric by use of a smooth Hermitian metric which is more differential geometric. Note that the theorems above prove that the conjecture of Kobayashi is true for the Thullen domains. Not only that, we would like to point out that the Wu metric is the only candidate for proving this conjecture among the known invariant metrics, since it is the only known (at least by far) distance-decreasing Hermitian invariant metric.

Remark 4.5 It was generally believed that the Euclidean convexity (up to biholomorphisms) of a bounded domain is closely related to the negativity of the curvature of invariant metrics. In this regard, the negativity of holomorphic curvature of the Wu metric on E_λ with $\lambda < 1/2$ should be remarked, since this domain is not biholomorphic to any bounded convex domains in $\mathbf{C}^2$. (The reason for nonconvexity up to biholomorphisms is as follows: notice that the Kobayashi indicatrix of E_λ at the origin is linearly equivalent to the domain itself, since it is completely circular. Consequently the indicatrix is not convex. Now, Lempert's work [14] implies that the Kobayashi indicatrix of a bounded convex domain is convex. Since the corresponding Kobayashi indicatrices of corresponding points should be linearly equivalent when the domains are biholomorphic to each other, one sees immediately that E_λ with $0 < \lambda < 1$ can never be mapped holomorphically to a bounded convex domain.)

5 Asymptotic Behavior

In this section, we investigate the asymptotic behavior of the Wu metric near the strongly pseudoconvex boundary point of a bounded domain of holomorphy. Resonant with our minimum ellipsoid viewpoint, we first study the asymptotic behavior of the Kobayashi metric following the earlier work of I. Graham [10] and present an improvement exploiting the scaling methods. Then we show that there is a Hermitian limit metric, which we call the *Levi metric*, for both the Kobayashi and the Wu metrics.

However, in order not to get into too much of intricate technicalities, we restrict ourselves in this article to a special case of complex dimension two and the "radial" approaching directions. This exposition is thus special, but it shows all the essential ideas suitable even for the general cases in all dimensions. Complete detailed arguments for the general case will appear elsewhere.

5.1 Graham's estimates by Scaling Methods

Let G be a bounded domain in $\mathbf{C}^2$, and let $p \in \partial G$ such that the boundary ∂G is C^2 smooth and strongly pseudoconvex in a neighborhood U_p of p in $\mathbf{C}^2$.

Denote by $\tilde{G} = G \cap U_p$. Shrinking U_p if necessary, we may have a holomorphic embedding $\varphi : U_p \to \mathbf{C}^2$ such that

- $\varphi(0) = 0$,
- $\varphi(\tilde{G}) = \{(z, w) \in \mathbf{C}^2 \mid \operatorname{Re} z + |w|^2 + o(|z| + |w|^2) < 0\}$,

- $\varphi(U_p \cap \partial G) = \{(z, w) \in \mathbf{C}^2 \mid \operatorname{Re} z + |w|^2 + o(|z| + |w|^2) = 0\}$, and
- $\varphi(\tilde{G}) \subset E$ where $E = \{(z, w) \in \mathbf{C}^2 \mid \operatorname{Re} z + |w|^2/4 < 0\}$.

Note that $\varphi(U_p \cap \partial G)$ and ∂E are tangent to each other.

Now we take the crucial step of scaling along the non-tangential *radial curve*. With the mapping φ defined above, we define the following:

$$\Lambda = \{(-t, 0) \mid 0 < t < r\}$$
$$\Lambda_\varphi = \varphi^{-1}(\Lambda) \subset \tilde{G}$$

where $r > 0$ is chosen so that $\Lambda \subset \varphi(\tilde{G})$.

For each $\xi = \varphi^{-1}(-t, 0) \in \Lambda_\varphi$, we define the "scaling map" by

$$L_\xi(z, w) = \left(\frac{z}{t}, \frac{w}{\sqrt{t}}\right)$$

Note that

$$L_\xi(E) = E, \quad \forall \xi \in \Lambda_\varphi$$

If we denote by

$$E_0 = \{(z, w) \in \mathbf{C}^2 \mid \operatorname{Re} z + |w|^2 < 0\}$$

then we observe that

$$\lim_{\Lambda_\varphi \ni \xi \to p} L_\xi(\tilde{G}) = E_0$$

where the set convergence is in the local Hausdorff sense. In fact, the Hausdorff convergence is taken first for the closures of the domains in consideration. Then the interior of the limit set is considered as the limit domain. Of course this procedure is troublesome in general. However, such a consideration is valid in this exposition as one easily sees in what follows.

First notice that

$$L_\xi(\varphi(\xi)) = (-1, 0), \ \forall \xi \in \Lambda_\varphi,$$

and that *this point is bounded away from the boundaries of the domains* E_0, E *and* $L_\xi \circ \varphi(\tilde{G})$ *for all* $\xi \in \Lambda_\varphi$ *altogether.*

Note also that every domain $L_\xi \circ \varphi(\tilde{G})$ in the successive scaling above is contained in the "envelope" E, since $L_\xi \circ \varphi(E) = E$ for all $\xi \in \Lambda_\varphi$. Thus, if we apply the linear fractional mapping

$$\Phi(z, w) = \left(\frac{1+z}{1-z}, \frac{2w}{1-z}\right)$$

to all of these domains simultaneously, all the scaled domains as well as E and E_0 are simultaneously realized as bounded domains as follows:

$$\Phi(E) = B^2_\# \equiv \{(z,w) \mid |z|^2 + |w|^2/4 < 1\}$$
$$\Phi(E_0) = B^2 \equiv \{(z,w) \mid |z|^2 + |w|^2 < 1\}$$
$$\Phi \circ L_\xi \circ \varphi(\tilde{G}) \subset \Phi(E) = B^2_\#$$

We now have the global Hausdorff set convergence

$$\lim_{\Lambda_\varphi \ni \xi \to p} \Phi \circ L_\xi \circ \varphi(\tilde{G}) = B^2$$

by which we mean:

for every $\epsilon > 0$ there exists $\delta > 0$ such that

$$\frac{1}{1+\epsilon} B^2 \subset \Phi \circ L_\xi \circ \varphi(\tilde{G}) \subset \frac{1}{1-\epsilon} B^2 \tag{2}$$

whenever $\xi \in \Lambda_\varphi$ satisfies $|\xi - p| < \delta$.

Notice also that

$$\Phi \circ L_\xi \circ \varphi(\xi) = 0, \ \forall \xi \in \Lambda_\varphi$$

Now we introduce the key estimate of the Kobayashi metric near the boundary. By (2) and the distance decreasing property of the Kobayashi metric, we immediately obtain

$$(1-\epsilon)\|d(\Phi \circ L_\xi \circ \varphi)_\xi(v)\| \le k^{\tilde{G}}_\xi(v) \le (1+\epsilon)\|d(\Phi \circ L_\xi \circ \varphi)_\xi(v)\| \tag{3}$$

for every $v \in \mathbf{C}^2$. Here, $\mathbf{C}^2$ in fact represents the fiber of the trivial bundle $T\tilde{G} \cong \tilde{G} \times \mathbf{C}^2$, under the obvious identification.

Consequently, we have

$$\lim_{\Lambda_\varphi \ni \xi \to p} \frac{k^{\tilde{G}}_\xi(v)}{\|d(\Phi \circ L_\xi \circ \varphi)_\xi(v)\|} = 1, \ \ \forall v \in \mathbf{C}^2 \setminus 0$$

and the convergence is *uniform* on $v \in \mathbf{C}^2$ with $\|v\| = 1$.

The remaining steps are an appropriate interpretation of the meaning of the estimate (3) that will eventually yield both Graham's results and the asymptotic behavior of the Kobayashi and Wu metrics.

Computation of $d(\Phi \circ L_\xi \circ \varphi)_\xi$

Recall the definitions of Φ and L_ξ and the fact that $L_\xi \circ \varphi(\xi) = (-1, 0)$. We immediately obtain

$$d\Phi_{(-1,0)} = \begin{pmatrix} \frac{1}{2} & 0 \\ 0 & 1 \end{pmatrix}$$

and

$$dL_\xi = \begin{pmatrix} \frac{1}{t} & 0 \\ 0 & \frac{1}{\sqrt{t}} \end{pmatrix}$$

In order to compute $d\varphi_\xi$, we remark here that the normalization map φ is constructed in the following four steps:

- Move p to the origin, by translation by $-p$.
- Apply a unitary map, say T, so that $\operatorname{Re} z$ axis becomes normal to the boundary and that $\{\operatorname{Re} z = 0\}$ becomes the tangent plane to the boundary at the origin. The transformed domain at this stage takes a local defining function near the origin of the form

  $$\operatorname{Re} z + aw^2 + \bar{a}\bar{w}^2 + r|w|^2 + o(|z| + |w|^2) < 0$$

  where $r > 0$.
- Apply $(z, w) \mapsto (z + 2aw^2, w)$, which transforms the domain near the origin to the one defined by the inequality

  $$\operatorname{Re} z + r|w|^2 + o(|z| + |w|^2) < 0.$$

- Apply finally the map $(z, w) \mapsto (z, rw)$.

Thus, we have

$$d\varphi_\xi = \begin{pmatrix} 1 & 0 \\ 0 & r \end{pmatrix} \cdot T$$

Notice that $r =$ (the Levi form)/2, according to Graham's choice for the Levi form.[10]

Graham's estimate for complex normal vectors

Let $v \in \mathbf{C}^2$ be a complex normal vector to ∂G at p. Then, it follows that

$$d\varphi_\xi(v) \to (\hat{v}, 0) \in \mathbf{C}^2, \text{ as } \Lambda_\varphi \ni \xi \to p.$$

Observe $||d\varphi_\xi(v)|| = |\hat{v}|$. Hence, we arrive at

$$||d(\Phi \circ L_\xi \circ \varphi)_\xi(v)|| = \frac{|\hat{v}|}{2t}.$$

Therefore (3) implies

$$\lim_{\Lambda_\varphi \ni \xi \to p} k_\xi^{\tilde{G}}(v) \cdot t = \frac{1}{2}|\hat{v}|.$$

Recollecting the definitions of $d\varphi_\xi$ and L_ξ, we easily obtain Graham's estimate of the Kobayashi metric along the complex normal vector v as follows:

$$\lim_{\Lambda_\varphi \ni \xi \to p} k_\xi^{\tilde{G}}(v) \cdot dis(\xi, \partial G) = \frac{1}{2}||v||.$$

Graham's estimate along complex tangent vectors

Let $w \in \mathbf{C}^2$ be a complex tangent vector to ∂G at p. Then we obtain

$$d\varphi_\xi(w) \to (0, r\hat{w}) \in \mathbf{C}^2, \text{ as } \Lambda_\varphi \ni \xi \to p.$$

Notice also that $||d\varphi_\xi(w)|| \to r|\hat{w}| = r||w||$. Moreover,

$$||d(\Phi \circ L_\xi \circ \varphi_\xi)(w)|| = \frac{r|\hat{w}|}{\sqrt{t}} = \frac{\sqrt{\mathcal{L}_p(w, \bar{w})}}{\sqrt{2t}}$$

Here, $\mathcal{L}_p$ denote the Levi form in Graham's notation with the restriction that the gradient of the defining function has length 1 at p. Therefore, one obtains

$$\lim_{\Lambda_\varphi \ni \xi \to p} (k_\xi^{\tilde{G}}(w))^2 \cdot dis(\xi, \partial G) = \frac{1}{2}\mathcal{L}_p(w, \bar{w})$$

which is the remaining estimate of Graham on the Kobayashi metric along the complex tangent vector w to ∂G at p.

5.2 A uniform asymptotic estimate of the Wu metric

We are going to exploit the full strength of the estimate (3). We re-write (3) as:

For every $\epsilon > 0$ there exists $\delta > 0$ such that

$$(1 - \epsilon) \le \frac{k_\xi^{\tilde{G}}(v)}{||d(\Phi \circ L_\xi \circ \varphi)_\xi(v)||} \le (1 + \epsilon)$$

whenever $\xi = \varphi^{-1}(-t, 0)$ satisfies $0 < t < \delta$.

Now, notice that the map $d(\Phi \circ L_\xi \circ \varphi)_\xi$ is canonically determined by the local defining function for ∂G near p. Thus at each $\xi \in \Lambda_\varphi$, we may define by

$$\mathcal{E}_\xi = \{v \in \mathbf{C}^2 \mid \|d(\Phi \circ L_\xi \circ \varphi)_\xi(v)\| \leq 1\}$$

and we call it *the Levi infinitesimal ball at* ξ.

We now cite the uniform localization result of Graham[10], which is:

For every $\epsilon > 0$ *there exists* $\delta > 0$ *such that*

$$1 \leq \frac{k_\xi^{\tilde{G}}(v)}{k_\xi^G(v)} \leq 1 + \epsilon$$

for all $\xi = \varphi^{-1}(-t, 0) \in \Lambda_\varphi$ *with* $0 < t < \delta$.

This in particular makes all the estimates above for $k_\xi^{\tilde{G}}(v)$ valid (with possibly a smaller δ) for $k_\xi^G(v)$.

Now let h_ξ denote the Wu metric at ξ for the domain G. Let

$$\mathcal{W}_\xi = \{v \in \mathbf{C}^2 \mid h_\xi(v, v) \leq 1\}$$

and let

$$\mathcal{K}_\xi = \{v \in \mathbf{C}^2 \mid k_\xi^G(v) \leq 1\}.$$

Notice that we can construct a Hermitian metric at ξ explicitly by

$$\lambda_\xi(v, w) = \langle d(\Phi \circ L_\xi \circ \varphi)_\xi(v), d(\Phi \circ L_\xi \circ \varphi)_\xi(w) \rangle$$

where $\langle\ ,\ \rangle$ denote the standard Hermitian inner product for $\mathbf{C}^2$.

For convenience, denote by

$$A_\xi = d(\Phi \circ L_\xi \circ \varphi)_\xi$$

which is understood as a linear map from $\mathbf{C}^2$ to $\mathbf{C}^2$. Then we obtain that $1 - 2\epsilon \leq \|A_\xi(w)\| \leq 1 + 2\epsilon$ for every $w \in \mathcal{K}_\xi$. In other words,

$$(1 - 2\epsilon)B^2 \subset A_\xi(\mathcal{K}_\xi) \subset (1 + 2\epsilon)B^2$$

where B^2 denotes the unit ball in $\mathbf{C}^2$.

Since the best fitting complex ellipsoid is transformed to another best fitting ellipsoid by a complex linear isomorphism (due to its uniqueness) and since $\mathcal{W}_\xi$ is the best fitting ellipsoid of $\mathcal{K}_\xi$, it follows that $A_\xi(\mathcal{W}_\xi)$ is the best fitting ellipsoid of $A_\xi(\mathcal{K}_\xi)$. By continuity of the choice of best fitting ellipsoid and by (19), there exists $0 < \delta_1 < \delta$ such that

$$(1 - 3\epsilon)B^2 \subset A_\xi(\mathcal{W}_\xi) \subset (1 + 3\epsilon)B^2$$

whenever $\xi = \varphi^{-1}(-t, 0)$ satisfies $0 < t < \delta_1$. Notice that

$$A_\xi^{-1}(B^2) = \mathcal{E}_\xi.$$

This implies that

$$(1 - 3\epsilon)\mathcal{E}_\xi \subset \mathcal{W}_\xi \subset (1 + 3\epsilon)\mathcal{E}_\xi$$

which in turn implies the uniform estimate

$$(1 - 6\epsilon)^2 \leq \frac{h_\xi(v, v)}{\lambda_\xi(v, v)} \leq (1 + 6\epsilon)^2, \quad \forall v \in \mathbf{C}^2 \setminus 0$$

or simply

$$\lim_{\Lambda_\varphi \ni \xi \to p} \frac{h_\xi(v, v)}{\lambda_\xi(v, v)} = 1$$

and the convergence is uniform on $v \in \mathbf{C}^2$ with $\|v\| = 1$.

Finally, the preceding arguments summarize into:

Theorem 5.1 *For every bounded pseudoconvex domain in* $\mathbf{C}^2$ *with a strongly pseudoconvex boundary point* p*, the Kobayashi and the Wu metrics are uniformly asymptotically equivalent to the Levi metric at* p *which depends only upon the strongly pseudoconvex boundary near* p.

Acknowledgments

The core of this work was presented in The Third Pacific Rim Geometry Conference held in December, 1996. We would like to express our thanks to the organizers for their warm invitation.

Research of the author is supported in part by the KOSEF Grant 961-0104-021-2, by BSRI-97-1429, and by GARC of Seoul National University.

References

1. L. Ahlfors, *An extension of Schwarz's lemma,* Trans. Amer. Math. Soc. **43** (1938), 359-364.
2. K. Azukawa and M. Suzuki, *The Bergman metric on a Thullen domain,* Nagoya Math. J. **89** (1983), 1-11.

3. T.J. Barth, *The Kobayashi distance induces the standard topology,* Proc. Amer. Math. Soc. **35** (1972), 439-441.
4. S. Bergman, *The kernel function and conformal mapping,* (2nd ed.), Math. Surveys, No. 5, Amer. Math. Soc., Providence, R.I., 1970.
5. J. Bland, *The Einstein-Kähler metric on* $\{|z|^2 + |w|^{2p} < 1\}$, Michigan Math. J. **33** (1986), 209-220.
6. S.Y. Cheng and S.T. Yau, *On the existence of a complete Kähler metric on noncompact complex manifolds and the regularity of Fefferman's equation,* Comm. Pure Appl. Math. **33** (1980), 507-544.
7. S.S. Chern, *On holomorphic mappings of hermitian manifolds of the same dimension,* Proc. Symp. Pure Math. **11** (1966), 157-170.
8. C.K. Cheung and K.T. Kim, *Analysis of the Wu metric I: The case of convex Thullen domains,* Trans. Amer. math. Soc. **348** (1996), 1429-1457.
9. C.K. Cheung and K.T. Kim, *Analysis of the Wu metric II: The case of nonconvex Thullen domains,* Proc. Amer. Math. Soc. **125** (1997), 1131-1142.
10. I. Graham, *Boundary behavior of the Carathéodory and Kobayashi metrics on strongly pseudoconvex domains in* $\mathbf{C}^n$ *with smooth boundary,* Trans. Amer. Math. Soc. **207** (1975), 219-240.
11. F. John, *Extremum problems with inequalities as subsidiary conditions,* Studies and Essays presented to Richard Courant, Interscience, New York (1948), 187-204.
12. K.T. Kim and J. Yu, *Boundary behavior of the Bergman curvature in strictly pseudoconvex polyhedral domains,* Pacific J. Math. **176** (1996), 141-163.
13. S. Kobayashi, *Hyperbolic manifolds and holomorphic mappings,* Marcel-Dekker, New York, 1970.
14. L. Lempert, *La métrique de Kobayashi et la representation des domain sur la boule, Bull. Soc. Math. France* **109** (1981), 427-474.
15. H.L. Royden, *Remarks on the Kobayashi metric,* Several Complex Variables II, Maryland 1970, Springer-Verlag, 1971, 125-137.
16. T. Sunada, *Holomorphic equivalence problem for bounded Reinhardt domains,* Math. Ann. **235** (1978), 111-128.
17. H. Wu, *Old and new invariant metrics,* Several Complex Variables: Proc. of Mittag-Leffler Inst. 1987-88 (J.E. Fornaess ed.), Math Notes, Princeton Univ. Press **38** (1993), 640-682.
18. S.T. Yau, *A general Schwarz lemma for Kähler manifolds,* American J. math. **100** (1978), 197-203.

STRUCTURE OF THE IMAGE OF ANALYTIC DISCS ATTACHED TO TOTALLY REAL SUBMANIFOLDS

DAESUNG KWON
Department of Mathematics, Seoul National University,
Seoul 151–742, KOREA
E-mail:dskwon@math.snu.ac.kr

YONG-GEUN OH
Department of Mathematics, University of Wisconsin,
Madison, WI 53706, USA
E-mail:oh@math.wisc.edu

We prove that, allowing multiplicity, the image of a holomorphic map from a disc attached to a (maximally) totally real submanifold in $\mathbf{C}^n$ can be decomposed into a finite union of the images of *simple* discs with boundaries in the same totally real submanifold.

1 Introduction

Local structure of analytic discs attached to a given *smooth* totally real submanifold (in $\mathbf{C}^n$) is rather complex when the submanifold is not real-analytic.[3] In Kwon and Oh[6] and Oh[8,9], we study the local structure of such maps in the context of more general Gromov's pseudo-holomorphic discs and prove a general structure theorem of the images of pseudo-holomorphic discs, which seems to be new even in the classical context.

From now on, we will call this paper[6] by [KwO].

This paper is motivated by the second author's discussion with J. P. Rosay. Rosay pointed out that the proof in (Corollary 2.3, Oh[9]) of the unique continuation result is considerably simpler than that in (Theorem 1, Alinhac, Baouendi and Rothschild[1]) (see Remarks 2 of the paper) of a similar continuation result: *The proof in Oh[9] is purely soft, i.e., does not involve any hard analysis unlike the one in Alinhac, Baouendi and Rothschild[1].*

Because the main results from [KwO][6] seem to be equally interesting in the context of several complex variables, we decided to present them in this article. This paper is a simplified version of [KwO][6] restricted to $\mathbf{C}^n$. We hope that main ideas of the proofs in [KwO][6] become more accessible to the several complex variable community.

Although properties of the local structure of analytic discs continuously (and hence smoothly) attached to smooth totally real submanifolds are well-known,[3] the main global arguments we used in [KwO][6] do not seem to be simplified even in the classical context, and we repeat most of them here.

Let $D^2 \subset \mathbf{C}$ be the standard closed unit disc and $\Delta = D^2 \setminus \partial D^2$.

Theorem 1.1 (Stout [10]) *Let $w : \Delta \to \mathbf{C}^n$ be a non-constant bounded holomorphic map which is proper onto its image. Then $w(\Delta)$ is analytic subvariety of $\mathbf{C}^n$ and if $(\widetilde{V}, \widetilde{w})$ is a normalization of $w(\Delta)$, then $\widetilde{V}$ is biholomorphic to Δ and the map $w : \Delta \to w(\Delta)$ lifts to a proper holomorphic map $\phi : \Delta \to \widetilde{V}$ with*

$$w = \widetilde{w} \circ \phi. \tag{1.1}$$

Moreover, if we identify $\widetilde{V}$ with Δ, then ϕ is a finite Blaschke product, i.e., of the form with $s \geq 1$

$$\phi(z) = e^{ic} \prod_{k=1}^{s} \frac{z - \alpha_k}{1 - \overline{\alpha}_k z}$$

where c is a real constant and α_k are complex constants with $|\alpha_k| < 1$.

From the definition of a normalization, $\widetilde{w}$ is one-to-one away from discrete points in Δ. Following Stout [10] and Oh [9], we call the map $\widetilde{w}$ a "simple" map.

Boundedness and properness of the map in Theorem 1.1 can be replaced by the condition (1.2) below in the following theorem if we require that the map is smoothly extended to the boundary and the boundary of the image is on a given totally real submanifold. Following the notations from Globevnik and Stout [5] and Oh [9], we denote

$$A = \text{Image of } w|_{D^2}$$
$$bA = \text{Image of } w|_{\partial D^2}.$$

Theorem 1.2 (Theorem I, Oh [9]) *Let R be a totally real smooth submanifold of almost complex manifold (M, J) and $w : (D^2, \partial D^2) \to (M, R)$ be a non-constant holomorphic map such that it extends smoothly to the boundary and*

$$\text{“ } Int\ D^2 \backslash w^{-1}(bA) \text{ is connected ”.} \tag{1.2}$$

Then we can write

$$w = \widetilde{w} \circ b \tag{1.3}$$

where $\widetilde{w} : (D^2, \partial D^2) \to (M, R)$ is a simple holomorphic disc with smooth extension to the boundary, and $b : D^2 \to D^2$ is a finite Blaschke product.

It is easy to check that (1.2) is equivalent to

$$Int\ D^2 \cap w^{-1}(bA) \text{ is discrete.}$$

In the above theorem, the hypothesis (1.2) cannot be dropped as shown by the following example given in [Remark 3.3, Oh [8]]: Consider the holomorphic map

$$w : \mathbf{H} \to \mathbf{C}, \quad w(z) = z^k \tag{1.4}$$

where $\mathbf{H} \subset \mathbf{C}$ is the upper half-plane. Identify $\mathbf{C} \cup \{\infty\} = S^2$ and $\mathbf{H} \cup \{\infty\} = D^2 \subset \mathbf{C}$ and $\partial\mathbf{H} = \mathbf{R} \subset \mathbf{C}$. If k is odd, it cannot be written as a multiple covering of a simple disc since the multiplicity is not constant. To formulate a structure theorem without (1.2), we consider connected domains in $\Delta \setminus w^{-1}(bA)$. The finiteness of the number of components of $\Delta \setminus w^{-1}(bA)$ can be shown easily (see [KwO] [6] for details of the proof). We denote

$$\Delta\backslash w^{-1}(bA) = \bigcup_{j=1}^{\ell} E_j. \tag{1.5}$$

and by $\widetilde{E}_j$ the domain obtained by adding a discrete set of points in Int $\overline{E_j}$ so that $\widetilde{E}_j$ has no discrete holes. On general almost-complex manifold (M, J), $\widetilde{E}_j$ need not to be simply connected (see [KwO] [6] for such an example). But on $\mathbf{C}^n$ we have

Theorem I *Let w be an analytic disc attached to a totally real submanifold R in $\mathbf{C}^n$. Let $\{\widetilde{E}_j\}_{j=1}^{\ell}$ be the connected domains in (1.5). Then all $\widetilde{E}_j$'s are simply connected.*

Essentially the same proof of Theorem I applies as long as the ambient manifold (M, J) does not admit a non-constant J-holomorphic sphere. [6]

Since $w|_{\overline{\widetilde{E}_j}}$ satisfies (1.2), combining Theorem 1.2 and Theorem I, we obtain the following main structure theorem.

Theorem II (Main Structure Theorem) *The image of an analytic disc attached to a smooth totally real submanifold in $\mathbf{C}^n$ can be decomposed into a finite union of simple discs (with multiplicity) attached to the same totally real submanifold.*

If R is real-analytic, we can improve the above to the following stronger result by modifying the proof of Theorem I combined with the local reflection of the image along its boundary:

Theorem III *Let R be a real-analytic totally real submanifold in $\mathbf{C}^n$. Then for any holomorphic disc $w : (D^2, \partial D^2) \to (\mathbf{C}^n, R)$, $\Delta \setminus w^{-1}(bA)$ is connected and so w allows the factorization (1.3).*

Rosay's example in Section 6 shows that Theorem III fails for general smooth totally real boundary condition.

2 Preliminaries

In this section, we organize various theorems on the local structure of holomorphic discs which are proved in the theory of J-holomorphic discs. We borrow them mainly from the McDuff *et al* [7], Floer *et al* [4] and Oh [9]. Let R be a (maximally) smooth totally real submanifold of $\mathbf{C}^n$, i.e.,

$$\dim R = n \text{ and } T_x R \cap i\, T_x R = \{0\} \text{ for all } x \in R.$$

A smooth map

$$w : (D^2, \partial D^2) \to (\mathbf{C}^n, R)$$

is called an *analytic disc attached to* R if w satisfies

$$\begin{cases} i \circ dw = dw \circ i \\ w(\partial D^2) \subset R. \end{cases} \tag{2.1}$$

Note that by the regularity theorem, any continuous map which satisfies (2.1) is smooth (see e.g., Chirka [3]). Hence we always assume the map is smooth. We also denote by $B_\epsilon(0)$ the open disc centered at 0 with radius $\epsilon > 0$ and by $D_\epsilon(0)$ the semi-disc centered at 0 with radius $\epsilon > 0$ and with boundary

$$\partial D_\epsilon(0) = (-\epsilon, \epsilon) \subset \mathbf{R} \subset \mathbf{C}.$$

Following the notations from Globevnik *et al* [5] and Oh [9], we denote

$$\begin{aligned} A &= \text{Image of } w|_{D^2} \\ bA &= \text{Image of } w|_{\partial D^2} \end{aligned}$$

and introduce the *multiplicity function* $\kappa : A \to \mathbf{Z}_+$.

Definition 2.1 We define the integer valued function $\kappa : A \to \mathbf{Z}_+$ by

$$\kappa(a) = \#(w^{-1}(a)), \quad a \in A$$

and call it *the multiplicity* of a. If R is real-analytic, then for any $x \in R$, we can find neighborhoods U of x, V of $0 \in \mathbf{C}^n$ and a biholomorphic map $\phi : U \to V$ such that

$$\phi(U \cap R) \subset \mathbf{R}^n.$$

However, when R is not real-analytic, we can only require ϕ is a diffeomorphism. In this case, $\phi_* i := \phi_* \circ i \circ (\phi^{-1})_*$ is an integrable almost complex structure which is *not standard* and $\mathbf{R}^n$ is totally real with respect to $\phi_* i$ on V. We call such a local diffeomorphism *boundary flattening map*. Therefore we are naturally led to the realm of general pseudo-holomorphic discs when we (locally) flatten smooth totally real submanifold R. In this sense, it will be natural and useful to use the general theory from Oh [8,9] and Kwon *et al* [6] of pseudo-holomorphic discs even in the study of structure of classical analytic discs attached to totally real smooth submanifold.

We now quote several results on the local structure of J-holomorphic discs (with boundary) from Floer *et al* [4] or McDuff *et al* [7], and Oh [9]. We denote either by $z = (x, y)$ or by $z = x + iy$ the standard coordinates in $\mathbf{R}^2$ or $\mathbf{C}$. Let J be an almost complex structure on $\mathbf{R}^{2n}$ and R be a (maximally) totally real submanifold with respect to J, i.e., $\dim R = n$ and $T_xR \cap JT_xR = \{0\}$ for all $x \in R$. A smooth map $w : (D^2, \partial D^2) \to (\mathbf{R}^{2n}, R)$ is called a *J-holomorphic disc with the boundary condition* R if it satisfies

$$\begin{cases} J \circ dw = dw \circ i \\ w(\partial D^2) \subset R. \end{cases} \tag{2.2}$$

We denote by $\mathcal{T}_J(n)$ the set of *totally real* planes in $\mathbf{R}^{2n}$ with respect to the complex structure J, i.e. an automorphism $J : \mathbf{R}^{2n} \to \mathbf{R}^{2n}$ with $J^2 = -Id$. Maps whose first derivatives lie in L^p are called $H^{1,p}$-maps. Let $z \mapsto J(z)$ be a $H^{1,p}$-map associating to $z \in D_\epsilon(0)$ a complex structure $J(z)$ on $\mathbf{R}^{2n}$, and let $z \mapsto C(z)$ be a map in $L^p(D_\epsilon(0), \mathcal{L}_{\mathbf{R}}(\mathbf{R}^{2n}))$ where $\mathcal{L}_{\mathbf{R}}(\mathbf{R}^{2n})$ is the set of endomorphisms on $\mathbf{R}^{2n}$. Now, for each given smooth path $\tau(x) \in \mathcal{T}_{J(x)}(n)$, $x \in (-\epsilon, \epsilon)$, we consider the first-order elliptic partial differential equation

$$\begin{cases} \frac{\partial \xi}{\partial x}(z) + J(z)\frac{\partial \xi}{\partial y}(z) + C(z)\xi(z) = 0, \ \xi(0) = 0 \\ \xi(x, 0) \in \tau(x) \quad \text{for} \quad x \in (-\epsilon, \epsilon). \end{cases} \tag{2.3}$$

One important useful result in the theory of pseudo-holomorphic curves is the following *Similarity Principle* which is a boundary version of the one from Floer *et al* [4]. This tells us that the behavior of a solution of (2.3) is similar to that of classical cases. We get the perturbed equation (2.3) either by differentiating (2.2) or by flattening R locally. After locally flattening R , we study the boundary condition

$$\xi(x, 0) \in \mathbf{R}^n. \tag{2.4}$$

The main reason why we flatten the boundary is similar to that of why we do so in the general theory of elliptic boundary value problems: *We need to study estimates for the equation with variable coefficients (with flat boundary) even for that of constant coefficients for general boundary value problems.*

Theorem 2.2 (Theorem 2.1, Oh [9]) *Let $p \in (2, \infty)$ and assume $\xi \in H^{1,p}(D_\epsilon(0), \mathbf{R}^{2n})$ solves (2.3). Then there exists $\delta \in (0, \epsilon)$ and maps*

$\Phi \in H^{1,p}(D_\delta(0), \mathcal{L}_{\mathbf{R}}(\mathbf{R}^{2n}))$ *such that* $\Phi(z)$ *is invertible for all* $z \in D_\delta(0)$, *and* $\sigma \in C^\infty(D_\delta(0), \mathbf{R}^{2n})$ *such that*

$$\xi(z) = \Phi(z)\sigma(z), \; J(\dot{z})\Phi(z) = \Phi(z)i \tag{2.5}$$

and if we consider σ *as a map from* $D_\delta(0)$ *to* $\mathbf{C}^n$, *it satisfies*

$$\begin{cases} \overline{\partial}\sigma = 0 \\ \sigma(x,0) \in \mathbf{R}^n \subset \mathbf{C}^n. \end{cases} \tag{2.6}$$

From this theorem, we get following corollaries.

Corollary 2.3 (Corollary 2.2, Oh [9]) *Let* R *be a totally real submanifold. Let* $w : (D_\epsilon(0), \partial D_\epsilon(0)) \to (\mathbf{R}^{2n}, R)$ *be a smooth solution of*

$$\begin{cases} \frac{\partial w}{\partial x} + i\frac{\partial w}{\partial y} = 0 \\ w(x,0) \in R \quad \text{for } x \in (-\epsilon, \epsilon) = \partial D_\epsilon(0). \end{cases} \tag{2.7}$$

Then there exists $\delta > 0$ *such that* $dw(z) \neq 0$ *for all* z *with* $0 < |z| < \delta$.
Corollary 2.3 implies that the set of critical points of a non-constant J-holomorphic disc is discrete. In the following Corollary 2.4 and Lemma 2.5, E_ϵ denotes either $B_\epsilon(0)$ or $D_\epsilon(0)$.

Corollary 2.4 (Corollary 2.3, Oh [9]) *Let* R *as above and let* $w_1, w_2 : E_\epsilon \to \mathbf{R}^{2n}$ *be smooth solutions of (2.7). If the* ∞*-jets of* w_1 *and* w_2 *coincide with* 0, *then*

$$w_1 \equiv w_2.$$

If $dw(0) \neq 0$, then we may choose a better boundary flattening map. We mainly apply the following Lemma 2.5 for the complex structure $\phi_* i$.

Lemma 2.5 (Lemma 2.2.2, McDuff and Salamon [7]) *Let* $w : E_\epsilon \to \mathbf{R}^{2n}$ *be* J*-holomorphic with* $dw(0) \neq 0$. *Then there is a local chart* $\alpha : U \to \mathbf{C}^n$ *near* $w(0)$ *such that*

$$\alpha \circ w(z) = (z, 0, \ldots, 0)$$
$$d\alpha(w(z)) \circ J(w(z)) = id\alpha(u(z))$$

for $z \in E \cap w^{-1}(U)$.

Lemma 2.6 *Let* $u, v : B_\epsilon(0) \to \mathbf{R}^{2n}$ *be non-constant holomorphic maps with* $u(0) = v(0)$. *Assume that there are sequences* z_μ, $\xi_\mu \in B_\epsilon(0)$ *such that*

$$u(z_\mu) = v(\xi_\mu), \; \lim_{\mu\to\infty} z_\mu = \lim_{\mu\to\infty} \xi_\mu = 0, \; z_\mu \neq 0 \neq \xi_\mu.$$

Then there is a holomorphic map $\phi : B_\delta(0) \to B_\epsilon(0)$ *with* $0 < \delta < \epsilon$, $\phi(0) = 0$ *and*

$$v = u \circ \phi$$

on $B_\delta(0)$.

We cannot expect the obvious analogue of Lemma 2.6 for the semi-disc case. This is because the images can be in the "different" sides of the totally real submanifold and because they can also branch off as Rosay's example in Section 6 illustrates. For the semi-disc case, we could expect only the following. The classical proof for Lemma 2.6 cannot be applied to the boundary point *because general holomorphic disc cannot be extended across its boundary as a smooth holomorphic map,* unless R is real analytic near the point. We refer readers to the proof in Oh [9] for this point.

Lemma 2.7 (Corollary 2.4, Oh [9]) *Let* $w_i : (D_{\epsilon_i}(0), \partial D_{\epsilon_i}(0)) \to (\mathbf{C}^n, R)$ *be holomorphic with*

$$w_1(0) = w_2(0), \quad dw_1(0) \neq 0.$$

Assume that

$$\text{Im } w_1|_{\partial D_{\epsilon_1}(0)} = \text{Im } w_2|_{\partial D_{\epsilon_2}(0)}.$$

Then the following alternative holds:
(i) Im w_1 *and* Im w_2 *overlap on some open neighborhoods of* 0 *or*
(ii) $dw_2(0) \neq 0$ *and* Im $w_1 \cup$ Im w_2 *smoothly matches up along the boundary* $\partial D_\epsilon(0)$ *near* 0.

One special case in Lemma 2.7 is worthwhile to separately mention and will be very useful for later discussions: Suppose that for $i = 1, 2$, $w_i : (D_{\epsilon_i}(0), \partial D_{\epsilon_i}(0)) \to (\mathbf{C}^n, R)$ be holomorphic with

$$w_1(0) = w_2(0), \quad dw_1(0) \neq 0 \neq dw_2(0)$$

such that there are some neighborhoods U_1 and U_2 of 0 in $D_\epsilon(0)$ respectively and

$$\text{Im } w_1\big|_{\partial D_{\epsilon_1}(0) \cap U_1} = \text{Im } w_2\big|_{\partial D_{\epsilon_2}(0) \cap U_2}.$$

We assume that $w_i : U_i \to \mathbf{R}^{2n}$ are one-one, which can be always achieved by choosing smaller U_i's if necessary. Denote

$$\partial D_{\epsilon_1}(0) \cap U_1 = (-\delta_1^1, \delta_2^1) =: \Lambda_1$$
$$\partial D_{\epsilon_2}(0) \cap U_2 = (-\delta_1^2, \delta_2^2) =: \Lambda_2$$

for some $\delta_j^i > 0$. Now consider the diffeomorphism

$$\left(w_2\big|_{\Lambda_2}\right)^{-1} \circ w_1\big|_{\Lambda_1} : \Lambda_1 \to \Lambda_2. \tag{2.8}$$

We provide the boundary orientation to $\Lambda_i \subset \partial D_{\epsilon_i}(0)$ induced from the complex orientation in $D_{\epsilon_i}(0) \subset \mathbf{C}$.

Corollary 2.8 *Let w_i's, Λ_i's as above. Then the following alternative holds:*
(i) if the diffeomorphism (2.8) preserves the orientation, then Lemma 2.7 (i) must hold.
(ii) if the diffeomorphism (2.8) reverses the orientation, then Lemma 2.7 (ii) must hold.

3 Local structure of piercing

As we mentioned in the introduction, it is essential to analyze how bA pierces through A for the purpose of understanding structure of the images of holomorphic discs. We study this piercing by analyzing the "net" on Δ provided by $w^{-1}(bA) \cap \Delta$. In this section, we derive several results concerning the local structure of this piercing based on the results from Section 2. We will always assume that all (local) holomorphic maps appearing below have a finite number of critical points, and a discrete set of isolated self-intersection points (away from the critical values). For example, all the results in this section will apply to the maps obtained by restricting the domain of any given holomorphic map $w : (D^2, \partial D^2) \to (\mathbf{C}^n, R)$. The following notion of the collar neighborhood turns out to be useful for this purpose.

Definition 3.1 For a point $z \in D^2$, we say that $U \subset D^2$ is a *collar neighborhood* of z (or λ) if $\partial U = \lambda$ is an embedded piecewise smooth arc containing z.

The following lemma is a variation of Lemma 2.6, which shows that if the boundary of holomorphic disc is contained in the *interior* image of another such disc near a regular point of it, images of the two discs should overlap in a collar neighborhood of the boundary of the first disc. Because of the *branching phenomenon* as we mentioned before, we cannot expect this if we replace interior image by *boundary* image of the second disc.

Lemma 3.2 *Let $u : (D_\epsilon(0), \partial D_\epsilon(0)) \to (\mathbf{C}^n, R)$ and $v : B_\epsilon(0) \to \mathbf{C}^n$ be holomorphic. Let $\{z_\mu\} \subset \partial D^2_\epsilon$ and $\{w_\mu\} \subset B^2_\epsilon$ be sequences with*

$$u(z_\mu) = v(w_\mu) \qquad \text{and}$$
$$\lim_{\mu\to\infty} z_\mu = 0,\ z_\mu \neq 0, \quad \lim_{\mu\to\infty} w_\mu = 0,\ w_\mu \neq 0.$$

Suppose in addition that $dv(0) \neq 0$. Then there exists a collar neighborhood $U \subset D_\epsilon(0)$ of z with $\lambda = \partial U$, an arc contained in ∂D^2 such that

$$u(U) \subset v(B_\epsilon(0)).$$

Proof Let ϕ be a local diffeomorphism of an open neighborhood U of $u(0) = v(0)$ onto an open neighborhood V of 0 such that $\phi(U \cap R) \subset \mathbf{R}^n$. We may assume, by Lemma 2.5, that

$$\begin{aligned}\tilde{v}(z) &= \phi \circ v(z) = (z, 0, \ldots, 0)\\ \phi_* i(z, 0, \ldots, 0) &= i\end{aligned}$$

for all $z \in B_\epsilon(0)$. Let $\tilde{u}(z) = \phi \circ u(z)$ and write

$$u(z) = (a(z), b(z)), \;\; a(z) \in \mathbf{C}, \; b(z) \in \mathbf{C}^{n-1}$$

for $z \in D_\epsilon(0)$ such that $u(\partial D_\epsilon(0)) \subset \mathbf{R}^n$ and $b(z_\mu) = 0$. If ∞-jet of $b(z)$ at 0 does not vanish, then there is an non-negative integer ℓ such that $b(z) = O(|z|^\ell)$ and $b(z) \neq O(|z|^{\ell+1})$ and hence $\phi_* i(u(z)) = i + O(|z|^\ell)$. Consider the Taylor expansion up to order $\ell - 1$ of the equation

$$\frac{\partial u}{\partial x} + \phi_* i(u(z)) \frac{\partial u}{\partial y} = 0$$

on $D_\epsilon(0)$. Then we have

$$\frac{\partial u_\ell}{\partial x} + i \frac{\partial u_\ell}{\partial y} = 0, \qquad \text{for } z \in D_\epsilon(0)$$

where u_ℓ is the Taylor expansion of u up to order ℓ. So, there exists a nonzero a such that

$$b(z) = az^\ell + O(|z|^{\ell+1}) \qquad \text{for } z \in D_\epsilon(0)$$

and hence $b(z) \neq 0$ in some open neighborhood of 0. This then implies that the images of v and u do not intersect away from 0 in the neighborhood, which contradicts the assumption in the lemma. Therefore ∞-jet of $b(z)$ at 0 must vanish. Since $\phi_* i(z, 0, \ldots, 0) \equiv i$ for $z \in B_\epsilon(0)$, $b(z)$ satisfies the equation

$$\begin{aligned}0 &= \tfrac{\partial b}{\partial x} + pr_2(\phi_* i(a, b) \tfrac{\partial u}{\partial y})\\ &= \tfrac{\partial b}{\partial x} + pr_2\Big(\big(i + \textstyle\int_0^1 D_b(\phi_* i(a, b)) d\tau \cdot v_2\big) \tfrac{\partial u}{\partial y}\Big)\\ &= \tfrac{\partial b}{\partial x} + i \tfrac{\partial b}{\partial y} + (Ab)(z)\end{aligned}$$

in a neighborhood of $\partial D_\epsilon(0)$ in $D_\epsilon(0)$ where $pr_2 : \mathbf{C}^n \to \mathbf{C}^{n-1}$ is the projection and the operator A is defined by

$$(A\xi)(z) := pr_2\Big(\int_0^1 D_b(\phi_* i(a, \tau b) d\tau) \cdot \xi \frac{\partial u}{\partial y}\Big)(z).$$

Since $b(x, 0) \in \mathbf{R}^{n-1}$ and the ∞-jet of b at 0 vanishes, By Lemma 2.3,

$$b(z) \equiv 0.$$

This finishes the proof. □

An immediate corollary, when combined with following Proposition 3.5 and Proposition 3.7, is

Corollary 3.3 *Let* $w : (D^2, \partial D^2) \to (\mathbf{C}^n, R)$ *be a holomorphic disc with* $dw(z') \neq 0$ *for some* $z' \in w^{-1}(w(z)) \cap \Delta$ *where* $z \in \partial D^2$. *Suppose that* $\lambda_0 \subset \partial D^2$ *is an arc with* z *as one end point and that there exists a collar neighborhood* V *of* λ_0 *such that*

$$w(V) \subset w(\Delta).$$

Then there exists a collar neighborhood U *of an open arc* $\lambda = \partial U$ *containing* z *in its interior with*

$$w(U) \subset w(\Delta).$$

This corollary shows that the following branching picture of the images cannot occur.

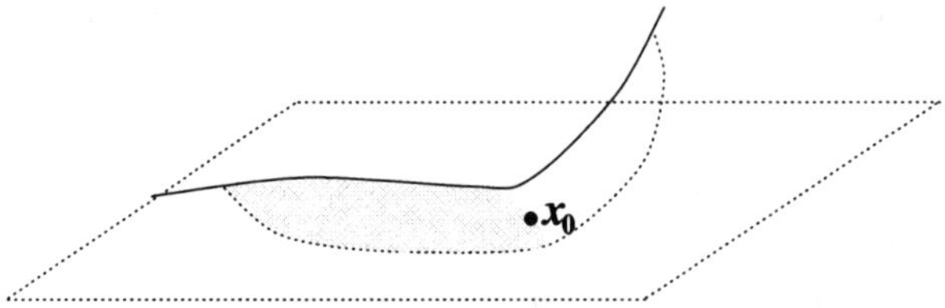

Figure 1

The following objects appear often in later discussions and so we assign a special name to them. They describe the structure of $B_\epsilon(z) \cap w^{-1}(bA)$ at a critical point z of w with $w(z) \in bA$.

Definition 3.4 The union of k embedded arcs $\lambda_1, \ldots, \lambda_k$ in the plane that share one end point z from each as drawn in Figure 2, is called a *k-armed asterisk*. An asterisk $\cup_{j=1}^k \lambda_j \subset D_\epsilon(z)$ with

$$\lambda_1 \cup \lambda_k \subset \partial D_\epsilon(z)$$

as drawn in Figure 2, is called a *semi-asterisk*. We call each λ_j in the asterisk an *arm* and the shared end point the *center* of the asterisk. We always order the arms either clockwise or counterclockwise in the plane.

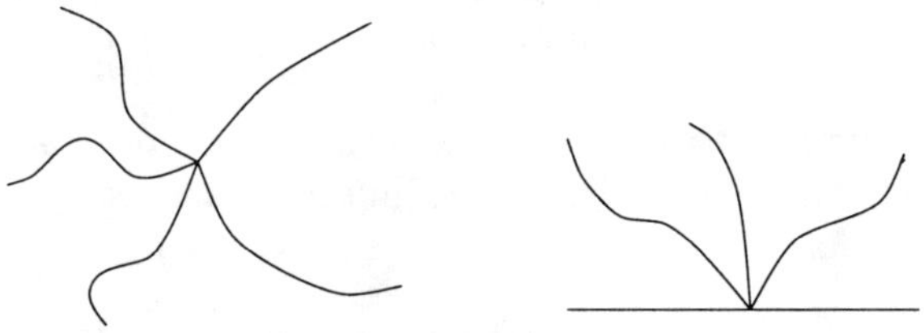

Figure 2

The following proposition is the crucial tool in the proof of the main theorem. Roughly speaking, it shows that as long as the boundary of one disc does not intersect with the interior of the other, "overlap" of the images of two discs uniquely propagates *along the boundary of discs.* Recall that it is a well-known and an immediate consequence of Lemma 2.3 that overlap of the interior images of the two discs uniquely propagates.

Proposition 3.5 (Unique continuation of the image) *Let* $w_i : (D_{\epsilon_i}(0), \partial D_{\epsilon_i}(0)) \to (\mathbf{C}^n, R)$ *for* $i = 1, 2$ *be two holomorphic maps satisfying the following properties:*

(i) they can be smoothly extended to a neighborhood $D_{\epsilon_i'}(0)$ *with* $\epsilon_i' > \epsilon_i$ *of* $\overline{D_{\epsilon_i}(0)}$ *in* $\mathbf{H} \subset \mathbf{C}$.

(ii) $w_1(0) = w_2(0)$

(iii) $dw_i(z) \neq 0$, $w_i(z) \neq w_i(0)$ *for any* z *with* $0 < |z| < \epsilon_i'$ *respectively for* $i = 1, 2$.

(iv) Int $D_{\epsilon_i}(0) \cap w_i^{-1}(w_j(\partial D_{\epsilon_j'}(0)))$ *is discrete in* $\overline{D_{\epsilon_i}(0)} \setminus \{0\}$, *for all* $i, j = 1, 2$.

(v) there exists some $d > 0$ *such that*

$$\text{dist}\,(p, w_i(\overline{D_{\epsilon_i}(0)} \setminus D_{\epsilon_i}(0))) > d > 0$$

where $p = w_1(0) = w_2(0)$.

If there exist collar neighborhoods U_i *of* $(-\epsilon_i, 0)$ *in* $D_{\epsilon_i}(0)$ *with* $\partial U_i = (-\epsilon_i, 0)$ *for* $i = 1, 2$ *and with* $w_1(U_1) = w_2(U_2)$, $w_1(\partial U_1) = w_2(\partial U_2)$, *then we can find collar neighborhoods* V_i *of* $(-\epsilon_i, \delta_i)$ *such that* $\delta_i > 0$ *and*

$$w_1(V_1) = w_2(V_2).$$

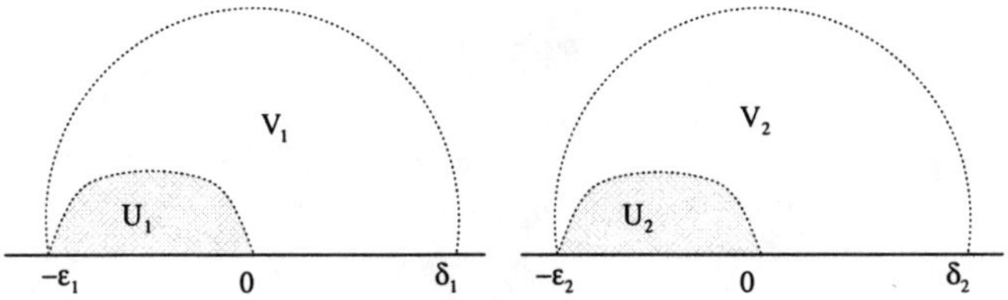

Figure 3

Remark 3.6 (i) The hypothesis (v) can be always obtained by choosing ϵ_i's smaller if necessary.
(ii) This proposition fails to be true if we drop the hypothesis (iv), i.e., allow the image of the boundary to pierce through that of the interior. For example, consider the two maps

$$w_1(z) = z, \quad w_2(z) = z^2 : (D_\epsilon(0), \partial D_\epsilon(0)) \to (\mathbf{C}, \mathbf{R})$$

and start with the intervals $(0, \epsilon_i)$ for appropriate $0 < \epsilon_i < \epsilon$ for $i = 1$, 2.
(iii) This proposition rules out the following branching phenomenon under the hypotheses in the proposition, which otherwise could occur. Compare with Figure 1.

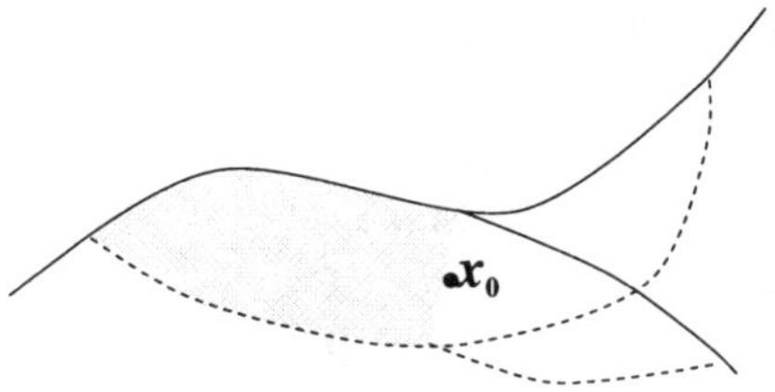

Figure 4

To prove this proposition, we start with two lemmas. The first lemma rules out folding of the image (See Figure 5). The second one is an continuation lemma of the collar neighborhood.

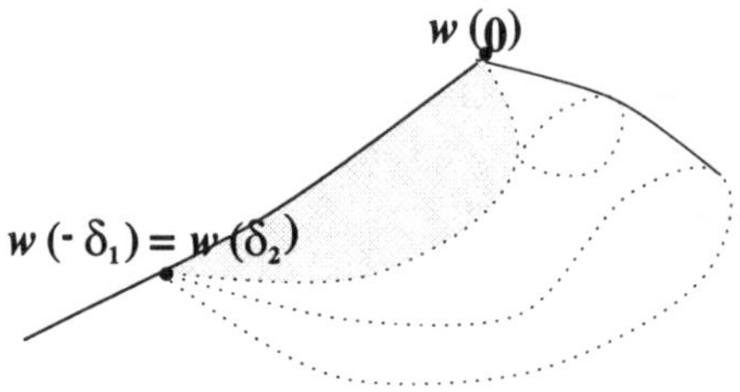

Figure 5

Lemma 3.5.1 *Under the hypotheses as in Proposition 3.5, there cannot be any pair of collar neighborhoods U of $(-\delta_1, 0)$ and V of $(0, \delta_2)$ for some $0 < \delta_1, \delta_2 < \min\{\epsilon_1, \epsilon_2\}$ such that*

$$w_i(U) = w_j(V)$$

for any $i \neq j$, $1 \leq i, j \leq 2$.

Proof Because of (iii), w_i are immersed on $(-\epsilon_i, \epsilon_i)\backslash\{0\}$ for $i = 1, 2$. Suppose the contrary that there exist such neighborhoods U, V and assume without loss of any generality that

$$w_1(U) = w_2(V).$$

Since self-intersection points can accumulate only at critical points, there exist some $0 < \delta_1' < \widetilde{\delta}_1 < \delta_1$ and $0 < \delta_2' < \widetilde{\delta}_2 < \delta_2$ and collar neighborhoods A, B of $(-\widetilde{\delta}_1, -\delta_1')$ and $(\delta_2', \widetilde{\delta}_2)$ in $D_{\epsilon_1}(0)$ and $D_{\epsilon_2}(0)$ respectively such that $w_1|_A$ and $w_2|_B$ are one-one and

$$w_1(A) = w_2(B), \;\; w_1(-\delta_1') = w_2(\delta_2'), \;\; w_1(-\widetilde{\delta}_1, -\delta_1') = w_2(\delta_2', \widetilde{\delta}_2).$$

Then we consider the map $(w_1|_A)^{-1} \circ w_2|_B : B \to A$. Because $(w_1|_A)^{-1} \circ w_2|_B : B \to A$ is holomorphic and so preserves the complex orientations of $B, A \subset \mathbf{C}$, the diffeomorphism

$$\left(w_1|_{(-\widetilde{\delta}_1, -\delta_1')}\right)^{-1} \circ w_2|_{(\delta_2', \widetilde{\delta}_2)} : (\delta_2', \widetilde{\delta}_2) \to (-\widetilde{\delta}_1, -\delta_1')$$

must preserve the induced (boundary) orientations which are nothing but the standard orientations on them. However this is impossible by Corollary 2.7 because we assume $w_1(-\delta_1') = w_2(\delta_2')$ from (ii) and hence this finishes the proof.□

Lemma 3.5.2 *Under the same hypotheses as in Proposition 3.5, there exist some $\delta_1, \delta_2 > 0$ such that*

$$w_1((0, \delta_1)) = w_2((0, \delta_2)).$$

Proof Suppose the contrary, say, that there exist a sequence of points $x_j \in (0, \epsilon_1)$ with $x_j \to 0$ such that

$$w_1(x_j) \notin w_2((0, \epsilon_2)). \tag{3.1}$$

By the continuity of w_1 and by (v) for a sufficiently large j_0, we can choose a smooth curve $c_0 : [0, 1] \to D_{\epsilon_1}(0)$ such that

$$\begin{aligned} &c_0(0) \in (-\epsilon_1, 0), \quad c_0(1) = x_{j_0} \\ &c_0(t) \in \text{Int } D_{\epsilon_1}(0) \setminus w_1^{-1}(w_2(\partial D_{\epsilon_2'}(0))) \;\text{ and} \\ &\text{dist }(p, w_1(c_0(t))) \leq \tfrac{d}{2} \quad \text{for } 0 < t < 1. \end{aligned} \tag{3.2}$$

It follows from (3.2) and (v) that

$$w_1(c_0) \cap w_2(\overline{D_{\epsilon_2}} \backslash D_{\epsilon_2}) = \emptyset. \tag{3.3}$$

Define

$$t_0 = \sup\{t \in [0,1] \mid w_1(c_0(t)) \in \text{Im } w_2\}.$$

From (iv), the main hypothesis on the collar neighborhoods U_1 , U_2 and hypothesis (3.1), it follows that

$$0 < t_0 < 1.$$

By the continuity of w_2 and from (3.3) we derive that

$$w(c_0(t_0)) = w_2(z) \quad \text{for some } z \in D_{\epsilon_2}(0).$$

From (3.2), $z \in \text{Int } D_{\epsilon_2'}(0)$. Since $dw_1(c_0(t_0)) \neq 0 \neq dw_2(z)$ from (iii), we can choose a small disc $B_\delta(z) \subset$ Int $D_{\epsilon_2}(0)$ by Lemma 2.5 such that $w_1 \circ \phi = w_2$ on $B_\delta(z)$ for a holomorphic map $\phi : B_\delta(z) \to \text{Int } B_{\epsilon_1}(0)$. Therefore we can find $t_1 > t_0$ such that $w_1(c_0(t_1)) \subset \text{Im } w_2$ which contradicts the definition of t_0. Hence, we have proved Lemma 3.5.2.□

Proof of Proposition 3.5 Again by choosing sufficiently small $\widetilde{\delta_i}$'s, we may assume that the map

$$\left(w_1\big|_{(0,\widetilde{\delta_1})}\right)^{-1} \circ w_2\big|_{(0,\widetilde{\delta_2})} : (0,\widetilde{\delta_2}) \to (0,\widetilde{\delta_1})$$

becomes a diffeomorphism. Since $w_1(0) = w_2(0)$ from (ii), this diffeomorphism must preserve the orientations on $(0,\widetilde{\delta_i})$'s. Therefore Proposition 3.5 immediately follows from Corollary 2.7 (ii) and Lemma 2.5, which finishes the proof. □

If R is real-analytic, then we can use the reflection principle along the boundary of the image, which immediately gives rise to Proposition 5.2. It basically rules out the *branching phenomenon along the boundary* in the real analytic case which complicates the image of general smooth cases as the map $h_\pm$ illustrate in Rosay's example in Section 6.

The next proposition shows that even when the boundary image of one disc penetrates into the interior image of the other, overlap of the image uniquely propagates across the arms of certain asterisk, at least near the center of the asterisk. We recommend readers to see Figure 6 before reading the statement of this proposition.

Proposition 3.7 *Let* $w : (D_\epsilon(0), \partial D_\epsilon(0)) \to (\mathbf{C}^n, R)$ *be a holomorphic map with* $dw(z) \neq 0$ *on* $D_\epsilon(0)\backslash\{0\}$, *and let* $C = \bigcup_{j=1}^{k+2} \lambda_j \subset D_\epsilon(0)$ *be a* $(k+2)$*-armed semi-asterisk in* $D_\epsilon(0)$ *centered at* 0 *such that* $w(C) \subset R$ *and* $w\big|_{C\backslash\{0\}}$ *is an embedding (away from a discrete set of points). And let* $u : \Delta \to \mathbf{C}^n$ *be a holomorphic map such that all the points in* $(w(C)\backslash w(0)) \cap \text{Im } u$*, if any, are* *regular values of u. Suppose that there exists a collar neighborhood U_1 of λ_1 and an arc $\delta_1 \subset \Delta$ and a collar neighborhood V_1 of δ_1 such that*

$$w(\lambda_1) = u(\delta_1) \text{ and } w(U_1) = u(V_1). \tag{3.4}$$

Then there exists a set of arcs $\delta_1, \delta_2, \ldots, \delta_{k+2}$ in Δ that share one of end points from each, and collar neighborhoods U_ℓ, V_ℓ of $\lambda'_\ell \subset \lambda_\ell$ and δ_ℓ respectively such that

$$w(\lambda'_\ell) = u(\delta_\ell) \text{ and } w(U_\ell) = u(V_\ell) \tag{3.5}$$

for all $1 \leq \ell \leq k+2$. Note that in particular the union $\bigcup_{j=1}^{k+2} \delta_j$ is connected, and that some of δ_j's could be overlapped with one another.

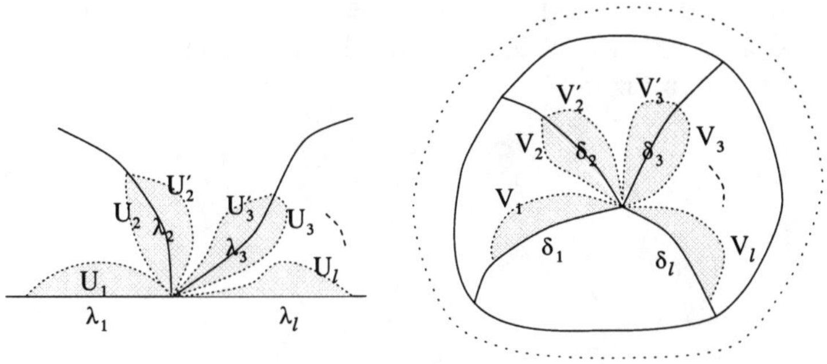

Figure 6

Proof When $k = 0$, $C = (-\epsilon_1, 0] \cup [0, \epsilon_2)$ for some $0 < \epsilon_i < \epsilon$. From (3.4) and Corollary 3.3, there exists $0 < \epsilon'_2 \leq \epsilon_2$ and a collar neighborhood N of $(-\epsilon_1, \epsilon'_2)$ in $D_\epsilon(0)$ such that

$$w(N) \subset \text{Im } u.$$

Consider the subset $u^{-1}(w(N)) \subset \Delta$. Certainly this contains the collar neighborhood $V_1 \cap u^{-1}(w(N))$ of δ_1 in Δ. Let $z_1 \in \Delta$ be one of the end points of δ_1 in Δ such that $w(0) = u(z_1)$. It will be enough to prove that the connected component of $u^{-1}(w(C \cap N))$ containing δ_1 contains a non-trivial connected arc δ_2 having z_1 as one end point. Then it will follow

$$u(\delta_2) = w((0, \epsilon''_2))$$

for some $0 < \epsilon''_2 < \epsilon'_2$ and so (3.5) for $\ell = 2$. However this immediately follows from the fact that $u : B_\delta(z_1)\backslash\{z_1\} \to \mathbf{C}^n$ is immersed for some $\delta > 0$, and so $u : B_\delta(z_1) \to \text{Im } u$ is a branched covering and an open map. This finishes the proof of Proposition 3.7 for $k = 0$. Now suppose that we have proven the proposition for $1 \le \ell \le k+1$, and consider the case $\ell = k+2$. Consider the asterisk $C' = C\backslash\{\lambda_{k+2}\}$ and apply the Riemann-mapping theorem to make C' into a $(k+1)$-armed semi-asterisk. Applying the induction hypothesis, there exists a sequence of $\{\delta'_\ell\}_{1\le\ell\le k+1}$ consisting of connected arcs sharing one of the end points from each such that there exist sub-arcs $0 \in \lambda'_\ell \subset \lambda_\ell$

$$w(\lambda'_\ell) = u(\delta'_\ell)$$

and collar neighborhoods U_ℓ, V_ℓ of λ'_ℓ and δ_ℓ respectively with

$$w(U_\ell) = u(V_\ell).$$

Finally consider the 2-armed asterisk $C'' = \lambda_{k+1} \cup \lambda_{k+2} \subset D_\epsilon(0)$. First note that the collar neighborhood U_{k+1} obtained above lie in the side towards the arm λ_k as drawn in Figure 7.

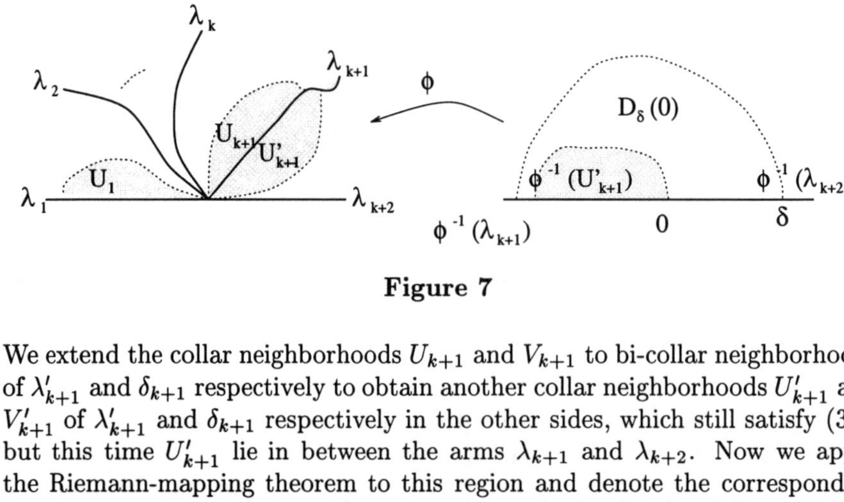

Figure 7

We extend the collar neighborhoods U_{k+1} and V_{k+1} to bi-collar neighborhoods of λ'_{k+1} and δ_{k+1} respectively to obtain another collar neighborhoods U'_{k+1} and V'_{k+1} of λ'_{k+1} and δ_{k+1} respectively in the other sides, which still satisfy (3.5) but this time U'_{k+1} lie in between the arms λ_{k+1} and λ_{k+2}. Now we apply the Riemann-mapping theorem to this region and denote the corresponding Riemann map by ϕ and then the composition $w\circ\phi : (D_\delta(0), \partial D_\delta(0)) \to (\mathbf{C}^n, R)$ and u satisfy the hypothesis of Proposition 3.5, and so we can choose δ_{k+2}, λ'_{k+2} and U_{k+2}, V_{k+2} satisfying all the requirements by applying Proposition 3.5. This finishes the proof. □

4 Proof of Theorem I and II

We analyze structure of the set $\Delta\backslash w^{-1}(bA)$ in this section. When $\Delta\backslash w^{-1}(bA)$ is connected, (Theorem I, Oh[9]) gives a complete answer and so we assume that

$\Delta\backslash w^{-1}(bA)$ has more than one connected components. As a first step, we prove that $\Delta\backslash w^{-1}(bA)$ has only finitely many connected components. The following lemma shows that if there is any accumulation point of self-intersections, the limit point must be a critical point. The case in which the limit point is in the interior is classical and so we will consider only the case where the limit point lies in the boundary ∂D^2 of D^2.

Lemma 4.1 *Let $w : (D^2, \partial D^2) \to (\mathbf{C}^n, R)$ be a holomorphic map, $z \in \partial D^2$ and λ be an open arc in ∂D^2 containing z, in its interior. We fix sufficiently small collar neighborhood U of λ. Assume that there is a sequence $\{x_k\} \subset w(\partial D^2) \cap w(\mathrm{Int}U)$ such that $\lim_{k\to\infty} x_k = w(z)$. Then we must have*

$$dw(z) = 0.$$

Proof Suppose the contrary that $dw(z) \neq 0$. By applying Lemma 2.5 and considering local pieces of w near z, we get a $\phi_* i$-holomorphic map $u : D_\epsilon(0) \to \mathbf{C}^n$ such that $u = \phi \circ w$ and

$$u(z) = (z, 0, \ldots, 0), \quad \phi_* i(z, 0, \ldots, 0) = i \tag{4.1}$$

for some boundary flattening local diffeomorphism ϕ from an open neighborhood U of z onto an open neighborhood V of 0. Furthermore, there is a sequence $\{z_\mu\} \subset \mathrm{Int}D_\epsilon(0)$ such that

$$u(z_\mu) \in \mathbf{R}^n, \quad \lim_{\mu\to\infty} z_\mu = 0, \quad z_\mu \in D_\epsilon(0)\backslash\partial D_\epsilon(0). \tag{4.2}$$

On the other hand it follows from (4.1) that $u(z) \notin \mathbf{R}^n$ for $z \notin \partial D_\epsilon(0)$. This gives rise to a contradiction. □

Unlike the case of closed curves, it is not likely that the number of self-intersection points of discs is finite in general. (See some examples in Chirka [3] from which we believe one can construct a counter example).

Denote by $X \subset D^2$ the set of critical points. We recall the multiplicity function

$$\kappa : A \to \mathbf{Z}_+$$

and denote

$$A_k = \{a \in A \mid \kappa(a) = k\} \text{ and}$$
$$S = D^2\backslash w^{-1}(A_1).$$

We denote by $S_0 \subset S$ the subset of *isolated self-intersection points* of at least two different branches of the image.

Lemma 4.1 together with Lemma 2.6 implies that $S_0\backslash X$ is a discrete subset of D^2. We next quote a useful result from Oh [9].

Lemma 4.2 (Lemma 4.4, Oh [9]) *There exists a positive integer $M > 0$ such that*

$$1 \leq \kappa(a) < M < \infty$$

for all $a \in bA$ (and so for all $a \in A$).

Since the "vertices" of the "net" $\Delta \backslash w^{-1}(bA)$ must be critical points of w in Δ and the number of different branches of $\Delta \cap w^{-1}(bA)$ at each vertex must be finite by Lemma 4.2, we have the following immediate corollary of the above discussion

Corollary 4.3 $\Delta \backslash w^{-1}(bA)$ *has only finite number of connected components.*

We now denote

$$\Delta \backslash w^{-1}(bA) = \coprod_{j=1}^{k} E_j$$

where $\{E_j\}_{j=1}^k$ the set of connected components. Note that the isolated points in $\cup_{j=1}^k \overline{E}_j \backslash E_j$ are exactly

$$S_0 \cap w^{-1}(bA) =: S_0'.$$

Here we denote by $\overline{E}_j$ the closure of E_j in D^2, not in Δ. We add the points in S_0' to E_j to define

$$\widetilde{E}_j = E_j \cup (S_0' \cap \operatorname{Int} \overline{E}_j) \subset \Delta.$$

Then we have

$$(\Delta \backslash w^{-1}(bA)) \cup S_0' = \coprod_{j=1}^{k} \widetilde{E}_j \tag{4.3}$$

where $\widetilde{E}_j$ are open subsets of D^2 which are domains in $\mathbf{C}$ without discrete holes in its interior.

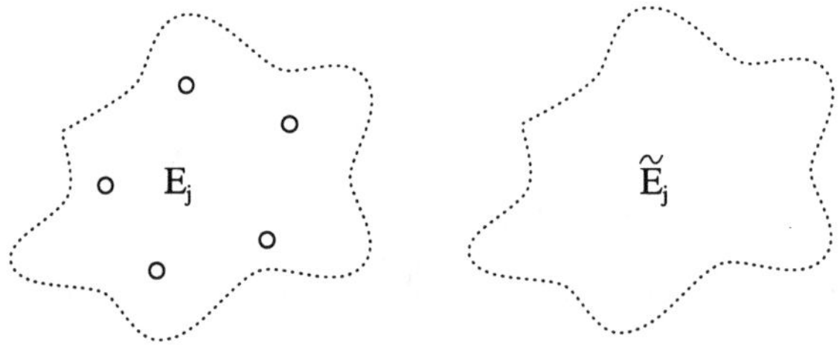

Figure 8

When the ambient manifold is not $\mathbf{C}^n$ and admits a non-constant holomorphic sphere, $\widetilde{E}_j$ could be non-simply connected as following example shows.

Example Consider the complex plane $\mathbf{C}$ with coordinates $z = x + iy$. We identify $S^2 = \mathbf{C} \cup \{\infty\}$. Let $B \subset S^2$ be the closed disc with the radius $1/2$ centered at $1 + i$ and let D be the complement to the interior of B in S^2. Then consider a map $F : S^2 \to S^2$, $F(z) = z^4$ and let w be its restriction to D. Denote by Γ the image of $w|_{\partial B}$. Note that Γ is a smooth embedded circle in S^2 and hence a totally real submanifold in S^2. For the map $w : (D, \partial D) \to (M, R)$ where $M = S^2$ and $R = \Gamma$, it is easy to check that $D \setminus w^{-1}(\Gamma)$ contains a connected component which is *non-simply connected* (even after adding a discrete set of points).

But in $\mathbf{C}^n$, we have the following theorem.

Theorem 4.4 *Let w be an analytic disc attached to totally real submanifold R in $\mathbf{C}^n$. And then in (4.3), each domain $\widetilde{E}_j$ is simply connected.*

Remark 4.5 We refer to [KwO] [6] for general structure theorem in the case when the ambient almost complex manifold (M, J) admits a non-constant J-holomorphic sphere.

Before giving the proof of this theorem, let us give the proof of Theorem II.

Proof of Theorem II (assuming Theorem 4.4) From Theorem 4.4, it follows that

$$\text{Image } w\big|_{\Delta} = \bigcup_{j=1}^{k} (\text{Image } w\big|_{\widetilde{E}_j}).$$

For each given $\widetilde{E}_j$, we choose a Riemann-map $\phi_j : D^2 \to \widetilde{E}_j$ and consider the composition $w_j := w \circ \phi_j : D^2 \to \mathbf{C}^n$ which is continuous up to the boundary. Since the Riemann-map $\phi_j : D^2 \to \widetilde{E}_j$ is proper i.e., maps ∂D^2 to ∂E_j and $w(\partial D^2) \subset R$, we have $w_j(\partial D^2) \subset R$. Then by the boundary regularity theorem (see Chirka [3]) for holomorphic discs with totally real boundary condition, $w_j : D^2 \to \mathbf{C}^n$ is smooth up to the boundary and defines a smooth holomorphic map

$$w_j : (D^2, \partial D^2) \to (\mathbf{C}^n, R).$$

Furthermore, $D^2 \backslash (w_j)^{-1}(w_j(\partial D^2))$ is connected by the construction of $\widetilde{E}_j$ and w_j. Now we apply Theorem 1.1 to w_j to conclude that there exists a Blaschke product $b_j : D^2 \to D^2$ such that

$$w_j = \widetilde{w}_j \circ b_j$$

with $\widetilde{w}_j : (D^2, \partial D^2) \to (\mathbf{C}^n, R)$ is a simple map. Combining the above discussion, we have finished the proof of Theorem II.□

Now, the remaining section will be spent to prove Theorem 4.4.

Suppose that at least one of $\widetilde{E}_j$'s is not simply connected. Then we can find $\widetilde{E}_{j_0}$, by taking an innermost one, such that its boundary $\partial\widetilde{E}_{j_0}$ is not connected. Then it will

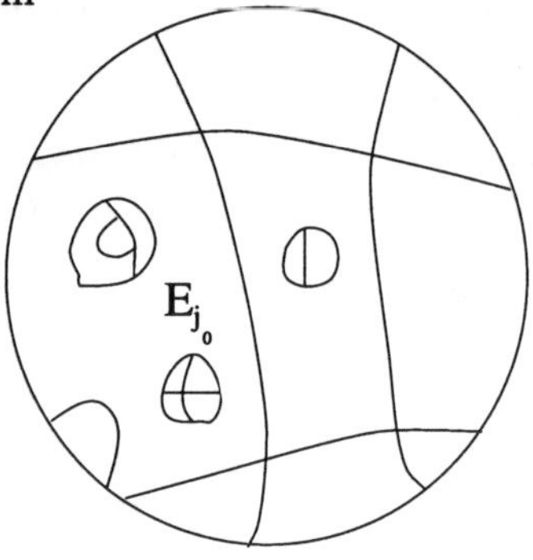

Figure 9

i.e., consist of one outer boundary and a finite number of inner boundaries contained in $\Delta$where the outer boundary is a piecewise smooth simple closed curve, and each inner boundary contains only simply connected $\widetilde{E}_k$'s in its interior. See Figure 9. Note that some inner boundary may not be equal to $\cup_k \partial\widetilde{E}_k$ where $\widetilde{E}_k$'s are the domains contained in the inside of the inner boundary. See Figure 9 again.

For the convenience of notation, we assume $E_1 = E_{j_0}$. Denote by B_1 the outer boundary and $\{B_2^i\}_{i=1}^m$ be the set of connected components of the inner boundary.

We take partitions

$$\partial D^2 \backslash X = \coprod_{i=1}^{s} \lambda_i, \; B_1 \backslash X = \coprod_{j=1}^{t} \mu_j, \; B_2^i \backslash X = \coprod_{k} \delta_k^i \tag{4.4}$$

where λ_i, μ_j and δ_k^i are connected open arcs without critical points of w and the orderings are taken clockwise. By refining the partitions further if necessary, we may assume that any pair from $w(\lambda_i)$'s, $w(\mu_i)$'s and $w(\delta_k^i)$'s is either identical or disjoint except possibly away from a discrete set of intersections. We have

$$\text{Interior of } B_2^i = \coprod_{j=1}^{k_i} \left(\overline{\widetilde{E}_{i,j}} \backslash B_2^i\right) \qquad (k_i \geq 0)$$

where $\widetilde{E}_{i,j}$'s are the domains inside B_2^i among the set of $\widetilde{E}_k$'s. We denote

$$F^i = B_2^i \cup \Big(\bigcup_{j=1}^{k_i} \partial \widetilde{E}_{i,j} \Big). \tag{4.5}$$

By the choice of E_1, it follows that each F^i must be connected.

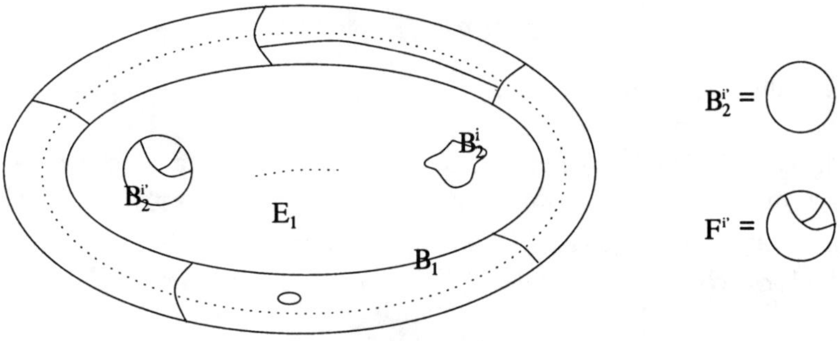

Figure 10

Now consider a (collar-)neighborhood U of ∂D^2 in D^2 and open neighborhoods V^i of $\overline{\bigcup_{j=1}^{k_i} \widetilde{E}_{i,j}}$ such that U and V^i are all disjoint and $\big(U \cap w^{-1}(bA)\big) \setminus S_0'$ consists of disjoint open arcs one or both of whose two ends lie in ∂D^2. See Figure 11 below.

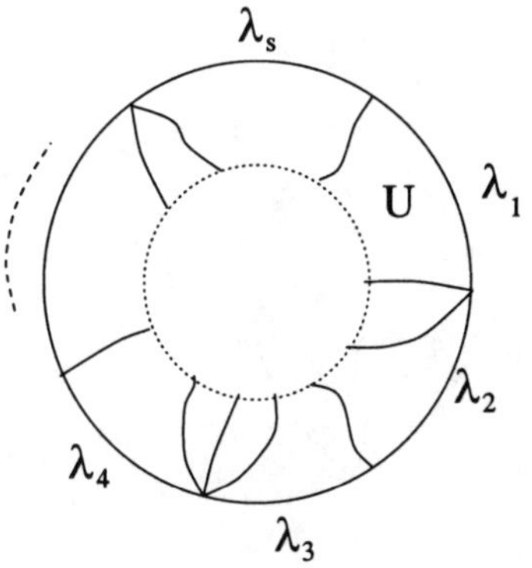

Figure 11

Since $w(B_2^i) \subset bA = w(\partial D^2)$, there exist $\lambda \in \{\lambda_j\}_{j=1}^s$, $\delta^i \in \{\delta_k^i\}$ and collar neighborhoods $U_1 \subset U$ and $V_1^i \subset V^i$ of λ and δ^i respectively such that they satisfy

$$w(U_1) = w(V_1^i),\ w(\lambda) = w(\delta^i) \subset w(B_2^i) \subset w(F^i). \tag{4.6}$$

We would like to note that the collar neighborhood V_1^i *may not be contained in* E_1 *but in* $\widetilde{E}_{i,j_i}$ *for some* $j_i \in \{1, 2, \cdots, k_i\}$.

Claim $bA = w(\partial D^2) = w(F^i)$.

Proof Obviously, $w(F^i) \subset bA$ and so it is enough to prove

$$bA(= w(\partial D^2)) \subset w(F^i).$$

By cyclically re-ordering $\{\lambda_j\}$, $\{\delta_j^i\}$ if necessary, we assume $\lambda = \lambda_1$ and $\delta^i = \delta_1^i$.□

The claim immediately follows from the following lemma which proves that the overlap in (4.6) propagates along all the components $\{\lambda_j\}_{j=1}^s$ through.

Lemma 4.6 *Let* i_0 *be fixed. For each* λ_k*, there is an arc* $\delta_k^{i_0} \subset F^{i_0}$ *and collar neighborhoods* U_k *and* V_k^i *of* λ_k *and* $\delta_k^{i_0}$ *respectively such that*

$$w(U_k) = w(V_k^{i_0}).$$

Proof We will prove this by induction. For $k = 1$, it follows from (4.6). Suppose we have proven it for any $1 \le \ell \le k$ and consider the case for $k+1$. We consider a small collar neighborhood W of $\overline{\lambda_k} \cup \overline{\lambda_{k+1}}$ in U and the neighborhood V^{i_0} chosen above.

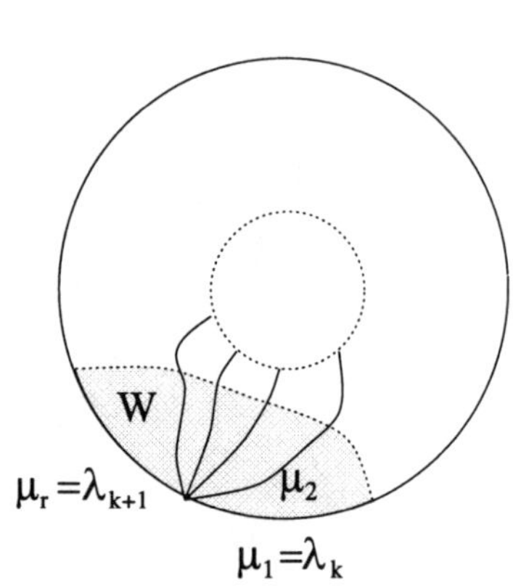

Figure 12

In general, $W \cap w^{-1}(bA)$ becomes an r-armed (semi)-asterisk C with λ_k and λ_{k+1} as the initial and the final arms respectively (see Figure 12). We order these arms $\{\mu_j\}_{j=1}^r$ counterclockwise with $\mu_1 = \lambda_k$, $\mu_r = \lambda_{k+1}$. We now apply Proposition 3.7 to the maps

$$w\big|_{(W,\partial W)} \quad \text{and} \quad w : V^{i_0} \to \mathbf{C}^n$$

and find a sequence of arcs, $\nu_1, \cdots, \nu_r \subset V^{i_0}$, sub-arcs $\mu'_j \subset \mu_j$ and collar neighborhoods V'_j and U'_j of ν_j and μ'_j respectively such that

$$w(\nu_j) = w(\mu'_j) \quad \text{and} \quad w(U'_j) = w(V'_j)$$

and ν_j's share one of the end points from each and in particular

$$\bigcup_{j=1}^{r} \nu_j \quad \text{is connected.}$$

Since $\nu_1 = \delta_1^{i_0} \subset F^{i_0}$ and F^{i_0} is a connected component of $w^{-1}(bA)$, it follows that

$$\bigcup_{j=1}^{r} \nu_j \subset F^{i_0}.$$

Note that due to the way how we partitioned in (4.4), we can extend μ'_r and ν_r correspondingly V'_r and U'_r until μ'_r becomes a full edge of $\mu_r = \lambda_{k+1}$ so that

$$w(\nu_r) = w(\lambda_{k+1}), \quad w(V'_r) = w(U'_r)$$

in F^{i_0}, we have proven the existence of $\gamma_{k+1}^{i_0}$, U_{k+1} and V_{k+1} required as in Lemma 4.6 by setting $V'_r = V_{k+1}$ and $U'_r = U_{k+1}$.□

Now consider the image of a holomorphic disc $w : (D^2, \partial D^2) \to (\mathbf{C}^n, R)$. In this case, by the Claim, for each i

$$w(\partial D^2) = w(F^i) \subset w(\Delta).$$

Hence the image becomes a closed analytic variety without boundary in $\mathbf{C}^n$, which gives rise to a contradiction. This finishes the proof of Theorem 4.4.

5 The case of real-analytic R

In this section we restrict to the classical case in which the totally real submanifold R is real-analytic. For the real-analytic R, we have the following theorem which seems to be new even in the classical context of several complex variables.

Theorem 5.1 *Let R be a real-analytic totally real submanifold in $\mathbf{C}^n$. Then for any holomorphic disc $w : (D^2, \partial D^2) \to (\mathbf{C}^n, R)$, $\Delta \setminus w^{-1}(bA)$ is connected and so allows the factorization (1.3).*

The remaining section will be spent to prove Theorem 5.1. Since R is real-analytic, we can apply locally reflection principle along the image of ∂D^2. Thus, we get the following proposition from which Theorem 5.1 follows.

Let $\overline{D}_\epsilon(0)$ be the semi-disk centered at 0 with boundary $(-\epsilon, \epsilon)$ in the lower half plane of $\mathbf{C}$. We will need to prove an analogue to Lemma 3.5.2 for the maps

$$w_1 : (D_\epsilon(0), \partial D_\epsilon(0)) \to (\mathbf{C}^n, R)$$
$$w_2 : (\overline{D}_\epsilon(0), \partial \overline{D}_\epsilon(0)) \to (\mathbf{C}^n, R)$$

satisfying all the hypothesis on Proposition 3.5.

Proposition 5.2 *Let*

$$w_1 : (D_{\epsilon_1}(0), \partial D_{\epsilon_1}(0)) \to (\mathbf{C}^n, R)$$
$$w_2 : (\overline{D}_{\epsilon_2}(0), \partial \overline{D}_{\epsilon_2}(0)) \to (\mathbf{C}^n, R)$$

be holomorphic maps satisfying $(i) - (v)$ in Proposition 3.5. Suppose that there exist collar neighborhoods U_i of $(-\epsilon_i, 0)$ in $D_{\epsilon_i}(0)$ with $\partial U_i = (-\epsilon_i, 0)$ for i=1,2 and $w_i|_{\partial U_i}$ is an embedding, and that the diffeomorphism $\left(w_1|_{(-\epsilon_1,0)}\right)^{-1} \circ w_2|_{(-\epsilon_2,0)} : (-\epsilon_2, 0) \to (-\epsilon_1, 0)$ preserves the orientations induced from the complex orientations on $D_{\epsilon_1}(0)$ and $\overline{D}_{\epsilon_2}(0)$ respectively. Then we can find collar neighborhoods V_i of $(-\epsilon_i, \delta_i)$ such that $\delta_i > 0$ and

$$w_1(\partial V_1) = w_2(\partial V_2).$$

In particular, by Corollary 2.7 (ii), the union $w_1(V_1) \cup w_2(V_2)$ forms a smooth surface possibly with a branched point at 0.

Proof Since R is real-analytic and $w_2(\partial \overline{D}_{\epsilon_2}) \subset R$, we can apply the classical reflection principle to w_2 to get a map $\widetilde{w}_2 : (D_{\epsilon_2}(0), \partial D_{\epsilon_2}(0)) \to (\mathbf{C}^n, R)$. We can check that w_1 and $\widetilde{w}_2$ satisfy the assumptions in Proposition 3.5. So, from Proposition 3.5, there are $\delta_i \in (0, \epsilon_i)$ and collar neighborhoods W_i of $(-\epsilon_i, \delta_i)$ for $i = 1, 2$ such that

$$w_1(W_1) = \widetilde{w}_2(W_2).$$

Since $w_2(-\epsilon_2, \delta_2) = \widetilde{w}_2(-\epsilon_2, \delta_2)$, This finishes the proof.□

Proof of Theorem 5.1 Assume that there are more than one components in the decomposition (4.3). We know by Theorem 4.4, $\widetilde{E}_j$'s are simply connected.

Assume that $\widetilde{E}_j$ and $\widetilde{E}_k$ are adjacent, i.e., there are arcs $s_1(\subset \partial\widetilde{E}_j \setminus X)$ and $t_1(\subset \partial\widetilde{E}_k \setminus X)$ such that the diffeomorphism $(w|_{t_1})^{-1} \circ w|_{s_1} : s_1 \to t_1$ preserves the orientation induces from the complex orientation. By applying Proposition 5.2, we can get arcs $s_2(\subset \partial\widetilde{E}_j \setminus X)$ and $t_2(\subset \partial\widetilde{E}_k \setminus X)$ such that $\overline{s}_1 \cap \overline{s}_2, \overline{t}_1 \cap \overline{t}_2$ are one points and for some collar neighborhoods U_2 of s_2 in $\widetilde{E}_j$ and V_2 of t_2 in $\widetilde{E}_k$ respectively such that $w(s_2) = w(t_2)$ and $w(U_2) \cup w(V_2)$ form a smooth surface.

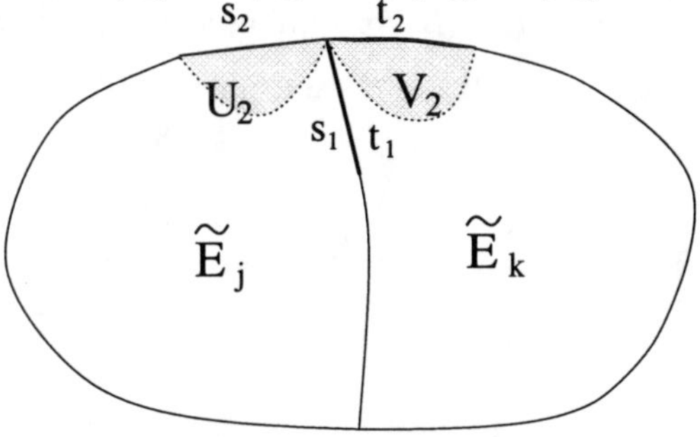

Figure 13

Since the diffeomorphism $(w|_{t_2})^{-1} \circ w|_{s_2} : s_2 \to t_2$ preserves the orientation induces from the complex orientation, by applying Proposition 5.2 inductively we conclude that $w(\overline{\widetilde{E}_j}) \cup w(\overline{\widetilde{E}_k})$ is compact and has no boundary. It is impossible. This finishes the proof. □

6 Rosay's example

The example in this section is given by Rosay and taken from the appendix of [KwO] [6].

We show that the hypothesis of real analyticity of R is needed in Theorem 5.1. The results do not generalize to the case when R is only smooth. By Δ we denote the open unit disk, $b\Delta$ is the unit circle and $\overline{\Delta} = \Delta \cup b\Delta$. A (smooth) *holomorphic disk* in $\mathbf{C}^2$ is a smooth map from $\overline{\Delta}$ into $\mathbf{C}^2$, whose restriction to Δ is holomorphic.

Proposition 6.1 *There exists a non-constant (smooth) holomorphic disk in $\mathbf{C}^2$, and M_2 a 2-dimensional totally real manifold in $\mathbf{C}^2$ such that $f(b\Delta) \subset M_2$ and $f^{-1}(f(b\Delta))$ disconnects Δ.*

Remark 6.2 This cannot happen, by Theorem 5.1, if M_2 is real analytic.

Construction of f and M_2:

1) Let D^+ and Ω^+ be smooth by bounded, simply connected, regions in $\{\text{Im } z > 0\}$ (in $\mathbf{C}$) such that

$$\begin{cases} bD^+ \cap \mathbf{R} = [-1, +1] \\ b\Omega^+ \cap \mathbf{R} = [-1/2, +1/2]. \end{cases}$$

Let h^+ be a conformal map from D^+ to Ω^+, normalized so that $h^+[-1/2, 1/2] = [-1/2, +1/2]$, ($h^+$ extending smoothly to bD^+). The main fact is that Im $h^+ = 0$ on $[-1/2, +1/2]$ and Im $h^+ > 0$ on $([-1, +1] - [-1/2, +1/2])$.

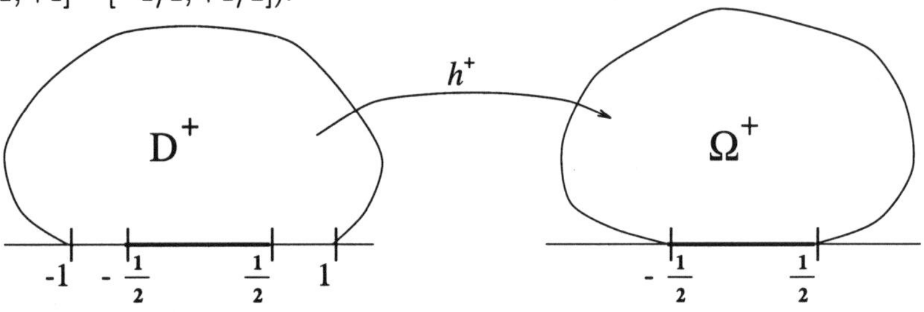

Figure 14

2) Define D^-, Ω^- by symmetry with respect to the real axis and f^- by Schwartz reflection:

$$\begin{cases} D^- = \{z \in \mathbf{C}, \overline{z} \in D^+\} \\ \Omega^- = \{z \in \mathbf{C}, \overline{z} \in \Omega^-\} \end{cases}$$

$$h^- : D^- \to \Omega^-, \quad h^-(z) \; = \; \overline{h^+(\overline{z})}.$$

3) Let $D = D^+ \cup D^- \cup (-1/2, +1/2)$. By Schwartz reflection, there exists h holomorphic on D such that $h|D^\pm = h^\pm$. For later use, set also $h(\pm 1/2) = h^+(\pm 1/2) = h^-(\pm 1/2)$.

4) *Construction of* M_2.

Claim: There exists a smooth totally real manifold M_2 in $\mathbf{C}^2$ such that $(z, h^+)(bD^+)$ and $(z, h^-)(bD^-) \subset M_2$. (Where $(z, h^\pm)$ denotes the maps $z \mapsto (z, h^\pm(z))$. The only difficulty is at the points $x = \pm 1/2$, where $(x, h^+(x))$ and $(x, h^-(x))$ "bifurcate". Let $M_2^0 = \{(x, h^+(x)) + i(0, t), \quad -1 < x < +1, \quad t \in \mathbf{R}\}$. This is a smooth totally real manifold. Taking $t = 0$, we see that it contains $(z, h^+)([-1, +1])$. Taking $t = -2$ Im $h(x)$, we see that it contains $(z, h^-(z))([-1, +1])$, since $h^+(x) - h^-(x)$ is purely imaginary, for $-1 \leq x \leq +1$.

Extending M_2^0 to a totally real manifold containing $(z, h^{\pm})(bD^{\pm})$ is now "soft". One can, for example, set, for ϵ small enough: $M_2 = M_2^0 \cup M_2^+ \cup M_2^-$.

$$M_2^{\pm} = \{(z, h^{\pm}(z)) + i(0,t), \ \ z \in bD^{\pm} - (-1,+1), \ \ -\epsilon < t < \epsilon\}.$$

Note: The above construction shows that the real analyticity of R is needed in Proposition 5.2.

5) *End of construction.*

Let G be a conformal map from Δ into D. For simplicity impose $G(0) = 0, G'(0) > 0$. This yields $G([0,1)) = [0,1/2)$, by symmetry. Let F be the holomorphic map from Δ into $\mathbf{C}^2$ defined by

$$F(z) = (G(z^2), h(G(z^2))).$$

Notice that F extends continuously to $b\Delta$ by $F(z) = (G(z^2), h^{\pm}(G(z^2)))$, $\pm$ depending whether Im $G(z_1^2) \geq 0$ or ≤ 0 for $z_1 \in \Delta$, z_1 close to z (if $G(z^2) = \pm 1/2$, $F(z) = (\pm 1/2, h(\pm 1/2))$. Since $z \in b\Delta$ implies that $G(z^2) \in bD, F(b\Delta) \subset M_2$. But also

$$F([-1,+1]) = \{(x, h(x)), 0 \leq x \leq 1/2\} \subset M_2.$$

Let $\Delta_1 = \Delta - [0,1]$. It follows from the above that $F(b\Delta_1) = F(b\Delta \cup [0,1]) \subset M_2$. but since F is even $F^{-1}F([0,1]) \supset [-1,+1]$. So $F^{-1}(F([0,1]))$ disconnects Δ_1. Let ψ be a conformal map from Δ onto Δ_1. Set $f = F \circ \psi$. The pair (f, M_2) gives the desired example.

Note: Although the smoothness of f could be checked by following the construction, it also follows immediately from the continuity of f and the fact that $f(b\Delta) \subset M_2$ (M_2 totally real, and smooth).

Rosay's example above however still has constant multiplicity 2 and allows the factorization. We can slightly modify the example to produce an analytic disc attached to M_2 that has points of two different multiplicities and so does not allow the factorization.

We consider the map

$$\chi : \Delta^+ \to \Delta, \quad \chi(z) = z^3$$

where $\Delta^+ = \{z \in \Delta \mid \text{Im } z > 0\}$, the upper semi-disc. Note that

$$\chi(b\Delta^+) = b\Delta \cup [-1,1].$$

Let M_2 be the totally real submanifold and G be the map from Δ into D in Appendix 1. Define the map $\widetilde{F}(z) : \Delta^+ \to \mathbf{C}^2$ by

$$\widetilde{F}(z) = \Big(G(\chi(z)), h(G(\chi(z)))\Big).$$

Since $\chi(b\Delta^+) = b\Delta \cup [-1, 1]$, we have correct boundary condition $\widetilde{F}(b\Delta^+) \subset M_2$. Then it follows that $\widetilde{F}$ has points of two different multiplicities (in fact, 1 and 2). Composing $\widetilde{F}$ with a conformal map $\phi : \Delta \to \Delta^+$, we have obtained the seeked example $g = \widetilde{F} \circ \phi : \Delta \to \mathbf{C}^2$ with $g(b\Delta) \subset M_2$.

Acknowledgments

The first author is partially supported by GARC, SRC and Research Institute of Basic sciences and the second author is partially supported by the NSF Grant # DMS-9504455 .

Moreover, the second author would like to thank Jean-Pierre Rosay for useful discussions and for providing the example in Section 6. The first author is grateful to Professor Hong-Jong Kim for helping to improve English.

References

1. S. Alinhac, M. S. Baouendi and L. P. Rothschild, *Duke Math. J.* **61**, 635 (1990).
2. E. Bedford and B. Gaveau, *Amer. J. Math.* **105**, 975 (1983).
3. E. M. Chirka, *Math. USSR Sbornik* **45**, 291 (1983).
4. A. Floer, H. Hofer and D. Salamon, *Duke Math. J.* **80**, 251 (1995).
5. J. Globevnik and E. Stout, *Ann. Math.* **129**, 380 (1988)
6. D. Kwon and Y.-G. Oh, (submitted).
7. D. McDuff and D. Salamon, *J-holomorphic Curves and Quantum Cohomology*, (Univ. Lec. Series **6**, Amer. Math. Soc. 1994).
8. Y.-G. Oh, *Math. Z.* **222**, 505 (1996).
9. Y.-G. Oh, *J. Geom. Anal.* (to appear).
10. E. Stout, *J. Analyse Math.* **10**, 239 (1975).

MAYER-VIETORIS FORMULA FOR DETERMINANTS OF ELLIPTIC OPERATORS OF LAPLACE-BELTRAMI TYPE. (AFTER BURGHELEA, FRIEDLANDER AND KAPPELER)

Y. LEE
Department of mathematics
Inha University
Inchon, 402-751, Korea
E-mail: ywlee@math.inha.ac.kr

The purpose of this note is to provide a short cut presentation of a Mayer-Vietoris formula due to Burghelea-Friedlander-Kappeler for the regularized determinant in the case of elliptic operators of Laplace Beltrami type in the form typically needed in applications to torsion.

1 Statement of Mayer-Vietoris Formula for Determinants

Let (M, g) be a closed oriented Riemannian manifold of dimension d and Γ be an oriented submanifold of codimension 1. We denote by ν the unit normal vector field along Γ. Let M_Γ be the compact manifold with boundary $\Gamma^+ \sqcup \Gamma^-$ obtained by cutting M along Γ, where Γ^+ and Γ^- are copies of Γ and denote by $p : M_\Gamma \to M$ the identification map. The vector field ν has the lift on M_Γ which we denote by ν again. Denote by Γ^+ the component of the boundary where the lift of ν points outward. Given a smooth vector bundle $E \to M$, denote by E_Γ the pull back of $E \to M$ to M_Γ by p. Let $A : C^\infty(E) \to C^\infty(E)$ be an elliptic, essentially self-adjoint, positive definite differential operator of Laplace-Beltrami type, where we say that A is of Laplace-Beltrami type if A is an operator of order 2 whose principal symbol is $\sigma_L(x, \xi) = \| \xi \|^2 \, Id_x$, $Id_x \in End_x(E_x, E_x)$. We denote by $A_\Gamma : C^\infty(E_\Gamma) \to C^\infty(E_\Gamma)$ the extension of A to smooth sections of E_Γ.

Consider Dirichlet and Neumann boundary conditions B, C on $\Gamma^+ \sqcup \Gamma^-$ defined as follows:

$$B : C^\infty(E_\Gamma) \to C^\infty(E_\Gamma \mid_{\Gamma^+ \sqcup \Gamma^-}), B(f) = f \mid_{\Gamma^+ \sqcup \Gamma^-}$$

$$C : C^\infty(E_\Gamma) \to C^\infty(E_\Gamma \mid_{\Gamma^+ \sqcup \Gamma^-}), C(f) = \nu(f) \mid_{\Gamma^+ \sqcup \Gamma^-} .$$

Consider $\tilde{A_{\Gamma,B}} = (A_\Gamma, B) : C^\infty(E_\Gamma) \to C^\infty(E_\Gamma) \oplus C^\infty(E_\Gamma \mid_{\Gamma^+ \sqcup \Gamma^-})$. ¿From the properties of A it follows that $\tilde{A_{\Gamma,B}}$ is invertible. Therefore we can define the corresponding Poisson operator P_B as the restriction of $\tilde{A_{\Gamma,B}}^{-1}$ to $0 \oplus C^\infty(E_\Gamma \mid_{\Gamma^+ \sqcup \Gamma^-})$. Denote by A_B the restriction of $\tilde{A_{\Gamma,B}}$ on $\{u \in C^\infty(E_\Gamma) \mid$

$B(u) = 0\}$. Then A_B is also essentially self-adjoint and positive definite (cf Lemma 3.1). This (using standard analytic continuation technique due to Seeley (cf.[3])) allows us to define

$$logDet(A) = -\frac{d}{ds}\,|_{s=0}\, tr\frac{1}{2\pi i}\int_\gamma \lambda^{-s}(\lambda - A)^{-1} d\lambda \tag{1}$$

$$logDet(A_\Gamma, B) = -\frac{d}{ds}\,|_{s=0}\, tr\frac{1}{2\pi i}\int_\gamma \lambda^{-s}(\lambda - A_B)^{-1} d\lambda, \tag{2}$$

where γ is a path around the negative real axis,

$$\{\rho e^{i\pi} \mid \infty > \rho \geq \epsilon\} \cup \{\epsilon e^{i\theta} \mid \pi \geq \theta \geq -\pi\} \cup \{\rho e^{-i\pi} \mid \epsilon \leq \rho < \infty\}$$

with $\epsilon > 0$ chosen sufficiently small to ensure that Γ does not separate the spectrum.

We define the Dirichlet to Neumann operator, associated to A, B and C,

$$R : C^\infty(E\,|_\Gamma) \to C^\infty(E\,|_\Gamma)$$

by the composition of the following maps

$$C^\infty(E\,|_\Gamma) \overset{\triangle_{ia}}{\to} C^\infty(E\,|_{\Gamma+}) \oplus C^\infty(E\,|_{\Gamma-}) \overset{P_B}{\to} C^\infty(E_\Gamma) \overset{C}{\to} C^\infty(E\,|_{\Gamma+}) \oplus C^\infty(E\,|_{\Gamma-})$$

$$\overset{\triangle_{if}}{\to} C^\infty(E\,|_\Gamma),$$

where $\triangle_{ia}(f) = (f, f)$ is the diagonal inclusion and $\triangle_{if}(f, g) = f - g$ is the difference operator. Then R is an essentially self-adjoint, positive definite, elliptic operator of order 1 (cf Corollary 3.5).

Theorem 1.1 (Mayer-Vietoris Type Formula for Determinants [1])

Let (M, g) be a closed oriented Riemannian manifold of dimension d and A be an elliptic, essentially self-adjoint, positive definite differential operator of Laplace-Beltrami type acting on smooth sections of a vector bundle $E \to M$. Then A_B and R are essentially self-adjoint, positive definite elliptic operators and

$$Det(A) = cDet(A_\Gamma, B)Det(R),$$

where c is a local quantity which can be computed in terms of the symbols of A, B and C along Γ.

Theorem 1.2 Let (M, g) be a closed oriented Riemannian manifold of dimension d with the same situation as above. If d is even, then $c = 1$ and therefore

$$Det(A) = Det(A_\Gamma, B)Det(R). \tag{3}$$

Remark The above result can be extended to manifolds with boundary. E.g. consider an oriented, compact, smooth manifold M whose boundary ∂M is a disjoint union of two components $\partial_+ M$ and $\partial_- M$ with $\Gamma \cap \partial M = \emptyset$, an operator A of Laplace-Beltrami type and differential elliptic boundary conditions B_+ respectively B_- for A on $\partial_+ M$ respectivley $\partial_- M$. Denote by $A^{(0)}$ the operator A with domain $\{u \in C^\infty(E) \mid B_+ u = 0, B_- u = 0\}$. Then Theorem 1.1 remains true with A replaced by $A^{(0)}$.

2 The Asymptotics of Determinants of Elliptic Pseudodifferential Operators with Parameter

Let V be an open angle in the complex λ-plane and P(λ), $\lambda \in$V, a family of ΨDO's of order m, m a positive integer, acting on smooth sections of a vector bundle $E \to M$ of rank ν, where M denotes a closed smooth Riemannian manifold of dimension d.

Definition 2.1 (cf.[4]) The family P(λ), $\lambda \in V$, is said to be a ΨDO with parameter of weight $\chi > 0$ if in any coordinate neighborhood U of M, not necessarily connected, and for an arbitarily fixed $\lambda \in V$, the complete symbol $p(\lambda; x, \xi)$ of P(λ) is in $C^\infty(U \times \mathbf{R}^d, End(\mathbf{C}^\nu))$ and, moreover, for any multiindices α and β, there exists a constant $C_{\alpha,\beta}$ such that $\mid \partial_\xi^\alpha \partial_x^\beta p(\lambda; x, \xi) \mid \leq C_{\alpha\beta}(1 + \mid \xi \mid + \mid \lambda \mid^{\frac{1}{\chi}})^{m-|\alpha|}$.

Definition 2.2 P(λ) is called classical if in any chart the complete symbol $p(\lambda; x, \xi)$ admits an expansion of the form

$$p(\lambda; x, \xi) \sim p_m(\lambda; x, \xi) + p_{m-1}(\lambda; x, \xi) + \cdots, \tag{4}$$

where $p_j(\tau^\chi \lambda; x, \tau\xi) = \tau^j p_j(\lambda; x, \xi)(\tau > 0, j \leq m)$. The family P($\lambda$) is said to be elliptic with parameter if $p_m(\lambda; x, \xi)$ is invertible for all $x \in M$, $\xi \in T_x^*(M)$ and $\lambda \in$V satisfying $\mid \xi \mid + \mid \lambda \mid^{\frac{1}{\chi}} \neq 0$.

Definition 2.3 Let Q be an elliptic ΨDO. The angle π is called an Agmon angle for Q if for some $\epsilon > 0$, $spec(Q) \cap \Lambda_\epsilon = \emptyset$, where $spec(Q)$ denotes the spectrum of Q and $\Lambda_\epsilon = \{z \in \mathbf{C} \mid \pi - \epsilon < arg(z) < \pi + \epsilon \text{ or } \mid z \mid < \epsilon\}$.

The following theorem can be found in the appendix of [1].

Theorem 2.4 Let $P(\lambda)$ be an essentially self-adjoint, positive definite, classical ΨDO of order $m \in N$ with parameter $\lambda \in V$ of weight $\chi > 0$ such that (i) $P(\lambda)$ is elliptic with parameter and
(ii) for each $\lambda \in V$, $P(\lambda)$ has π as an Agmon angle.

Then $logDetP(\lambda)$ admits an asymptotic expansion for $\lambda \in V$, $|\lambda| \to \infty$, of the form

$$logDetP(\lambda) \sim \sum_{j=-d}^{\infty} \pi_j \mid \lambda \mid^{-\frac{j}{\chi}} + \sum_{j=0}^{d} q_j \mid \lambda \mid^{\frac{j}{\chi}} log \mid \lambda \mid . \qquad (5)$$

The coefficients π_j and q_j can be evaluated in terms of the symbol of P and $\frac{\lambda}{|\lambda|}$. In particular, π_0 is independent of perturbations by lower order operators, whose orders differ at least by $d+1$ from the order of $P(\lambda)$.

The following result is due to Voros [5] and Friedlander [2].

Theorem 2.5 Let $\{\lambda_k\}_{k\geq 1}$ be a sequence in $V_{0,\frac{\pi}{2}} = \{z \in \mathbf{C} \mid -\frac{\pi}{2} + \epsilon < arg(z) < \frac{\pi}{2} - \epsilon\}$, possibly with multiplicities, arranged in such a way that $0 < Re\lambda_1 \leq Re\lambda_2 \leq \cdots$. Assume that the heat trace, $\theta(t) := \sum_{k=0}^{\infty} e^{-t\lambda_k} (t > 0)$ admits an asymptotic expansion for $t \to 0$ of the form

$$\theta(t) \sim \sum_{n \geq 0} c_{i_n} t^{i_n}, \qquad (6)$$

where $i_0 < 0$ and $i_0 < i_1 < i_2 < \cdots \to +\infty$. For λ with $Re\lambda > 0$, let $\zeta(s,\lambda) = \sum_{k=0}^{\infty} (\lambda_k + \lambda)^{-s}$. Then $\pi_0 = 0$, where π_0 is the constant term in the asymptotic expansion of $-\frac{d}{ds}\zeta(s,\lambda) \mid_{s=0}$ for $|\lambda| \to +\infty$.

3 Auxiliary Results for the Proof of Theorem 1.1

We begin by collecting a number of results about operators related to A and Γ. Denote by $H^s(E_\Gamma)$ the Sobolev spaces of $E_\Gamma-$valued sections. Throughout section 3 and section 4 we assume that A satisfies the hypothesis of Theorem 1.1 and fix $\epsilon > 0$, so that the spectrum of A is bounded from below by ϵ.

Lemma 3.1 (i) The operator $A_B : \{u \in C^\infty(E_\Gamma) \mid B(u) = 0\} \to C^\infty(E_\Gamma)$ has a self-adjoint extension $\bar{A}_B$ with domain $D(\bar{A}_B) := \{u \in H^2(E_\Gamma) \mid B(u) = 0\}$. (ii) The operator $\bar{A}_B$ is positive definite and its spectrum is bounded below by ϵ. (iii) The operator

$$(A_\Gamma, B) : C^\infty(E_\Gamma) \to C^\infty(E_\Gamma) \oplus C^\infty(E_\Gamma \mid_{\Gamma^+ \sqcup \Gamma^-})$$

defined by $(A_\Gamma, B)(u) = (A_\Gamma(u), B(u))$ can be extended to an invertible operator $(A_\Gamma, B)^{\sim}$

$$(A_\Gamma, B)^{\sim} : H^2(E_\Gamma) \to L^2(E_\Gamma) \oplus H^{2-\frac{1}{2}}(E_\Gamma \mid_{\Gamma^+ \sqcup \Gamma^-}).$$

Set $\alpha_k = e^{i\frac{\pi+2k\pi}{d}}$ for $0 \le k \le d-1$, where $d = dim(M)$.

Lemma 3.2 The following operators are invertible for $0 \le k \le d-1$ and $t \ge 0$

$$(A_\Gamma - \alpha_k t, B) : C^\infty(E_\Gamma) \to C^\infty(E_\Gamma) \oplus C^\infty(E_\Gamma \mid_{\Gamma^+ \sqcup \Gamma^-}).$$

Since $(A_\Gamma - \alpha_k t, B)$ is invertible, we can define the Poisson operator $P(\alpha_k t)$ associated to $(A_\Gamma - \alpha_k t, B)$, $P(\alpha_k t) : C^\infty(E_\Gamma \mid_{\Gamma^+ \sqcup \Gamma^-}) \to C^\infty(E_\Gamma)$, i.e. for $\varphi \in C^\infty(E_\Gamma \mid_{\Gamma^+ \sqcup \Gamma^-})$, $u = P(\alpha_k t)\varphi$ is the solution in $C^\infty(E_\Gamma)$ of $(A_\Gamma - \alpha_k t)u = 0$ with boundary conditions $u \mid_{\Gamma^+ \sqcup \Gamma^-} = \varphi$.
Let $R(\alpha_k t) : C^\infty(E \mid_\Gamma) \to C^\infty(E \mid_\Gamma)$ be the Dirichlet to Neumann operator corresponding to $A_\Gamma - \alpha_k t, B$ and C. Then the following result holds:

Lemma 3.3 For $0 \le k \le d-1$, and $t \ge 0$, $R(\alpha_k t)$ is an invertible classical ΨDO of order 1, which is elliptic with parameter t of weight 1.

Proof In a sufficiently small collar neighborhood U of Γ, choose coordinates $x = (x', s)$ such that $(x', 0) \in \Gamma$ and $\frac{\partial}{\partial s} \mid_{(x',0)} = \nu_{(x',0)}$. Let $\xi = (\xi', \eta)$ be coordinates in the cotangent space corresponding to the coordinates (x', s). Let $D_s = \frac{1}{i}\frac{\partial}{\partial s}$ and write $(A - \alpha_k t) = A_2 D_s^2 + A_1 D_s + A_0$, where the A_j's are differential operators of order at most $2-j$. The A_j's induce, when restricted to Γ, differential operators, again denoted by A_j, $A_j : C^\infty(E \mid_\Gamma) \to C^\infty(E \mid_\Gamma)$. Since $\sigma_L(x, (\xi', \eta)) = \| (\xi', \eta) \|^2$ and since $\nu_{(x',0)}$ is the unit normal to Γ at $(x', 0)$, one has $A_2(x) = Id_x \in End_x(E_x, E_x)$ on Γ.

For any $\varphi \in C^\infty(E \mid_\Gamma)$ and $t \ge 0$, we can choose $u \in C^\infty(E_\Gamma) \cap C(E)$ such that $(A - \alpha_k t)u = 0$ on $M - \Gamma$ and $u \mid_{\Gamma^+} = \varphi = u \mid_{\Gamma^-}$. Then $\frac{\partial u}{\partial s}(x', s)$ has a jump across Γ, which is $-R(\alpha_k t)(\varphi)(x')$. Hence

$$\frac{\partial u}{\partial s}(x', s) = -R(\alpha_k t)(\varphi)(x')H(s) + v(x', s), \tag{7}$$

where $v(x', s) \in C^\infty(E_\Gamma \mid_U) \cap C(E \mid_U)$ and $H(s)$ is the Heavyside function. Therefore, on U,

$$(A - \alpha_k t)u = A_2 R(\alpha_k t)(\varphi) \otimes \delta_\Gamma - A_2 \frac{\partial v}{\partial s} + \frac{1}{i} A_1 \frac{\partial u}{\partial s} + A_0 u. \tag{8}$$

Since $(A - \alpha_k t)u = 0$ on $M - \Gamma$, we conclude that, on $U \cap (M \setminus \Gamma)$,

$$-A_2 \frac{\partial v}{\partial s} + \frac{1}{i} A_1 \frac{\partial u}{\partial s} + A_0 u = 0. \tag{9}$$

As $-A_2 \frac{\partial v}{\partial s} + \frac{1}{i} A_1 \frac{\partial u}{\partial s} + A_0 u \in L^2(E \mid_U)$, it follows that

$$(A - \alpha_k t)u = A_2 \cdot (\cdot \otimes \delta_\Gamma) \cdot R(\alpha_k t)\varphi. \tag{10}$$

Using that $A_2 = Id$ on Γ, one therefore obtains $Id = J \cdot (A - \alpha_k t)^{-1} \cdot (\cdot \otimes \delta_\Gamma) \cdot R(\alpha_k t)$ where J is the restriction operator to Γ. From this identity it follows that $R(\alpha_k t)$ is invertible. Moreover, setting $\phi = R(\alpha_k t)\varphi$,

$$R(\alpha_k t)^{-1}\phi = J \cdot (A - \alpha_k t)^{-1} \cdot (\phi \otimes \delta_\Gamma)$$

$$= J \cdot \int_{\mathbf{R}^{d-1}} \int_{\mathbf{R}} e^{i(x',s)\cdot(\xi',\eta)} [(A - \alpha_k t)^{-1}(\phi \otimes \delta_\Gamma)]\hat{}(\xi',\eta) d\eta d\xi'$$

$$\int_{\mathbf{R}^{d-1}} e^{ix'\cdot\xi'} \int_{\mathbf{R}} \sigma((A - \alpha_k t)^{-1})(x', 0, \xi', \eta)\hat{\phi}(\xi') \cdot \frac{1}{\sqrt{2\pi}} d\eta d\xi'. \tag{11}$$

Hence $R(\alpha_k t)^{-1}$ is a classical ΨDO of order -1 with symbol

$$\frac{1}{\sqrt{2\pi}} \int_{\mathbf{R}} \sigma((A - \alpha_k t)^{-1})(x', 0, \xi', \eta) d\eta, \tag{12}$$

and therefore $R(\alpha_k t)$ is a classical ΨDO of order 1 with parameter t of weight 1. The ellipticity with parameter of $R(\alpha_k t)$ follows from the explicit formula of the symbol. □

Lemma 3.4 For $\epsilon' < \frac{\pi}{d}$ sufficiently small and $0 \le k \le d-1$, the operator $R(\alpha_k t)$ does not have any eigenvalues in $\Lambda_{\epsilon'}$, where $\Lambda_{\epsilon'} = \{z \in C \mid \pi - \epsilon' < arg(z) < \pi + \epsilon' \text{ or } |z| < \epsilon'\}$. Hence $R(\alpha_k t)$ has π as an Agmon angle.

Using the above formula, one obtains as an immediate consequence the following result.

Corollary 3.5 The operator $R = R(0)$ is essentially self-adjoint and positive definite.

Next we are collecting a number of results about operators involving the $d-$th power of A and the submanifold Γ.

Consider the families of operators $A^d + t^d$ and $A_\Gamma^d + t^d$ for nonnegative real numbers t. Then $A^d + t^d$ and $A_\Gamma^d + t^d$ are elliptic differential operators with parameter, where the weight of t is 2. Note that

$$A_\Gamma^d + t^d = (A_\Gamma - te^{i\frac{\pi}{d}})(A_\Gamma - te^{i\frac{3\pi}{d}}) \cdots (A_\Gamma - te^{i\frac{\pi+2\pi(d-1)}{d}}). \tag{13}$$

Let us introduce the boundary conditions $B_d(t), C_d(t)$ by setting

$$B_d(t) = (B, B(A_\Gamma - \alpha_0 t), B(A_\Gamma - \alpha_1 t)(A_\Gamma - \alpha_0 t), \cdots, B(A_\Gamma - \alpha_{d-2} t) \cdots (A_\Gamma - \alpha_0 t)),$$

and

$$C_d(t) = (C, C(A_\Gamma - \alpha_0 t), C(A_\Gamma - \alpha_1 t)(A_\Gamma - \alpha_0 t), \cdots, C(A_\Gamma - \alpha_{d-2} t) \cdots (A_\Gamma - \alpha_0 t)).$$

It follows from Lemma 3.2 that the following operator is invertible

$$(A_\Gamma^d + t^d, B_d(t)) : C^\infty(E_\Gamma) \to C^\infty(E_\Gamma) \oplus (\oplus_d C^\infty(E_\Gamma\,|_{\Gamma^+ \sqcup \Gamma^-}))\,.$$

Therefore the corresponding Poisson operator $\tilde{P}_d(t) : \oplus_d C^\infty(E\,|_{\Gamma^+\sqcup\Gamma^-}) \to C^\infty(E_\Gamma)$ is well defined.

Lemma 3.6 The Poisson operator $\tilde{P}_d(t)$ associated to $(A_\Gamma^d + t^d, B_d(t))$ is given by

$$\tilde{P}_d(t)(\varphi_0, \cdots, \varphi_{d-1}) = P(\alpha_0 t)\varphi_0 + (A_\Gamma - \alpha_0 t)_B^{-1} P(\alpha_1 t)\varphi_1 + \cdots +$$

$$(A_\Gamma - \alpha_0 t)_B^{-1}(A_\Gamma - \alpha_1 t)_B^{-1} \cdots (A_\Gamma - \alpha_{d-2} t)_B^{-1} P(\alpha_{d-1} t)\varphi_{d-1}, \qquad (14)$$

where $(A_\Gamma - \alpha_k t)_B$ is the restriction of $A_\Gamma - \alpha_k t$ to $\{u \in C^\infty(E_\Gamma) \mid Bu = 0\}$.

Further let us consider the boundary conditions $B_d(t)$ and $C_d(t)$ for $t = 0$. Note that

$$B_d(0) = (B, BA_\Gamma, \cdots, BA_\Gamma^{d-1}); C_d(0) = (C, CA_\Gamma, \cdots, CA_\Gamma^{d-1}).$$

Let $\Omega(t)$ be the following lowertriangular $d \times d$ matrix

$$\Omega(t) = \begin{pmatrix} 1 & 0 & & \cdots 0 \\ \alpha_0 t & 1 & & \cdots 0 \\ \vdots & \vdots & & \ddots \vdots \\ \alpha_0^{d-1} t^{d-1} & t^{d-2} \sum_{k=0}^{d-2} \alpha_0^{d-2-k} \alpha_1^k & & \cdots 1 \end{pmatrix}. \qquad (15)$$

Then $B_d(0) = \Omega(t) B_d(t)$ as well as $C_d(0) = \Omega(t) C_d(t)$. Let $P_d(t) := \tilde{P}_d(t)\Omega(t)^{-1}$ and notice that $P_d(t)$ is the Poisson operator corresponding to $(A_\Gamma^d + t^d, B_d(0))$.

Consider the Dirichlet to Neumann operator $\tilde{R}_d(t) = \Delta_{if} \cdot C_d(t) \cdot \tilde{P}_d(t) \cdot \Delta_{ia}$ corresponding to $A_\Gamma^d + t^d, B_d(t)$ and $C_d(t)$. Then

$$\tilde{R}_d(t)(\varphi_0, \cdots, \varphi_{d-1}) = \qquad (16)$$

$$\Delta_{if} \cdot (C, C(A_\Gamma - \alpha_0 t), \cdots, C(A_\Gamma - \alpha_{d-2} t) \cdots (A_\Gamma - \alpha_0 t)) \cdot$$
$$(P(\alpha_0 t)\, \Delta_{ia}\, \varphi_0 + (A_\Gamma - \alpha_0 t)_B^{-1} P(\alpha_1 t)\, \Delta_{ia}\, \varphi_1 + \cdots +$$
$$(A_\Gamma - \alpha_0 t)_B^{-1}(A_\Gamma - \alpha_1 t)_B^{-1} \cdots (A_\Gamma - \alpha_{d-2} t)_B^{-1} P(\alpha_{d-1} t)\, \Delta_{ia}\, \varphi_{d-1}).$$

Thus $\tilde{R}_d(t) : \oplus_d C^\infty(E\,|_\Gamma) \to \oplus_d C^\infty(E\,|_\Gamma)$ can be represented by a $d \times d$ matrix of upper triangular form,

$$\begin{pmatrix} R(\alpha_0 t) & * & \cdots & \Delta_{if} C(A_\Gamma - \alpha_0 t)_B^{-1} \cdots (A_\Gamma - \alpha_{d-2} t)_B^{-1} P(\alpha_{d-1} t) \Delta_{ia} \\ 0 & R(\alpha_1 t) & \cdots & \Delta_{if} C(A_\Gamma - \alpha_1 t)_B^{-1} \cdots (A_\Gamma - \alpha_{d-2} t)_B^{-1} P(\alpha_{d-1} t) \Delta_{ia} \\ \vdots & \vdots & \ddots & \vdots \\ 0 & 0 & \cdots & R(\alpha_{d-1} t) \end{pmatrix}, \tag{17}$$

where $*$ is $\Delta_{if} C(A_\Gamma - \alpha_0 t)_B^{-1} P(\alpha_1 t) \Delta_{ia}$ and $R(\alpha_k t)$ is the Dirichlet to Neumann operator corresponding to $A_\Gamma - \alpha_k t$, B and C defined earlier. In particular, we conclude that $\tilde{R_d}(t)$ is invertible and has π as an Agmon angle.

Finally introduce the Dirichlet to Neumann operator $R_d(t)$ associated to $A_\Gamma^d + t^d, B_d(0)$ and $C_d(0)$. Then

$$\begin{aligned} \tilde{R_d}(t) &= \Delta_{if} \cdot C_d(t) \cdot \tilde{P_d}(t) \cdot \Delta_{ia} = \Delta_{if} \cdot \Omega(t)^{-1} \cdot C_d(0) \cdot P_d(t) \cdot \Omega(t) \cdot \Delta_{ia} \\ &= \Omega(t)^{-1} \cdot \Delta_{if} \cdot C_d(0) \cdot P_d(t) \cdot \Delta_{ia} \cdot \Omega(t) = \Omega(t)^{-1} \cdot R_d(t) \cdot \Omega(t). \end{aligned} \tag{18}$$

As a consequence, $R_d(t)$ has the same spectrum as $\tilde{R_d}(t)$ and therefore, $R_d(t)$ is invertible, has π as an Agmon angle and satisfies $\log Det(R_d(t)) = \log Det(\tilde{R_d}(t))$. In view of the fact that $\tilde{R_d}(t)$ is of upper triangular form one has

$$logDet(\tilde{R_d}(t)) = \sum_{k=0}^{d-1} logDet(R(\alpha_k t)). \tag{19}$$

As A is positive and essentially selfadjoint, the operator $A^d + t^d : C^\infty(E) \to C^\infty(E)$ is invertible for $t \geq 0$. Using the kernel $k_t(x, y)$ of $(A^d + t^d)^{-1}$ this operator can be extended to $C^\infty(E_\Gamma)$ by setting ($u \in C^\infty(E_\Gamma)$)

$$((A^d + t^d)^{-1})_\Gamma u(x) = \int_{M_\Gamma} k_t(x, y) u(y) dy. \tag{20}$$

It follows from Lemma 3.2 that

$$(A_\Gamma^d + t^d, B_d(t)) : C^\infty(E_\Gamma) \to C^\infty(E_\Gamma) \oplus (\oplus_d C^\infty(E_\Gamma |_{\Gamma^+ \sqcup \Gamma^-})) .$$

is invertible. Thus, since $B_d(0) = \Omega(t) B_d(t)$, we conclude that $(A_\Gamma^d + t^d, B_d(0))$ is invertible as well. Denote by $(A_\Gamma^d + t^d)_{B_d(0)}$ the restriction of $A_\Gamma^d + t^d$ to $\{u \in C^\infty(E_\Gamma) \mid B_d(0) u = 0\}$ and let $(A_\Gamma^d + t^d)_{B_d(0)}^{-1}$ be its inverse.

Lemma 3.7

$$(A_\Gamma^d + t^d)_{B_d(0)}^{-1} = ((A^d + t^d)^{-1})_\Gamma - P_d(t) \cdot B_d(0) \cdot ((A^d + t^d)^{-1})_\Gamma . \tag{21}$$

Lemma 3.8

$$\text{(i)}\frac{d}{dt}P_d(t) = -dt^{d-1}(A_\Gamma^d + t^d)^{-1}_{B_d(0)} \cdot P_d(t). \tag{22}$$

$$\text{(ii)}R_d(t)^{-1} \cdot \frac{d}{dt}R_d(t) = -dt^{d-1}R_d(t)^{-1} \cdot \Delta_{if} \cdot C_d(0) \cdot (A_\Gamma^d + t^d)^{-1}_{B_d(0)} \cdot P_d(t) \cdot \Delta_{ia}. \tag{23}$$

In particular, d being the dimension of M, $R_d(t)^{-1} \cdot \frac{d}{dt}R_d(t)$ is of trace class.

Taking into account that $\Delta_{ia}(\oplus_d C^\infty(E\,|_\Gamma)) = \{(\varphi,\varphi) \mid \varphi \in \oplus_d C^\infty(E\,|_\Gamma)\}$ we may define $Pr_\Gamma : \Delta_{ia}(\oplus_d C^\infty(E\,|_\Gamma)) \to \oplus_d C^\infty(E\,|_\Gamma)$ by $Pr_\Gamma(\varphi,\varphi) = \varphi$.

Corollary 3.9

$$R_d(t)^{-1} \cdot \frac{d}{dt}R_d(t) = dt^{d-1}Pr_\Gamma \cdot B_d(0) \cdot ((A^d + t^d)^{-1})_\Gamma \cdot P_d(t) \cdot \Delta_{ia}. \tag{24}$$

As $(A^d + t^d)^{-1}, (A_\Gamma^d + t^d)^{-1}_{B_d(0)}$ and $R_d(t)^{-1}\frac{d}{dt}R_d(t)$ are of trace class we can apply the well known variational formula for regularized determinants:

Lemma 3.10 Let $Q(t)$ denote any of the operators $A^d + t^d$, $(A^d + t^d)_{B_d(0)}$ or $R_d(t)$. Then, for any $t \geq 0$,

$$\frac{d}{dt}logDetQ(t) = tr(Q(t)^{-1}\frac{d}{dt}Q(t)). \tag{25}$$

4 Proof of the Theorem 1.1 and 1.2

In the case where the operator A^{-1} is of trace class, the proof of Theorem 1.1 is considerably simpler. Unfortunately, this is only the case if the dimension d of M is equal to 1. Our strategy is to first prove a version of Theorem 1.1 for A^d (Lemma 4.1), using the fact that $(A^d)^{-1}$ is of trace class. Together with the auxiliary results of section 3 and the asymptotic expansion derived in section 2, the proof of Theorem 1.1 is then completed.

Lemma 4.1 Let $A^d + t^d$ and $(A_\Gamma^d + t^d, B_d(0))$ be as above. Then, for $t \geq 0$,

$$\frac{d}{dt}\left(logDet(A^d + t^d) - logDet(A_\Gamma^d + t^d, B_d(0))\right) = \frac{d}{dt}logDetR_d(t). \tag{26}$$

Proof Define $w(t) := \frac{d}{dt}(logDet(A^d + t^d) - logDet((A_\Gamma^d + t^d), B_d(0)))$. By Lemma 3.10 and Lemma 3.7

$$w(t) = tr(\frac{d}{dt}(A^d + t^d) \cdot (A^d + t^d)^{-1} - \frac{d}{dt}(A_\Gamma^d + t^d)_{B_d(0)} \cdot (A_\Gamma^d + t^d)^{-1}_{B_d(0)})$$

$$= dt^{d-1} tr((A^d + t^d)^{-1} - (A_\Gamma^d + t^d)^{-1}_{B_d(0)})$$
$$= dt^{d-1} tr(((A^d + t^d)^{-1})_\Gamma - (A_\Gamma^d + t^d)^{-1}_{B_d(0)})$$
$$= dt^{d-1} tr(P_d(t) \cdot B_d(0) \cdot ((A^d + t^d)^{-1})_\Gamma). \qquad (27)$$

On the other hand, by Lemma 3.10, Corollary 3.9 and the commutativity of the trace,

$$\frac{d}{dt} logDetR_d(t) = tr(\frac{d}{dt} R_d(t) \cdot R_d(t)^{-1})$$
$$= dt^{d-1} tr(Pr_\Gamma \cdot B_d(0) \cdot ((A^d + t^d)^{-1})_\Gamma \cdot P_d(t) \cdot \Delta_{ia})$$
$$= dt^{d-1} tr(P_d(t) \cdot \Delta_{ia} \cdot Pr_\Gamma \cdot B_d(0) \cdot ((A^d + t^d)^{-1})_\Gamma)$$
$$= dt^{d-1} tr(P_d(t) \cdot B_d(0) \cdot ((A^d + t^d)^{-1})_\Gamma). \qquad (28)$$

Combining the above two identities shows that

$$w(t) = \frac{d}{dt} logDetR_d(t). \qquad (29)$$

□

Since $logDetR_d(t) = \sum_{k=0}^{d-1} logDetR(\alpha_k t)$, we conclude from Lemma 4.1 that

$$logDet(A^d + t^d) - logDet((A_\Gamma^d + t^d), B_d(0)) = \tilde{c} + \sum_{k=0}^{d-1} logDetR(\alpha_k t), \qquad (30)$$

where $\tilde{c}$ is independent of t.

Note that $logDet(A^d + t^d)$, $logDet(A_\Gamma^d + t^d, B_d(0))$ and $logDetR(\alpha_k t)(0 \leq k \leq d-1)$ have asymptotic expansions as $t \to +\infty$. Since the eigenvalues of $A^d + t^d$ and $(A_\Gamma^d + t^d)_{B_d(0)}$ satisfy the condition in Theorem 2.5, the constant terms in the asymptotic expansions of $logDet(A^d + t^d)$ and $logDet((A_\Gamma^d + t^d), B_d(0))$ are zero. Let $\pi_0(R(\alpha_k t))$ be the constant term in the asymptotic expansion of $logDet(R(\alpha_k t))$. Then $\tilde{c} = -\sum_{k=0}^{d-1} \pi_0(R(\alpha_k t))$, which is computable in terms of the symbol of $R(\alpha_k t)$ by Theorem 2.4.

Lemma 4.2 (i) $Det(A_\Gamma^d, B_d(0)) = (Det(A_\Gamma, B))^d$; (ii) $Det(A^d) = (DetA)^d$.

Proof Statement (i) follows from the fact that λ is an eigenvalue of A_B if and only if λ^d is an eigenvalue of $(A_\Gamma^d)_{B_d(0)}$ and (ii) is proved in the same way. □

Proof of Theorem 1.1 Setting $t = 0$ in (30), one obtains

$$logDetA^d - logDet(A_\Gamma^d, B_d(0)) = \tilde{c} + logDetR_d(0). \qquad (31)$$

By Lemma 4.2, $log(DetA)^d - log(Det(A_\Gamma, B))^d = \tilde{c} + log(DetR)^d$. Hence $logDetA = log(c) + logDet(A_\Gamma, B) + logDetR$. Here

$$log(c) = -\frac{1}{d}\sum_{k=0}^{d-1} \pi_0(R(\alpha_k t)). \tag{32}$$

Using the result of Theorem 2.4, Theorem 1.1 follows. □

Proof of Theorem 1.2 ¿From (32) we get $logc = -\frac{1}{d}\sum_{k=0}^{d-1}\pi_0(R(\alpha_k t))$. We are going to prove that for each k, $\pi_0(R(\lambda)(\alpha_k t)) = 0$. Let $\sigma(R(\alpha_k t)) \sim p_1 + p_0 + p_{-1} + \cdots$ be the asymptotic symbols of $R(\lambda)(\alpha_k t)$. ¿From [1], we get a formula for π_0 as follows.

$$\pi_0(R(\alpha_k)) = \tag{33}$$

$$\frac{\partial}{\partial s}\Big|_{s=0} \frac{1}{(2\pi)^{d-1}}\int_\Gamma\int_{\mathbf{R}^{d-1}}\frac{1}{2\pi i}\int_\gamma \mu^{-s} r_{-1-(d-1)}(\mu, \alpha_k; x', \xi')d\mu d\xi' dvol(\Gamma).$$

Here r_{-1-j} is the homogeneous part of the asymptotic symbol of the resolvent $(\mu - R(\lambda))^{-1}$, i.e.

$$r_{-1} = (\mu - p_1(\lambda, x', \xi'))^{-1}, \tag{34}$$

$$r_{-1-j} = -(\mu - p_1)^{-1}\sum_{k=0}^{j-1}\sum_{\alpha,|\alpha|+l=j-k}\frac{1}{\alpha!}\partial_\xi^\alpha p_{1-l}(\lambda, x', \xi')D_{\xi'}^\alpha r_{-1-k}(\mu, \lambda, x', \xi').$$

The proof of the Theorem 1.2 reduces to the verification of the following equation

$$p_{1-j}(x', -\xi', \lambda) = (-1)^j p_{1-j}(x', \xi', \lambda). \tag{35}$$

Then

$$r_{-1-j}(\mu, \frac{\lambda}{|\lambda|}, x', -\xi') = (-1)^j r_{-1-j}(\mu, \frac{\lambda}{|\lambda|}, x', \xi'), \tag{36}$$

hence when d is even, $r_{-1-(d-1)}(\mu, \frac{\lambda}{|\lambda|}, x', \xi')$ is odd with respect to ξ'. Thus the integrand of R^{d-1} is zero and $\pi_0(R(\lambda)(\alpha_k t)) = 0$.

Let $\sigma(R(\lambda)(\alpha_k t)^{-1}) \sim q_{-1} + q_{-2} + \cdots$. ¿From the Lemma 3.3, we get a formula

$$q_{1-j} = \frac{1}{2\pi}\int_{-\infty}^{\infty}\beta_{2-j}(x', 0, \xi', \eta; \lambda, t)d\eta, \tag{37}$$

where $\sigma((\Delta + \lambda - \alpha_k t)^{-1}) \sim \sum_{j=0}^{\infty}\beta_{2-j}(x', s, \xi', \eta; \lambda, t)$.

Set $\sigma((\Delta + \lambda - \alpha_k t)) \sim w_2 + w_1 + w_0$, where $w_2(x, \xi; \lambda, t) = \sigma_L(\Delta + \lambda) - \alpha_k t$. Then $w_{2-j}(x, -\xi; \lambda, t) = (-1)^j w_{2-j}(x, \xi; \lambda, t)$ and hence $\beta_{-2-j}(x, -\xi; \lambda, t) = (-1)^j\beta_{2-j}(x, \xi; \lambda, t)$, where $x = (x', s)$ and $\xi = (\xi', \eta)$. ¿From these facts, we can conclude that $q_{1-j}(x', -\xi'; \lambda, t) = (-1)^j q_{1-j}(x', \xi'; \lambda, t)$ and hence $p_{1-j}(x', -\xi'; \lambda, t) = (-1)^j p_{1-j}(x', \xi'; \lambda, t)$. □

5 Mayer-Vietoris formula for the Laplacian on an even dimensional manifolds

In this section, we will consider the special case of the Mayer-Vietoris formula proved in the previous section. Throughout this section, (M, g) is an even dimensional closed Riemannian manifold. Let $\wedge^p T^*M$ be the p-th exterior product of the cotangent bundle. Set $A = \Delta + \lambda$, where Δ is a Laplacian acting on differential p-forms and λ is a positive real number. Then from Theorem 1.2, we get

$$Det(\Delta + \lambda) = Det(\Delta + \lambda, B) \cdot Det(R(\lambda)), \tag{38}$$

where B is the Dirichlet boundary condition and $R(\lambda)$ is the Dirichlet to Neumann operator.

By Corollary 3.5, $R(\lambda)$ is a positive definite, self-adjoint, elliptic ΨDO. When $\lambda = 0$, both the Laplacian Δ and $R := R(0)$ have zero eigenvalues and so $Det(\Delta) = Det(R) = 0$. In this case we can define the modified determinants $Det^*\Delta$ and Det^*R to be the determinants of Δ and R respectively, when restricted to the orthogonal complement of the null space.

Let H_p be the space of harmonic p-forms equipped with the natural inner product $\langle\varphi, \psi\rangle = \int_M \varphi \wedge *\psi = \int_M (\varphi, \psi) dvol(M)$, where $(,)$ is a metric in $E = \wedge^p T^*M$ induced by the Riemannian metric g on M. Let $H_p \mid_\Gamma$ be the restriction of harmonic forms to Γ. Define an inner product on $H_p \mid_\Gamma$ by $\langle\alpha, \beta\rangle_\Gamma = \int_\Gamma (\alpha, \beta) d\mu_\Gamma$, where $d\mu_\Gamma$ is a volume element of Γ coming from g restricted to Γ.

Suppose $k = dim H_p$, and let $\psi_1, \cdots, \psi_k$ be an orthonormal basis of H_p and $\phi_1, \cdots, \phi_k$ be an orthonormal basis of $H_p \mid_\Gamma$. Let $J : H_p \to H_p \mid_\Gamma$ denote the restriction map. Let $J(\psi_i) = a_{ij}\phi_j$ and let $A = (a_{ij})_{1 \le i,j \le k}$.

Theorem 5.1 For an even dimensional manifold M,

$$Det^*\Delta = \frac{1}{(det A)^2} Det(\Delta, B) \cdot Det^*R. \tag{39}$$

Remark If $p = 0$, then $E = M \times \mathbf{R}$ and A is a 1×1 matrix. If ω and ϕ are orthonormal harmonic 0-forms on M and Γ respectively, then $\omega = \frac{1}{\sqrt{V}}$ and $\phi = \frac{1}{\sqrt{l}}$, where V and l are the volumes of M and Γ respectively. Consider $J : H_0 \to H_0 \mid_\Gamma$. Then $J(\omega) = \frac{1}{\sqrt{V}} 1_\Gamma = \frac{\sqrt{l}}{\sqrt{V}} \cdot \phi$. Hence $A = (\frac{\sqrt{l}}{\sqrt{V}})$ and we obtain

$$Det^*\Delta = \frac{V}{l} Det(\Delta, B) \cdot Det^*R, \tag{40}$$

which is setablished in [1].

Let $\{\psi_j\}_{j\geq 0}$ be the complete orthonormal system of eigenforms of Δ with eigenvalue λ_j in $L^2(E)$. Let $k = dimH_p$. Then $\psi_1, \cdots, \psi_k$ are harmonic forms and $\lambda_1 = \cdots = \lambda_k = 0$.

Lemma 5.2

$$logDet(\Delta + \epsilon) = kloge + logDet^*(\Delta) + o(\epsilon) \text{ as } \epsilon \to 0. \tag{41}$$

Proof For $Res \gg 0$, $\zeta_{\Delta+\epsilon}(s) = k\epsilon^{-s} + \sum_{\lambda_i>0}(\lambda_i + \epsilon)^{-s}$. Then

$$\zeta'_{\Delta+\epsilon}(s) = -kloge \cdot \epsilon^{-s} - \sum_{\lambda_i>0} log(\lambda_i + \epsilon)(\lambda_i + \epsilon)^{-s}. \tag{42}$$

Thus

$$-\zeta'_{\Delta+\epsilon}(s) - (kloge + \sum_{\lambda_i>0} log\lambda_i \cdot \lambda_i^{-s}) \tag{43}$$

$$= kloge(\epsilon^{-s} - 1) + \sum_{\lambda_i>0} \left(log(\lambda_i + \epsilon) \cdot (\lambda_i + \epsilon)^{-s} - log\lambda_i \cdot \lambda_i^{-s}\right).$$

Since $\sum_{\lambda_i>0} log\lambda_i \cdot \lambda_i^{-s}$ converges uniformly for $Res \gg 0$ and has a meromorphic continuation with 0 as a regular value,

$$\lim_{\epsilon\to 0} \sum_{\lambda_i>0} \left(log(\lambda_i + \epsilon) \cdot (\lambda_i + \epsilon)^{-s} - log\lambda_i \cdot \lambda_i^{-s}\right) = 0 \text{ at } s = 0. \tag{44}$$

Thus we get

$$logDet(\Delta + \epsilon) - (kloge + logDet^*\Delta) = o(\epsilon). \tag{45}$$

□

Proof of Theorem 5.1 Let $\mu_j(\epsilon)(j \geq 1)$ be the eigenvalues of $R(\epsilon)$ with $0 < \mu_1(\epsilon) \leq \cdots \leq \mu_k(\epsilon) < \mu_{k+1}(\epsilon) \leq \cdots$. It is clear that $\lim_{\epsilon\to 0} \mu_j(\epsilon) = 0$ for $1 \leq j \leq k$. Then by the same argument as Lemma 5.2,

$$logDetR(\epsilon) = log(\mu_1(\epsilon) \cdot \mu_k(\epsilon)) + logDet^*R + o(\epsilon) \text{ as } \epsilon \to 0. \tag{46}$$

Now we want calculate $\mu_1(\epsilon) \cdots \mu_k(\epsilon)$. Let $\phi_1(\epsilon), \cdots, \phi_k(\epsilon)$ be orthonormal eigenforms of $R(\epsilon)$ corresponding to eigenvalues $\mu_1(\epsilon), \cdots, \mu_k(\epsilon)$. Then $\phi_j(\epsilon) \to \phi_j$ as $\epsilon \to 0$, where ϕ_j is the restriction of a harmonic form to Γ with $\langle\phi_j, \phi_j\rangle_\Gamma = 1$. Let $a_{ij}(\epsilon) = \langle\psi_i, \phi_j(\epsilon)\rangle_\Gamma$ $(1 \leq i, j \leq k)$ and $A(\epsilon) = (a_{ij}(\epsilon))$. Note that $\psi_i \mid_\Gamma = a_{ij}(\epsilon)\phi_j(\epsilon) + \tilde{\psi_i}(\epsilon) \mid_\Gamma$ for some $\tilde{\psi_i}(\epsilon) \mid_\Gamma \in (span\{\phi_1(\epsilon), \cdots, \phi_k(\epsilon)\})^\perp$. Define

$$I : C^\infty(E \mid_\Gamma) \to C^\infty(E \mid_\Gamma) \text{ by}$$

$$\varphi \longmapsto \sum_{j=1}^{k} \int_{\Gamma} (\varphi, \psi_j) d\mu_{\Gamma} \cdot \psi_j \mid_{\Gamma} = \sum_{j=1}^{k} \langle \varphi, \psi_j \rangle_{\Gamma} \cdot \psi_j \mid_{\Gamma} .$$

Then

$$\langle I(\phi_i(\epsilon)), \phi_j(\epsilon) \rangle_{\Gamma} = \sum_{l=1}^{k} a_{li}(\epsilon) a_{lj}(\epsilon) = ({}^t A(\epsilon) A(\epsilon))_{ij}. \tag{47}$$

Define

$$G_\epsilon : C^\infty(E|_\Gamma) \to C^\infty(E|_\Gamma) \text{ by}$$

$$\varphi \longmapsto R(\epsilon)^{-1}(\varphi) - \frac{1}{\epsilon} I(\varphi).$$

Then G_ϵ is a bounded operator with $\| G_\epsilon \| = \frac{1}{\mu_{k+1}(\epsilon)}$, which converges to $\frac{1}{\mu_{k+1}(0)}$ as $\epsilon \to 0$. Clearly

$$\langle R(\epsilon)^{-1} \phi_i(\epsilon), \phi_j(\epsilon) \rangle = \frac{1}{\epsilon} \langle I(\phi_i(\epsilon), \phi_j(\epsilon) \rangle + \langle G_\epsilon(\phi_i(\epsilon)), \phi_j(\epsilon) \rangle. \tag{48}$$

For $1 \leq j \leq k$,

$$\begin{aligned} \frac{1}{\mu_j(\epsilon)} &= \langle R(\epsilon)^{-1} \phi_j(\epsilon), \phi_j(\epsilon) \rangle \\ &= \frac{1}{\epsilon} \langle I(\phi_j(\epsilon)), \phi_j(\epsilon) \rangle + \langle G_\epsilon(\phi_j(\epsilon)), \phi_j(\epsilon) \rangle \\ &= \frac{1}{\epsilon} ({}^t A(\epsilon) A(\epsilon))_{jj} + N_j(\epsilon), \end{aligned} \tag{49}$$

where $N_j(\epsilon) := \langle G_\epsilon(\phi_i(\epsilon)), \phi_j(\epsilon) \rangle_\Gamma$ is bounded as $\epsilon \to 0$. For $i \neq j, 1 \leq i, j \leq k$,

$$\begin{aligned} 0 &= \langle R(\epsilon)^{-1}(\phi_i(\epsilon)), \phi_j(\epsilon) \rangle \\ &= \frac{1}{\epsilon} \langle I(\phi_i(\epsilon)), \phi_j(\epsilon) \rangle + \langle G_\epsilon(\phi_i(\epsilon)), \phi_j(\epsilon) \rangle \\ &= \frac{1}{\epsilon} ({}^t A(\epsilon) A(\epsilon))_{ij} + \langle G_\epsilon(\phi_i(\epsilon)), \phi_j(\epsilon) \rangle. \end{aligned} \tag{50}$$

Since $({}^t A(\epsilon) A(\epsilon))_{ij}$ and $\langle G_\epsilon(\phi_i(\epsilon)), \phi_j(\epsilon) \rangle$ are bounded, $({}^t A(\epsilon) A(\epsilon))_{ij} \to 0$. Thus

$$\frac{1}{\mu_1(\epsilon) \cdots \mu_k(\epsilon)} = \left(\frac{1}{\epsilon} ({}^t A(\epsilon) A(\epsilon))_{11} + N_1(\epsilon) \right) \cdots \left(\frac{1}{\epsilon} ({}^t A(\epsilon) A(\epsilon))_{kk} + N_k(\epsilon) \right) = \tag{51}$$

$$\frac{1}{\epsilon^k}(detA(\epsilon))^2\left(\frac{({}^tA(\epsilon)A(\epsilon))_{11}({}^tA(\epsilon)A(\epsilon))_{22}\cdots({}^tA(\epsilon)A(\epsilon))_{kk}}{(detA(\epsilon))^2}+\epsilon\cdot\frac{\tilde{N}(\epsilon)}{(detA(\epsilon))^2}\right),$$

where $\tilde{N}(\epsilon)$ is bounded as $\epsilon \to 0$. Hence

$$logDetR(\epsilon) = klog\epsilon - log(detA(\epsilon))^2 + logDet^*R + o(\epsilon) \text{ as } \epsilon \to 0. \quad (52)$$

From the Lemma 5.2 and Theorem 1.2, we get

$$logDet(\Delta + \epsilon) = klog\epsilon + logDet^*(\Delta) + o(\epsilon) \quad (53)$$

$$logDet(\Delta + \epsilon) = logDet(\Delta, B) + logDet(R(\epsilon)). \quad (54)$$

From these three equations, we get

$$logDet^*\Delta = -log(detA)^2 + logDet^*R + logDet(\Delta, B). \quad (55)$$

□

References

1. D.Burghelea, L.Friedlander, T.Kappeler, Journal of Functional Analysis **107**, 34 (1992)
2. L.Friedlander, Proceedings of American Mathematical Society **107**, 169 (1989).
3. R.Seeley, *Complex Powers of Elliptic Operators* (Proceedings of Symposia on Singular Integrals, Amer. Math. Soc., Providence, RI. 10, 288-307, 1967.
4. M.A.Shubin, *Pseudodifferential Operators and Spectral Theory*, Springer-Verlag, Berlin/New York, 1985.
5. A.Voros, Communications on Mathematical Physics **110**, 439 (1987)

CERTAIN SUBMANIFOLDS OF A KENMOTSU MANIFOLD

K. MATSUMOTO

Department of Mathematics, Faculty of Education,
Yamagata University, Yamagata, 990-8560, Japan

I. MIHAI

Faculty of Mathematics, Bucharest University,
Str. Academiei 14, 70109 Bucharest, Romania

M. H. SHAHID

Department of Mathematics, Faculty of Natural Sciences,
Jamia Millia Islamia, New Delhi-25, India

1 Preliminaries.

Let $(\tilde{M}, <,>)$ be a Hermitian manifold and denote by J the canonical almost conplex structure on $\tilde{M}$. According to the behavior of the tangent bundle TM with respect to the action of J, we may distinguish two special classes of submanifolds M in $\tilde{M}$.

a) complex submanifolds, i.e. $J(T_xM) = T_xM, \forall x \in M$.

b) totally real submanifolds, i.e. $J(T_xM) \subset T_x^{\perp}M, \forall x \in M$, where T_xM (resp. $T_x^{\perp}M$) is the tangent (resp. the normal) vector space of M at x. Such submanifolds were defined and studied by B.Y.Chen and K.Ogiue[7]. B.Y.Chen[5] introduced a new class of submanifolds in Hermitian manifolds, called *slant* submanifolds, which contains both complex submanifolds and totally real submanifolds.

A submanifold M of a Hermitian manifold $\tilde{M}$ is said to be a *slant* submanifold if for any tangent vector X to M, JX makes a constant angle with the tangent space $T_{\pi(X)}M$, where $\pi : TM \to M$ denotes the canonical projection.

Interesting examples of slant submanifolds were provided in 5.

On the other hand, K.Yano and S.Ishihara[19] considered a submanifold M whose tangent bundle TM splits into a complex subbundle $\mathcal{D}$ and a totally real subbundle $\mathcal{D}^{\perp}$. Later, such a submanifold was called a CR-submanifold[2,3].

D.E. Blair and B.Y.Chen[2] proved that a CR-submanifold of a locally conformal Kaehler manifold is a Cauchy-Riemann manifold in the sence of S.Greenfield.

The first main result on CR-submanifolds was obtained by A.Bejancu[1] and B.Y.Chen[3]: any CR-submanifold of a Kaehler manifold is foliated by totally real submanifolds (i.e. the totally real subbundle is involutive).

As non-trivial examples of CR-submanifolds, we can mention the (real) hypersurfaces of Hermitian manifolds[1,22].

B.Y.Chen defined the generic submanifold which is the generalization of the CR-submanifold.

A submanifold M of a Hermitian manifold $(\tilde{M}, J, <,>)$ is said to be a *generic* submanifold if its maximal holomorphic subspaces

$$\mathcal{D}_x = T_xM \cap J(T_xM), \quad x \in M$$

in T_xM define a differentiable subbundle of TM.

The Riemannian product of a complex submanifold and a slant submanifold of a Hermitian manifold is a generic submanifold.

It should be noticed that every real submanifold of a complex manifold is the closure of a union of some generic open submanifolds.

We want to mention that any generic submanifold of a Hermitian manifold is a Cauchy-Riemann manifold (see 13). But, on a generic submanifold, generally the purely real subbundle $\mathcal{D}^{\perp}$ is not involutive.

Recently, one of the present authors defined a new class of submanifolds of a Kaehler manifold, foliated by totally real submanifolds.

Definition[12]. A submanifold M of a Hermitian manifold $(\tilde{M}, J, <,>)$ is said to be a *generalized CR-submanifold* if its maximal totally real subspaces

$$\mathcal{D}_x^{\perp} = T_xM \cap J(T_x^{\perp}M), \quad x \in M,$$

in T_xM define a differentiable subbundle of TM.

It is easily seen that this class contains both CR-submanifolds and slant submanifolds. The Riemannian product of a slant submanifold and a totally real submanifold of a Hermitian manifold is a generalized CR-submanifold. And, moreover, the Riemannian submanifold of a Hermitian manifold is a generalized CR-submanifold. And, moreover, the Riemannian product of a complex submanifold and a totally real submanifold of a Hermitian manifold is a generalized CR-submanifold, too.

2 Generalized contact CR-submanifolds of a Kenmotsu manifold.

S.Tanno[16] has classified, into 3 classes, the connected almost contact Riemannian manifolds whose automorphisms group have the maximum dimensions:

1. homogeneous normal contact Riemannian manifolds with constant ϕ-holomorphic sectional curvature;

2. global Riemannian products of a line or circle and a Kaehlerian space form.

3. warped product spaces $L \times_f F$, where L is a line, F a Keahlerian manifold and f is a differentiable function on F.

K.Kenmotsu[8] studied the third class and characterized it by tensor equations. Later, such a manifold was called a *Kenmotsu* manifold.

A $(2m+1)$-dimensional Riemannian manifold $(\tilde{M}, g)$ is said to be a *Kenmotsu* manifold if it admits an endomorphism ϕ of its tangent bundle $T\tilde{M}$, a vector field ξ and a 1-form η, which satisfy:

$$\begin{cases} \phi^2 = -Id + \eta \otimes \xi, \quad \eta(\xi) = 1, \quad \phi\xi = 0, \quad \eta \circ \phi = 0, \\ g(\phi X, \phi Y) = g(X, Y) - \eta(X)\eta(Y), \quad \eta(X) = g(\xi, X), \\ (\tilde{\nabla}_X \phi)Y = -g(X, \phi Y)\xi - \eta(Y)\phi X, \quad \tilde{\nabla}_X \xi = X - \eta(X)\xi \end{cases} \tag{1}$$

for any vector fields X, Y on $\tilde{M}$, where $\tilde{\nabla}$ denotes the Riemannian connection with respect to g.

We denote by ω the fundamental 2-form of $\tilde{M}$, i.e. $\omega(X, Y) = g(\phi X, Y), \forall X, Y \in \Gamma(T\tilde{M})$. It was proved that the pairing (ω, η) defines a locally conformal cosymplectic structure, i.e.

$$d\omega = 2\omega \wedge \eta, \quad d\eta = 0. \tag{2}$$

A Kemnotsu manifold with constant ϕ-holomorphic sectional curvature c is called a *Kenmotsu space form* and it is denoted by $\tilde{M}(c)$. Then its curvature tensor $\tilde{R}$ is expressed by[9]

$$4\tilde{R}(X, Y)Z = (c-3)\{g(Y, Z)X - g(X, Z)Y\} \tag{3}$$

$$+(c+1)[\{\eta(X)Y - \eta(Y)X\}\eta(Z) + \{g(X, Z)\eta(Y) - g(Y, Z)\eta(X)\}\xi$$

$$+\omega(Y, Z)\phi X - \omega(X, Z)\phi Y - 2\omega(X, Y)\phi Z].$$

By analogy with submanifolds in a Kaehler manifold, different classes of submanifolds in Kaehlerian and Sasakian-like manifolds, respectively, were considered (see 9,10,11,14.15, 17,18, etc.).

In the present paper, we define a new class of submanifolds in a Kenmotsu manifold, called a *generalized contact CR-submanifold.*

Definition. Let $\tilde{M}$ be a Kenmotsu manifold. An $(n+1)$-dimensional submanifold M tangent to ξ in $\tilde{M}$ is called a *generalized contact CR-submanifold* if the maximal anti-invariant subspaces by ϕ

$$\mathcal{D}_x^{\perp} = T_x M \cap \phi(T_x^{\perp} M), \quad x \in M$$

in $T_x M$ define a differentiable subbundle of TM.

We will denote by $\mathcal{D}$ the complementary orthogonal subbundle to $\mathcal{D}^{\perp} \oplus \{\xi\}$ in TM, where $\{\xi\}$ is the 1-dimensional subbundle spanned by ξ in TM. For any $X \in \Gamma(\mathcal{D})$, ϕX has a non-zero component.

If $\mathcal{D}$ is invariant by ϕ, then M is called a contact CR-submanifold[21] and if $\mathcal{D} = \{0\}$, then M is anti-invariant.

For any vector field X tangent to M, we put

$$\phi X = PX + FX, \tag{4}$$

where PX (resp. FX) denotes the tangential (resp. normal) component of ϕX. Then P is an endomorphism of tangent bundle TM and F is a normal bundle valued 1-form on TM.

For any vector field V normal to M, we put

$$\phi V = tV + fV, \tag{5}$$

where tV (resp. fV) denotes the tangential (resp. normal) component of ϕV. Then f is an endomorphism of the normal bundle $T^{\perp}M$ and t is a tangent bundle-valued 1-form on $T^{\perp}M$.

On a generalized contact CR-submanifold, one has

$$P(\mathcal{D}_x) = \mathcal{D}_x, \quad x \in M,$$

$$t(F(\mathcal{D})) \subset \mathcal{D}, \quad t(\phi(\mathcal{D}^{\perp})) = \mathcal{D}^{\perp},$$

$$T^{\perp}M = F(\mathcal{D}) \oplus \phi(\mathcal{D}^{\perp}) \oplus \nu,$$

where ν denotes the maximal invariant subbundle by ϕ in $T^{\perp}M$,i.e.

$$\nu_x = T_x^{\perp}M \cap \phi(T_x^{\perp}M), \quad x \in M.$$

Necessary and sufficient conditions for a generalized contact CR-submanifold to be a contact CR-submanifold are given by the following;

Proposition 2.1. *Let M be a generalized contact CR-submanifold of a Kenmotsu manifold. Then the following assertions are equivarent to each other:*

i) *M is a contact CR-submanifold;*

ii) *the endomorphism P defines a f-structure in the tangent bundle, i.e.* $P^3 + P = 0$;

iii) *the endomorphism f defines a f-structure in the normal bundle, i.e.* $f^3 + f = 0$.

Proof. It is easily seen that one has

$$P^2 + tF = -Id + \eta \otimes \xi.$$

Then we get

$$P^3 + P = -tFP + (\eta \circ P) \otimes \xi.$$

If M is a contact CR-submanifold, then $P^3 + P = 0$.

Conversely, if P defines a f-structure in the tangent bundle TM, it follows that $tFP = 0$ on $\{\xi\}^{\perp}$. This implies $F(\mathcal{D}) = 0$, i.e. $\mathcal{D}$ is an invariant subbundle. Thus M is a contact CR-submanifold.

For f, the proof is similar with the case of P. So, we omit it.

A generalized contact CR-submanifold with non-trivial subbundles $\mathcal{D}$ and $\mathcal{D}^{\perp}$ is said to be *proper.*

An obstruction to the existence of a curvature invariant proper generalized contact CR-submanifold is given by the following result.

Proposition 2.2. *Let M be a proper generalized contact CR-submanifold of a Kenmotsu space form $\tilde{M}(c)$. Then M is a curvature invariant submanifold if and only if $c = -1$.*

Proof. Let M be a curvature invariant proper generalized contact CR-submanifold. For any $X, Y \in \Gamma(\mathcal{D})$ and $Z \in \Gamma(\mathcal{D}^{\perp})$, one has

$$\tilde{R}(X, Y)Z = \frac{c+1}{2} g(X, \phi Y)\phi Z \in \Gamma(TM).$$

Then $c = -1$.

The converse is obvious.

3 Involutivity of subbundles and geometry of leaves.

In this section, we will prove first that any generalized contact CR-submanifolds of a Kenmotsu manifold is foliated by anti-invariant submanifolds.

Theorem 3.1. *Let M be a generalized contact CR-submanifold of a Kenmotsu manifold $\tilde{M}$. Then the anti-invariant subbundle $\mathcal{D}^{\perp}$ is always involutive.*

Proof. Let $Z, W \in \Gamma(\mathcal{D}^{\perp})$ and $X \in \Gamma(TM)$. Since $\tilde{M}$ is a Kenmotsu manifold, using (2), we have

$$0 = 3d\omega(X, Z, W) = \omega(X, [Z, W]) = -g(X, \phi[Z, W]),$$

where [,] means the Lie bracket. The above equation means that $\phi[Z,W] \in \Gamma(T^{\perp}M)$, or equivalently, $[Z,W] \in \Gamma(\mathcal{D}^{\perp})$. Thus the subbundle $\mathcal{D}^{\perp}$ is involutive.

Corollary 3.2. *On a generalized contact CR-submanifold M of a Kenmotsu manifold, the subbundle $\mathcal{D}^{\perp} \oplus \{\xi\}$ is always invoutive.*

For the invariant subbundle $\mathcal{D}$, we have the following:

Theorem 3.3. *Let M be a generalized contact CR-submanifold of a Kenmotsu manifold $\tilde{M}$. Then the subbundle $\mathcal{D} \oplus \{\xi\}$ is involutive if and only if*

$$g(h(X,PY) + \nabla^{\perp}_{X}FY - h(PX,Y) - \nabla^{\perp}_{Y}FX, \phi Z) = 0$$

for any $X,Y \in \Gamma(\mathcal{D})$ and $Z \in \Gamma(\mathcal{D}^{\perp})$, where h and $\nabla^{\perp}$ denote the second fundamental form and the normal connection, respectively.

Proof. Let $X,Y \in \Gamma(\mathcal{D})$ and $Z \in \Gamma(\mathcal{D}^{\perp})$. Then we have

$$g([X,Y],Z) = g(h(X,PY) + \nabla^{\perp}_{X}FY - h(PX,Y) - \nabla^{\perp}_{Y}FX, \phi Z).$$

The above relation is equivalent to

$$h(X,PY) + \nabla^{\perp}_{X}FY - h(PX,Y) - \nabla^{\perp}_{X}FX \in \Gamma(F(\mathcal{D}))$$

for any $X,Y \in \Gamma(\mathcal{D})$.

Remark. Since η is a closed 1-form on a generalized contact CR-submanifold of a Kenmotsu manifold, the subbundle $\{\xi\}^{\perp} = \mathcal{D} \oplus \mathcal{D}^{\perp}$ is always involutive.

On the geometry of leaves $\mathcal{D}$ and $\mathcal{D}^{\perp}$, we may state the following results.

Proposition 3.4. *Let M be a generalized contact CR-submanifold of a Kenmotsu manifold $\tilde{M}$. If $\mathcal{D} \oplus \{\xi\}$ is involutive and its leaves are totally geodesic, then we have*

$$g(h(X,PY),\phi Z) = g(\nabla^{\perp}_{X}\phi Z, FY)$$

for any $X,Y \in \Gamma(\mathcal{D})$ and $Z \in \Gamma(\mathcal{D}^{\perp})$.

Proposition 3.5. *Let M be a generalized contact CR-submanifold of a Kenmotsu manifold $\tilde{M}$. If the leaves $\mathcal{D}^{\perp}$ are totally geodesic, then we have*

$$g(h(PX,Z),\phi W) = g(\nabla^{\perp}_{Z}\phi W, FX)$$

for any $X \in \Gamma(\mathcal{D})$ *and* $Z, W \in \Gamma(\mathcal{D}^{\perp})$.

4 B.Y.Chen's inequality. Recently, B.Y.Chen[6] has defined an invariant δ_M on a Riemannian manifold M and obtained a sharp inequality of this Riemannian invariant for submanifolds in real space forms.

In the present section, we will state an analogous result for contact CR-submanifolds in Kenmotsu space forms.

Recall the definition of B.Y.Chen's invariant[6].

Let (M, g) be an n-dimensional Riemannian manifold. Denote by $K(\pi)$ the sectional curvature of the plane section $\pi \subset T_pM, p \in M$. For any orthonormal basis $\{e_1, e_2, , , e_n\}$ of the tangent space T_pM, the scalar curvature τ at p is defined by

$$\tau(p) = \sum_{i \leq j \leq n} K(e_i \wedge e_j).$$

For each point $p \in M$, we put

$$(\inf K)(p) = \inf\{K(\pi); \pi \subset T_pM, \dim \pi = 2\}.$$

Let δ_M denote the difference between scalar curvature and $\inf K$, i.e.

$$\delta_M = \tau - \inf K.$$

δ_M is a well-defined Riemannian invariant, which we call *B.Y.Chen's invariant.*

Using the same method as in 8, we state the B.Y.Chen's inequality for contact CR-submanifolds in Kenmotsu space forms.

Theorem 4.1. *Let M be an $(n+1)$-dimensional contact CR-submanifold tangent to ξ in a Kenmotsu space form $\tilde{M}(c)$. Then we have*

$$\delta_M \leq \frac{n-1}{2}[\frac{(n+1)^2}{n}|H|^2 + \frac{1}{4}\{(n+2)(c-3) - \frac{2(n-3l)(c+1)}{(n-1)}\}],$$

where H is the mean curvature vector field of M and $2l = \dim \mathcal{D}$.

Proof. From (3) and the Gauss equation, the curvature tensor R with respect to the induced metric g is written as

$$4R(X,Y,Z,W) = (c-3)\{g(Y,Z)g(X,W) - g(X,Z)g(Y,W)\}$$

$$+(c+1)[\{g(\xi,X)g(Y,W) - g(\xi,Y)g(X,W)\}g(\xi,Z) + \{g(\xi,Y)g(X,Z)$$

$$-g(\xi,X)g(Y,Z)\}g(\xi,W) + g(\phi Y,Z)g(\phi X,W) - g(\phi X,Z)g(\phi Y,W)$$

$$-2g(\phi X, Y)g(\phi Z, W)] + 4g(h(X,W), h(Y,Z)) - 4g(h(X,Z), h(Y,W))$$

for any $X, Y, Z, W \in \Gamma(TM)$.

Now, let $\{e_1, e_2, ..., e_l, e_{l+1} = \phi e_1, e_{l+2} = \phi e_2, ..., e_{2l} = \phi e_l, e_{2l+1} = \xi\}$ and $\{e_{2l+2}, ...e_{n+1}\}$ be the local orthnormal basis of $\mathcal{D} \oplus \{\xi\}$ and $\mathcal{D}^\perp$, respectively.

Then, the sectional curvature $K(e_i \wedge e_j) = R(e_i, e_j, e_j, e_i)$ of a plane defined by e_i and e_j is written as

$$K(e_i \wedge e_j) = \frac{c-3}{4}(1 - \delta_{ji}) - \frac{3(c+1)}{4}\{2\delta_{2l+1,i}\delta_{2l+1,j} - \delta_{2l+1,j} - \delta_{2l+1,i}$$

$$+3g(\phi e_i, e_j)g(\phi e_i, e_j)\} + g(h(e_i, e_i), h(e_j, e_j)) - g(h(e_i, e_j), h(e_i, e_j)).$$

Using this, we obtain

$$2\tau = \frac{1}{4}\{n(n+1)(c-3) - 2(n-3l)(c+1)\} + (n+1)^2|H|^2 - ||h||^2.$$

Now, we put

$$\varepsilon = 2\tau - \frac{(n+1)^2(n-1)}{n}|H|^2 + \frac{1}{4}\{2(c+1)(n-3l) - (c-3)(n+2)(n-1)\}.$$

From the above two equations, we get

$$(n+1)^2|H|^2 = n\{||h||^2 - \frac{c-3}{2} + \varepsilon\}.$$

Next, let $\pi = \{e_1, e_2\}$ and e_{n+2} be a unit vector of the H-direction. Then, by virtue of Chen's lemma[6], we obtain

$$2K(\pi) \geq \varepsilon = 2\tau - \frac{(n+1)^2(n-1)}{n}|H|^2$$

$$+\frac{1}{4}\{2(c+1)(n-3l) - (c-3)(n-1)(n+2)\}$$

From the above inequality, we get our theorem.

By virtue of the above theorem, we can easily seen the following results;

Corollary 4.2. *Let M be an $(n+1)$-dimensional anti-invariant submanifold, that is $l = 0$[YK1], tangent to ξ in a Kenmotsu space form $\tilde{M}$. Then we have*

$$\delta_M \leq \frac{n-1}{2}[\frac{(n+1)^2}{n}|H|^2 + \frac{1}{4}\{(n+2)(c-3) - \frac{2n(c+1)}{n-1}\}].$$

Corollary 4.3. *Let M be a minimal anti-invariant $(n+1)$-dimensional submanifold of a Kenmotsu space form $\tilde{M}(c)$. Then*

$$\delta_M \leq \frac{n-1}{8}\{(n+2)(c-3) - \frac{2n(c+1)}{n-1}\}.$$

As a consequence of the above corollary, we obtain an obstruction to minimal immersions in a Kenmotsu space form.

Corollary 4.4. *Let (M, g) be an $(n+1)$-dimensional Riemanian manifold and $p \in M$. If there exists a point $p \in M$ such that*

$$\delta_M > \frac{n-1}{8}\{(n+2)(c-3) - \frac{2n(c+1)}{n-1}\}$$

then M admits no anti-invariant immersion into any m-dimensional Kenmotsu space form $\tilde{M}(c)$.

References

1. A. Bejancu, *CR-submanifolds of a Kaehler manifold* I, II, Proc. Amer. Math. Soc. **69**(1978),134-142, Trans. Amer. Math. Soc. **250**(1979).

2. D.E.Blair and B.Y.Chen, *On CR-submanifolds of Hermitian manifolds,* Israel J. Math.,**34**(1979).

3. B.Y. Chen, *CR-submanifolds of a Kaehler manifold,* J. Differential Geom., **16**(1981).

4. B.Y. Chen, *Geometry of Submanifolds and Its Applications,* Sci. Univ. Tokyo, 1981.

5. B.Y. Chen, *Geomety of Slant Submanifolds,* Kath. Univ. Leuven, 1990.

6. B.Y. Chen, *Some pinching and classification theorems for minimal submanifolds,* Archiv Math., **60**(1993).

7. B.Y. Chen and K. Ogiue, *On totally real submanifolds,* Trans. Amer. Math. Soc., **193**(1974).

8. F. Defever, I. Mihai and L. Verstraelen, *B.Y. Chen's inequality for C-*

totally real submanifolds in Sasakian space forms, Boll. Un. Mat. Ital., to appear.

9. K. Kenmotsu, *A class of almost contact Riemannian manifolds,* Tôhoku Math. J., **24**(1972).

10. K. Matsumoto, I. Mihai and R. Roşca, *A certain locally conformal cosymplectic manifolds and its submanifolds,* Tensor N. S., **51**(1992).

11. K. Matsumoto, I. Mihai and M.H. Shahid, *Semi-invariant submanifolds of certain almost contact manifolds,* Bull. Yamagata Univ., (Nat. Sci.) **13**(1994).

12. I. Mihai, *On generic submanifolds of a framed f-manifold,* Proc. Nat. Conf. Geom. Top., 153-158, Timişoara, 1984.

13. I. Mihai, *Certain submanifolds of a Kaehler manifold,* Geometry and Topology of Submanifolds, **7**, World Scientific, Singapore, 1995.

14. B. Opozda, *Generic submanifolds in almost Hermitian manifolds,* Ann. Polonic Math., **49**(1988).

15. R. Roşca, *CR-sous-variétés co-isotropes d'une veriété para-Kählerienne,* C. R. Acad. Sci. Paris, **298**(1984),149-151.

16. M.H. Shahid and I. Mihai, *Generic submanifolds of trans-Sasakian manifolds,* Publicationes Math. Debrecen, **47**(1995).

17. S. Tanno, *The automorphisms groups of almost contact Riemannian manifolds,* Tôhoku Math. J. **21**(1969).

18. P. Verheyen, *Generic submanifolds of Sasakian manifolds,* Med. Wisk. Inst., K.U. Leuven, 1982.

19. L. Verstraelen and L. Vrancken, *Pinching theorems for C-totally real submanifolds of Sasakian space forms,* J. Geom., **33**(1988).

20. K. Yano and S. Ishihara, *The f-structures induced on submanifolds of complex and almost complex spaces,* Kodai Math. Sem. Rep., **18**(1966).

21. K. Yano and M. Kon, *Anti-invariant submanifolds,* M. Dekker, New York, 1976.

22. K. Yano and M. Kon, *Differential Geometry of CR-submanifolds,* Geom.
Dedicata, **10**(1981).

23. K. Yano and M. Kon, *CR Submanifolds of Kaehlerian and Sasakian Manifolds,* Birkhäuser, Boston, 1983.

GAUGE-THEORETIC EQUATIONS FOR HARMONIC MAPS INTO SYMMETRIC SPACES

MARIKO MUKAI† and YOSHIHIRO OHNITA‡
Department of Mathematics, Graduate School of Science
Tokyo Metropolitan University
Minami-Ohsawa 1-1, Hachioji, Tokyo 192-0397, Japan
† *E-mail:mmukai@math.metro-u.ac.jp*
‡ *E-mail:ohnita@comp.metro-u.ac.jp*

The characteristic of harmonic maps of Riemann surfaces into symmetric spaces is that their equations have a gauge-theoretic formulation. In this paper we shall discuss the structure of the moduli space of solutions to the gauge-theoretic equations. Moreover we shall give a removability theorem for isolated singularity of solutions to the gauge-theoretic equations.

Introduction

The harmonic map equations of a Riemann surface M to a Lie group G have a gauge-theoretic formulation. It is known that each harmonic map of M into G produces a solution (A, ϕ) to the equations

$$(*)_G \qquad \begin{cases} F(A) + \frac{1}{2}[\phi \wedge \phi] = 0, \\ d_A\phi = d_A * \phi = 0 \end{cases}$$

in a principal bundle P over M with the structure group G, which has Lie algebra $\mathfrak{g}$, where A is a connection on P, $F(A)$ is the curvature form of the connection A and ϕ is a 1-form on M with values in the adjoint bundle $\mathfrak{g}_P$. The equations $(*)_G$ are invariant under the action of the group $\mathcal{G}$ of gauge transformations of P and are conformal invariant gauge-theoretic equations on a Riemann surface.

The gauge-theoretic equations were previously investigated by Hitchin [6,8] and Valli [17,18]. Hitchin [8] has provided a profound theory concerning moduli of harmonic maps from tori to the 3-sphere. He constructed a moduli space of solutions to $(*)_G$ for $G = SU(2)$ on tori in terms of geometry of compact algebraic curves, so called *spectral curves.* Recently the first named author [11] has discussed the fundamental properties, the deformation theory of solutions to $(*)_G$ and the local structure of moduli spaces of the solutions. We can observe that several important properties of such harmonic maps come from the character of the gauge-theoretic equations. In order to understand the structure of such harmonic maps, it is natural to study the general theory of the moduli spaces of solutions to the gauge-theoretic equations.

The purpose of this paper is to define the gauge-theoretic equations for harmonic maps of a Riemann surface into a general Riemannian symmetric space, which is an extension of the equations $(*)_G$, and to discuss the moduli spaces of solutions to the equations. It can be considered as a version of the work of the first named author [11] extended to the case of compact symmetric spaces which are not necessarily compact Lie groups. In this paper we shall discuss the following subjects: (1) gauge-theoretic formulation of harmonic map equations, (2) construction of moduli spaces, (3) deformations of solutions, (4) local structures of moduli spaces, (5) weak compactness theorems on moduli spaces.

1 Gauge-theoretic formulation of harmonic map equations

In this section we shall formulate gauge-theoretically the differential equations of harmonic maps into symmetric spaces. Let $\varphi : M \longrightarrow G/K$ be a smooth map from a Riemann surface to a symmetric space G/K. Let $\pi : G \longrightarrow G/K$ be the standard principal K-bundle over G/K and $P = \varphi^{-1}G$ be the pullback of G by φ, which is a principal K-bundle over M .

$$\begin{array}{ccc} P = \varphi^{-1}G & \xrightarrow{\tilde{\varphi}} & G \\ \pi\big\downarrow & & \pi\big\downarrow K \\ M & \xrightarrow[\varphi]{} & G/K \end{array}$$

Let $\mathfrak{g} = \mathfrak{k} + \mathfrak{m}$ be the canonical decomposition of $\mathfrak{g}$ with respect to a symmteric space G/K. Since $\mathfrak{k}$ and $\mathfrak{m}$ are invariant under the adjoint action of K, we can define associated vector bundles over M:

$$\begin{aligned} \mathfrak{k}_P &:= P \times_K \mathfrak{k}, \\ \mathfrak{m}_P &:= P \times_K \mathfrak{m}, \\ \mathfrak{g}_P &:= P \times_K \mathfrak{g}. \end{aligned}$$

Note that we have $\mathfrak{g}_P = \mathfrak{k}_P \oplus \mathfrak{m}_P$. Set $\alpha = \tilde{\varphi}^* \mu_G = \tilde{\varphi}^{-1} d\tilde{\varphi}$, where μ_G is the left invariant Maurer-Cartan form of G. Since α is a $\mathfrak{g}$-valued 1-form on M, we have a type decomposition $\alpha = \alpha_{\mathfrak{k}} + \alpha_{\mathfrak{m}}$ with $\mathfrak{k}$-valued part $\alpha_{\mathfrak{k}}$ and $\mathfrak{m}$-valued part $\alpha_{\mathfrak{m}}$. By the definitions we see that the $\mathfrak{k}$-valued 1-form $\alpha_{\mathfrak{k}}$ on P satisfies

$$\begin{aligned} &(1) \quad R_a^* \alpha_{\mathfrak{k}} = \mathrm{Ad}(a^{-1}) \alpha_{\mathfrak{k}} \quad \text{for each } a \in K, \\ &(2) \quad \alpha_{\mathfrak{k}}(X^*) = X \quad \text{for each } X \in \mathfrak{k} \end{aligned}$$

and the $\mathfrak{m}$-valued 1-form $\alpha_{\mathfrak{m}}$ on P satisfies

$$\begin{aligned} &(3) \quad R_a^* \alpha_{\mathfrak{m}} = \mathrm{Ad}(a^{-1}) \alpha_{\mathfrak{m}} \quad \text{for each } a \in K, \\ &(4) \quad \alpha_{\mathfrak{m}}(X^*) = 0 \quad \text{for each } X \in \mathfrak{k}. \end{aligned}$$

Thus $\alpha_{\mathfrak{k}}$ defines a connection A on P and $\alpha_{\mathfrak{m}}$ defines a 1-form ϕ with values in the vector bundle $\mathfrak{m}_P$. Then we obtain the following (cf. Dorfmeister, Pedit and Wu [1], Guest and Ohnita [4]).

Proposition 1.1 *If $\varphi : M \longrightarrow G/K$ is a smooth map, then (A, ϕ) satisfies the gauge-theoretic equations*

$$F(A) + \frac{1}{2}[\phi \wedge \phi] = 0, \tag{1.1}$$

$$d_A\phi = 0. \tag{1.2}$$

The map φ is a harmonic map if and only if

$$d_A * \phi = 0. \tag{1.3}$$

2 Construction of moduli spaces of harmonic maps

We shall begin by defining the gauge-theoretic equations over a Riemann surface for each symmetric space in an abstract setting. Let M be a Riemann surface and $N = G/K$ be a symmetric space. Let $\mathfrak{g} = \mathfrak{k} + \mathfrak{m}$ be the canonical decomposition of $\mathfrak{g}$ with respect to a symmetric space G/K. Let P be a principal K-bundle over M . Define associated vector bundles over M as $\mathfrak{k}_P := P \times_K \mathfrak{k}$, $\mathfrak{m}_P := P \times_K \mathfrak{m}$, $\mathfrak{g}_P := P \times_K \mathfrak{g}$. Note that we have $\mathfrak{g}_P = \mathfrak{k}_P \oplus \mathfrak{m}_P$. Let $\mathcal{A}_P$ denote the space of all smooth connections in P. For each $i = 0, 1, 2$, let $\Omega^i(\mathfrak{k}_P)$ and $\Omega^i(\mathfrak{m}_P)$ denote the vector spaces of all smooth i-forms with values in $\mathfrak{k}_P$ and $\mathfrak{m}_P$ on M, respectively. The space $\mathcal{A}_P$ is an infinite dimensional affine space associated to the vector space $\Omega^1(\mathfrak{k}_P)$. A connection $A \in \mathcal{A}_P$ is called *irreducible* if the holonomy algebra of the connection A is equal to $\mathfrak{k}$. For $(A, \phi) \in \mathcal{A}_P \times \Omega^1(\mathfrak{m}_P)$, consider the gauge-theoretic equations

$$(*)_{G/K} \qquad \begin{cases} F(A) + \frac{1}{2}[\phi \wedge \phi] = 0, \\ d_A\phi = d_A * \phi = 0. \end{cases}$$

A solution (A, ϕ) to the equations $(*)_{G/K}$ with $\phi \not\equiv 0$ is called *nontrivial* .

Let $\widetilde{\mathcal{N}}(M, G/K)$ be the set of all triplets (P, A, ϕ) of a principal K-bundle P over M and $(A, \phi) \in \mathcal{A}_P \times \Omega^1(\mathfrak{m}_P)$ and let $\widetilde{\mathcal{M}}(M, G/K)$ be the set of all triplets (P, A, ϕ) of principal K-bundles P over M and solutions $(A, \phi) \in \mathcal{A}_P \times \Omega^1(\mathfrak{m}_P)$ to the equations $(*)$. We define an equivalence relation $\simeq$ on $\widetilde{\mathcal{N}}$ as follows: For $(P, A, \phi), (P', A', \phi') \in \widetilde{\mathcal{N}}$,

$$(P, A, \phi) \simeq (P', A', \phi')$$

if and only if there exists a bundle isomorphism $F : P \longrightarrow P'$ such that $F^*(A', \phi') = (A, \phi)$. We see that if $(P, A, \phi) \simeq (P', A', \phi')$ and (A, ϕ) is a solution to $(*)_{G/K}$, then (A', ϕ') is also a solution to $(*)_{G/K}$. Let $\mathcal{N}(M, G/K)$ denote the quotient set $\widetilde{\mathcal{N}}/\simeq$ of $\widetilde{\mathcal{N}}(M, G/K)$ with respect to the equivalence relation $\simeq$. Let $\mathcal{M}(M, G/K)$ denote the quotient set of $\widetilde{\mathcal{M}}(M, G/K)$ by the equivalence relation $\simeq$.

The natural left action of G on the space G/K induces a left action of G on the space $C^\infty(M, G/K)$ of smooth maps. We take the quotient space $C^\infty(M, G/K)/G$ by the left action of G. Set $\mathcal{S}(M, G/K) = C^\infty(M, G/K)/G$. Similarly, set $\mathcal{M}_{HM}(M, G/K) = \mathrm{Harm}(M, G/K)/G$.

Proposition 2.1

$$\begin{array}{ccc} \mathcal{S}(M, G/K) & \xrightarrow{\text{injective}} & \mathcal{N}(M, G/K) \\ {\scriptstyle\text{inclusion}} \uparrow & & \uparrow {\scriptstyle\text{inclusion}} \\ \mathcal{M}_{HM}(M, G/K) & \xrightarrow{\text{injective}} & \mathcal{M}(M, G/K) \end{array}$$

Proof. For $\varphi, \varphi' \in C^\infty(M, G/K)$, we assume that $\varphi' = \tau_a \circ \varphi$ for each $a \in G$. We define a bundle isomorphism $F : P \longrightarrow P'$ by $F(z, c) = (z, ac)$ for each $(z, c) \in P$. Since $\tilde{\varphi}' \circ F = L_a \circ \tilde{\varphi}$, we have $(\tilde{\varphi}' \circ F)^*\theta_G = (L_a \circ \tilde{\varphi})^*\theta_G$ and thus $F^*(A', \phi') = (A, \phi)$. Hence $(P', A', \phi') \simeq (P, A, \phi)$. It induces maps $\mathcal{S}(M, G/K) \longrightarrow \mathcal{N}(M, G/K)$ and $\mathcal{M}_{HM}(M, G/K) \longrightarrow \mathcal{M}(M, G/K)$. If we assume that $(P, A, \phi) \simeq (P', A', \phi')$, then we obtain $\tilde{\phi}' \circ F = L_a \circ \tilde{\varphi}$ for some $a \in G$ and thus $\phi' = a\varphi$. Suppose that φ and φ' correspond to (P, A, ϕ) and (P', A', ϕ'), respectively. If $(P, A, \phi) \cong (P', A', \phi')$, then there is a bundle isomorphism $F : P \longrightarrow P'$ such that $F^*A' = A$ and $F^*\phi' = \phi$. Thus we see that the maps $\tilde{\varphi}$ and $\tilde{\varphi}' \circ F$ from P to G satisfy $\tilde{\varphi}^*\theta_G = (\tilde{\varphi}' \circ F)^*\theta_G$. Hence we have $\varphi' = a(\tilde{\varphi}' \circ F)$ for some $a \in G$. Therefore we obtain $\varphi' = a\varphi$. □

Remark. If M is contractible, then we have $\mathcal{M}_{HM}(M, G/K) = \mathcal{M}(M, G/K)$.

This construction is a generalization of one in the first named author [11], which treated the case of harmonic maps into a compact Lie group. In general the moduli space $\mathcal{M}$ includes and is much larger than the moduli space of harmonic maps of M into G/K.

Here we mention the case when N is a Lie group, that is, $N = (G \times G)/K \cong G$, where $K = \{(g, g) \in G \times G \mid g \in G\} \cong G$. The canonical decomposition of $\mathfrak{g} \oplus \mathfrak{g}$ is given by $\mathfrak{g} \oplus \mathfrak{g} = \mathfrak{k} + \mathfrak{m}$, where

$$\mathfrak{k} = \{(X, X) \in \mathfrak{g} \oplus \mathfrak{g} \mid X \in \mathfrak{g}\}$$

and

$$\mathfrak{m} = \{(X, -X) \in \mathfrak{g} \oplus \mathfrak{g} \mid X \in \mathfrak{g}\}.$$

Let P be a principal K-bundle over M. Note that there is a natural isomorphism $\mathfrak{k}_P \cong \mathfrak{m}_P$ as vector bundles. We denote by P_1 the principal G-bundle over M obtained by identify the structure group K of P with the Lie group G. Then there is a canonical bijective correspondence between solutions to the equations $(*)_{(G\times G)/K}$ on P and solutions to the equations $(*)_G$ on P_1.

Remark. If M is simply connected and $N = G/K$ is a Lie group, then we have $\mathcal{M}_{HM}(M, G/K) = \mathcal{M}(M, G/K)$.

3 The gauge-theoretic equations and duality of symmetric spaces

There are two types of gauge-theoretic equations for $(A, \phi) \in \mathcal{A}_P \times \Omega^1(\mathfrak{m}_P)$:

$$(*)_\pm \qquad \begin{cases} F(A) \pm \frac{1}{2}[\phi \wedge \phi] = 0, \\ d_A\phi = d_A * \phi = 0. \end{cases}$$

The equations $(*)_+$ are the gauge-theoretic equations $(*)_{G/K}$ for harmonic maps into a symmetric space G/K. We mention the relationship between these two equations $(*)_\pm$. We consider the symmetric space G^*/K dual to the symmetric space G/K. If G/K is a symmetric space with the symmetric Lie algebra $\mathfrak{g} = \mathfrak{k} + \mathfrak{m}$, then its dual symmetric space G^*/K is a symmetric space with the symmetric Lie algebra $\mathfrak{g}^* = \mathfrak{k} + \sqrt{-1}\mathfrak{m}$. Then we can clarify the relationship of solutions to the equations $(*)_-$ with solutions to the equations $(*)$ as follows.

Proposition 3.1 *$(A, \phi) \in \mathcal{A}_P \times \Omega^1(\mathfrak{m}_P)$ is a solution to $(*)_-$ for the symmetric space G/K if and only if $(A, \sqrt{-1}\phi) \in \mathcal{A}_P \times \Omega^1(\sqrt{-1}\mathfrak{m}_P)$ is a solution to $(*)_+$ for the dual symmetric space G^*/K.*

It is known that non-flat irreducible Riemannian symmetric spaces are classified as follows (cf. Helgason[5]):

(1) Type I : $N = G/K$ with a connected compact simple Lie group G.

(2) Type II : N is a connected compact simple Lie group.

(3) Type III : $N = G/K$ with a connected real simple Lie group G which has no compatible complex structure.

(4) Type IV : $N = G^{\mathbf{C}}/G$ with a connected compact simple Lie group G.

Type I and Type II are dual to Type III and Type IV, respectively.

By Proposition 3.1, the gauge-theoretic equations $(*)_+$ for a noncompact symmetric space $N \cong G^{\mathbf{C}}/G$ with compact G is converted to the self-duality equations $(*)_-$ for $N \cong G$, which was investigated by Hitchin[7]. The above observation was used by Donaldson[2] to prove the existence theorem of solutions to the self-duality equations $(*)_-$.

4 Infinitesimal deformations

From now on we assume that $N = G/K$ is a compact Riemannain symmetric space, that is, (G, K) is a Riemannian symmetric pair with compact G, and M is a compact Riemann surface. As we observed in Section 3, we may treat the equations $(*)_\pm$ for a compact Riemannian symmetric space $N = G/K$ instead of the equations $(*)$ for any Riemannian symmetric space. In Sections 4 and 5, we study infinitesimal deformations of solutions to the equations $(*)_\pm$ and the local structure of the moduli spaces by using the method due to Kuranishi. The formulas given here and their proofs are very similar to those in the first named author [11]. So we will omit the details here. We study infinitesimal deformations of solutions to the equations $(*)_\pm$. The linearized equations for $(*)_\pm$ are

$$(4.1) \qquad \begin{cases} d_A\alpha \pm [\phi \wedge \eta] = 0, \\ d_A\eta + [\phi \wedge \alpha] = 0, \\ d_A * \eta + [*\phi \wedge \alpha] = 0, \end{cases}$$

for $(\alpha, \eta) \in \Omega^1(\mathfrak{k}_P) \oplus \Omega^1(\mathfrak{m}_P)$. Set

$$\begin{aligned} \widetilde{\mathcal{M}}_\pm &= \widetilde{\mathcal{M}}_\pm(M, G/K) \\ &= \{(A, \phi) \in \mathcal{A}_P \times \Omega^1(\mathfrak{m}_P) \mid (A, \phi) \text{ is a solution to } (*)_\pm\}. \end{aligned}$$

Denote by $\mathcal{K}_P$ the group of gauge transformations of the principal K-bundle P. The right action of $\mathcal{K}_P$ on $\mathcal{A}_P \times \Omega^1(\mathfrak{m}_P)$ is defined by

$$k^*(A, \phi) = (k^*A, \mathrm{Ad}(k)^{-1}\phi).$$

The tangent space of the gauge orbit through (A, ϕ) is given by

$$\{(d_A\xi, [\phi, \xi]) \mid \xi \in \Omega^0(\mathfrak{k}_P)\}.$$

The moduli spaces of solutions to $(*)_\pm$ are defined by

$$\mathcal{M}_\pm = \mathcal{M}_\pm(M, G/K) = \widetilde{\mathcal{M}}_\pm/\mathcal{K}_P.$$

The elliptic cochain complex associated with deformations for solutions to the equations $(*)_\pm$ is, for $(A, \phi) \in \widetilde{\mathcal{M}}_\pm$,

$(C_{(A,\phi)})$

$$0 \longrightarrow \Omega^0(\mathfrak{k}_P) \xrightarrow{d_0} \Omega^1(\mathfrak{k}_P) \oplus \Omega^1(\mathfrak{m}_P) \xrightarrow{d_1} \Omega^2(\mathfrak{k}_P) \oplus \Omega^2(\mathfrak{m}_P) \oplus \Omega^2(\mathfrak{m}_P) \longrightarrow 0,$$

where the linear differential operators

$$d_0 = d_{0(A,\phi)} : \Omega^0(\mathfrak{k}_P) \longrightarrow \Omega^1(\mathfrak{k}_P) \oplus \Omega^1(\mathfrak{m}_P)$$

and

$$d_1 = d_{1(A,\phi)} : \Omega^1(\mathfrak{k}_P) \oplus \Omega^1(\mathfrak{m}_P) \longrightarrow \Omega^2(\mathfrak{k}_P) \oplus \Omega^2(\mathfrak{m}_P) \oplus \Omega^2(\mathfrak{m}_P)$$

are defined by

$$(4.2) \qquad d_{0(A,\phi)}(\xi) = (d_A\xi, [\phi, \xi]),$$

$$(4.3) \quad d_{1(A,\phi)}(\alpha, \eta) = (d_A\alpha \pm [\phi \wedge \eta], d_A\eta + [\phi \wedge \alpha], d_A * \eta + [*\phi \wedge \alpha]).$$

The formal adjoint operators d_0^* and d_1^* become

$$(4.4) \quad d_0^*(\alpha, \eta) = d_A^*\alpha - *[\phi \wedge *\eta],$$

$$(4.5) d_1^*(u, v, w) = (d_A^*u - *[\phi, *v] - *[*\phi, *w], \mp * [\phi, *u] + d_A^*v - d_A * w).$$

Set $P_0 = d_0^*d_0$, $P_1 = P_1^{\pm} = d_0d_0^* + d_1^*d_1$ and $P_2 = P_2^{\pm} = d_1d_1^*$. Then the i-th cohomology group $H^i_{(A,\phi)}$ of the cochain complex is isomorphic to $\mathrm{Ker}P_i$ for each $i = 0, 1, 2$.

Here we denote by $\mathfrak{c}(\mathfrak{g})$ the Lie algebra of the center $C(G)$ of Lie group G. We see that each element of $\mathfrak{c}(\mathfrak{g}) \cap \mathfrak{k}$ and $\mathfrak{c}(\mathfrak{g}) \cap \mathfrak{m}$ can be identified with a constant section of $\mathfrak{k}_P$ and $\mathfrak{m}_P$, respectively. Thus we have $\mathfrak{c}(\mathfrak{g}) \cap \mathfrak{k} \subset \Omega^0(\mathfrak{k}_P)$ and $\mathfrak{c}(\mathfrak{g}) \cap \mathfrak{m} \subset \Omega^0(\mathfrak{m}_P)$. Via the canonical isomorphisms $\Omega^0(\mathfrak{k}_P) \cong \Omega^2(\mathfrak{k}_P)$ and $\Omega^0(\mathfrak{m}_P) \cong \Omega^2(\mathfrak{m}_P)$ by the Hodge star operator, we have the inclusion

$$(\mathfrak{c}(\mathfrak{g}) \cap \mathfrak{k}) \oplus (\mathfrak{c}(\mathfrak{g}) \cap \mathfrak{m}) \oplus (\mathfrak{c}(\mathfrak{g}) \cap \mathfrak{m}) \subset \Omega^2(\mathfrak{k}_P) \oplus \Omega^2(\mathfrak{m}_P) \oplus \Omega^2(\mathfrak{m}_P).$$

Further we see that

$$\mathfrak{c}(\mathfrak{g}) \cap \mathfrak{k} \subset H^0_{(A,\phi)}$$

and

$$(\mathfrak{c}(\mathfrak{g}) \cap \mathfrak{k}) \oplus (\mathfrak{c}(\mathfrak{g}) \cap \mathfrak{m}) \oplus (\mathfrak{c}(\mathfrak{g}) \cap \mathfrak{m}) \subset H^2_{(A,\phi)}.$$

The two kinds of Jacobi differential operators

$$\mathcal{J}^{\mathfrak{k}}_{(A,\phi)} : \Omega^0(\mathfrak{k}_P) \longrightarrow \Omega^0(\mathfrak{k}_P)$$

and

$$\mathcal{J}^{\mathfrak{m}}_{(A,\phi)} : \Omega^0(\mathfrak{m}_P) \longrightarrow \Omega^0(\mathfrak{m}_P)$$

are defined as

$$(4.6) \qquad \mathcal{J}^{\mathfrak{k}}_{(A,\phi)}\xi = d_A^*d_A\xi \mp *[[\phi, \xi] \wedge *\phi]$$

for each $\xi \in \Omega^0(\mathfrak{k}_P)$ and

$$(4.7) \qquad \mathcal{J}^{\mathfrak{m}}_{(A,\phi)}\xi = d_A^*d_A\xi \mp *[[\phi, \xi] \wedge *\phi]$$

for each $\xi \in \Omega^0(\mathfrak{m}_P)$, respectively. In the case when N is a compact Lie group, $\mathcal{J}^{\mathfrak{k}}_{(A,\phi)}$ coincides with $\mathcal{J}^{\mathfrak{m}}_{(A,\phi)}$ via the natural isomorphism $\mathfrak{k}_P \cong \mathfrak{m}_P$.

The operators $P_2^{\pm}$ are given explicitly in terms of the Jacobi differential operators as follows:

$$P_2^+(u_1,u_2,u_3) = (* \mathcal{J}^{\mathfrak{k}}_{(A,\phi)}(*u_1) + 2[\phi \wedge \zeta_+], \\ * \mathcal{J}^{\mathfrak{m}}_{(A,\phi)}(*u_2) + 2[\phi \wedge \beta], \\ * \mathcal{J}^{\mathfrak{m}}_{(A,\phi)}(*u_3) + 2[*\phi \wedge \beta]) \tag{4.8}$$

and

$$P_2^-(u_1,u_2,u_3) = (*\mathcal{J}^{\mathfrak{k}}_{(A,\phi)}(*u_1), *\mathcal{J}^{\mathfrak{m}}_{(A,\phi)}(*u_2), *\mathcal{J}^{\mathfrak{m}}_{(A,\phi)}(*u_3)) \tag{4.9}$$

for each $(u_1,u_2,u_3) \in \Omega^2(\mathfrak{k}_P) \oplus \Omega^2(\mathfrak{m}_P) \oplus \Omega^2(\mathfrak{m}_P)$, where we set

$$\beta = d_A^* u_1 - *[\phi, *u_2] - *[*\phi, *u_3] \in \Omega^1(\mathfrak{k}_P), \tag{4.10}$$

$$\zeta_\pm = \mp * [\phi, *u_1] + d_A^* u_2 - d_A * u_3 \in \Omega^1(\mathfrak{m}_P). \tag{4.11}$$

Proposition 4.1 *If* $(u_1,u_2,u_3) \in Ker P_2^- \subset \Omega^2(\mathfrak{k}_P) \oplus \Omega^2(\mathfrak{m}_P) \oplus \Omega^2(\mathfrak{m}_P)$, *then*

$$\mathcal{J}^{\mathfrak{k}}_{(A,\phi)}(*u_1) = 0, \ \mathcal{J}^{\mathfrak{m}}_{(A,\phi)}(*u_2) = 0, \ \mathcal{J}^{\mathfrak{m}}_{(A,\phi)}(*u_3) = 0.$$

By the formula

$$\langle *[[\phi,\xi] \wedge *\phi], \xi\rangle * 1 = \langle [\phi,\xi], [\phi,\xi]\rangle * 1, \tag{4.12}$$

and Proposition 4.1 we obtain

Proposition 4.2 *Let* $(A,\phi) \in \mathcal{M}_-(M, G/K)$. *Assume that (1) G/K is a compact Lie group, or (2) G/K is of compact type and A is irreducible. If* $H^0_{(A,\phi)} = \mathfrak{c}(\mathfrak{g}) \cap \mathfrak{k}$, *then* $H^2_{(A,\phi)} = (\mathfrak{c}(\mathfrak{g}) \cap \mathfrak{k}) \oplus (\mathfrak{c}(\mathfrak{g}) \cap \mathfrak{m}) \oplus (\mathfrak{c}(\mathfrak{g}) \cap \mathfrak{m})$.

Proof. If $(u_1,u_2,u_3) \in H^2_{(A,\phi)} \cong \mathrm{Ker} P_2^- \subset \Omega^2(\mathfrak{k}_P) \oplus \Omega^2(\mathfrak{m}_P) \oplus \Omega^2(\mathfrak{m}_P)$, it follows from Proposition 4.1 and (4.12) that

$$d_A(*u_i) = 0 \text{ and } [\phi, *u_i] = 0$$

for $i = 1,2,3$. In the case of (1), via the natural isomorphism $\mathfrak{k}_P \cong \mathfrak{m}_P$, the condition that $H^0_{(A,\phi)} = \mathfrak{c}(\mathfrak{g}) \cap \mathfrak{k}$ implies that $*u_1 \in \mathfrak{c}(\mathfrak{g}) \cap \mathfrak{k}$ and $*u_i \in \mathfrak{c}(\mathfrak{g}) \cap \mathfrak{m}$ for $i = 2,3$. In the case of (2), note that $\mathfrak{c}(\mathfrak{g}) = \{0\}$, because G is semisimple. The condition that $H^0_{(A,\phi)} = \{0\}$ implies that $*u_1 = 0$ and hence $u_1 = 0$. The sections $*u_i$ $(i = 2,3)$ define $\mathfrak{k}$-fixed elements in $\mathfrak{m}$ because of the irreducibility of A. Since $\mathfrak{g}$ is compact and semisimple by the definition of G/K of compact type, it must be $*u_i = 0$ $(i = 2,3)$. □

5 Slice theorem and Kuranishi map

Set $\mathcal{S}_{(A,\phi)} = \{(\alpha,\eta) \in \Omega^1(\mathfrak{k}_P)\times\Omega^1(\mathfrak{m}_P) \mid d_0^*(\alpha,\eta) = 0\}$. The slice theorem says that the natural projection $\pi : \mathcal{A}_P \times \Omega^1(\mathfrak{m}_P) \longrightarrow (\mathcal{A}_P \times \Omega^1(\mathfrak{m}_P))/\mathcal{K}_P$ gives a homeomorphism of a neighborhood of the origin of $\mathcal{S}_{(A,\phi)}$ onto a neighborhood of $[A,\phi] \in (\mathcal{A}_P \times \Omega^1(\mathfrak{m}_P))/\mathcal{K}_P$.

Define a map

$$(5.1) \quad f = f_{(A,\phi)} : \Omega^1(\mathfrak{k}_P) \oplus \Omega^1(\mathfrak{m}_P) \longrightarrow \Omega^2(\mathfrak{k}_P) \oplus \Omega^2(\mathfrak{m}_P) \oplus \Omega^2(\mathfrak{m}_P)$$

by

$$(5.2) \quad \begin{aligned} f(\alpha,\eta) = (& \, d_A\alpha + \frac{1}{2}[\alpha\wedge\alpha] \pm [\phi\wedge\eta] \pm \frac{1}{2}[\eta\wedge\eta], \\ & d_A\eta + [\alpha\wedge\phi] + [\alpha\wedge\eta], d_A * \eta + [\alpha\wedge *\phi] + [\alpha\wedge *\eta]) \end{aligned}$$

for each $(\alpha,\eta) \in \Omega^1(\mathfrak{k}_P)\oplus\Omega^1(\mathfrak{m}_P)$. Then we have $(df)_{(0,0)} = d_1$. Let (A,ϕ) be a solution to $(*)_\pm$. Then, $(A+\alpha, \phi+\eta)$ is also a solution to $(*)_\pm$ if and only if $f(\alpha,\eta) = 0$.

The *Kuranishi map* $k = k_{(A,\phi)} : \Omega^1(\mathfrak{k}_P) \times \Omega^1(\mathfrak{m}_P) \longrightarrow \Omega^1(\mathfrak{k}_P) \times \Omega^1(\mathfrak{m}_P)$ is a nonlinear map defined by

$$(5.3) \qquad k(\alpha,\eta) = (\alpha,\eta) + d_1^* \circ G_2(f(\alpha,\eta) - d_1(\alpha,\eta)),$$

where G_2 denotes the Green operator associated to P_2 defined by $I = h_2 + P_2 \circ G_2$ and $h_2 : \Omega^2(\mathfrak{k}_P) \oplus \Omega^2(\mathfrak{m}_P) \oplus \Omega^2(\mathfrak{m}_P) \longrightarrow \mathrm{Ker}P_2$ is the Hermitian projection.

For $(A,\phi) \in \widetilde{\mathcal{M}}_\pm(M,G/K)$, we set

$$\Gamma_{(A,\phi)} = \{k \in \mathcal{K} \mid k^*(A,\phi) = (A,\phi)\}.$$

$(A,\phi) \in \widetilde{\mathcal{M}}_\pm(M,G/K)$ is called *regular* if $\Gamma_{(A,\phi)} = C(G) \cap K$, where $C(G)$ denotes the center of G. Set

$$\widetilde{\mathcal{M}}^*_\pm(M,G/K) = \{(A,\phi) \in \widetilde{\mathcal{M}}_\pm \mid (A,\phi) \text{ is regular}\}$$

and define $\mathcal{M}^*_\pm(M,G/K) = \widetilde{\mathcal{M}}^*_\pm(M,G/K)/\mathcal{K}$. Note that if $(A,\phi) \in \widetilde{\mathcal{M}}^*_\pm(M,G/K)$, then $H^0_{(A,\phi)} = \mathfrak{c}(\mathfrak{g}) \cap \mathfrak{k}$. The local structure of the moduli spaces are described in terms of a nonlinear real analytic map $\psi_{(A,\phi)}$ defined as

$$\psi_{(A,\phi)}(\beta,\zeta) = h_2(f(k^{-1}(\beta,\zeta)) - d_1(k^{-1}(\beta,\zeta))).$$

We can show that the image of $\psi_{(A,\phi)}$ is perpendicular to the subspace

$$(\mathfrak{c}(\mathfrak{g}) \cap \mathfrak{k}) \oplus (\mathfrak{c}(\mathfrak{g}) \cap \mathfrak{m}) \oplus (\mathfrak{c}(\mathfrak{g}) \cap \mathfrak{m}) \subset H^2_{(A,\phi)}$$

with respect to the L^2-inner product.

Theorem 5.1 *For each* $(A,\phi) \in \widetilde{\mathcal{M}}^*_{\pm}(M,G/K)$*, there exists a real analytic map* $\psi_{(A,\phi)} : U \longrightarrow H^2_{(A,\phi)}$ *defined on a neighborhood of the origin in* $H^1_{(A,\phi)}$ *with* $\psi_{(A,\phi)}(0) = 0$ *and* $(d\psi_{(A,\phi)})_0 = 0$ *such that a neighborhood of* $[A,\phi]$ *in* $\mathcal{M}_{\pm}(M,G/K) = \widetilde{\mathcal{M}}_{\pm}/\mathcal{K}_P$ *is homeomorphic to*

$$\psi^{-1}_{(A,\phi)}(0) = \{(\beta,\zeta) \in U \mid \psi_{(A,\phi)}(\beta,\zeta) = 0\}.$$

In general, we obtain the following.

Theorem 5.2 *For each* $(A,\phi) \in \widetilde{\mathcal{M}}_{\pm}(M,G/K)$*, there exists a real analytic* $\Gamma_{(A,\phi)}$*-equivariant map* $\psi_{(A,\phi)} : U \longrightarrow H^2_{(A,\phi)}$ *defined on a neighborhood of the origin in* $H^1_{(A,\phi)}$ *with* $\psi_{(A,\phi)}(0) = 0$ *and* $(d\psi_{(A,\phi)})_0 = 0$ *such that a neighborhood of* $[A,\phi]$ *in* $\mathcal{M}_{\pm}(M,G/K)$ *is homeomorphic to* $\psi^{-1}_{(A,\phi)}(0)/\Gamma_{(A,\phi)}$.

By combining Proposition 4.2 and Theorem 5.2, we obtain

Proposition 5.3 *(1) If* G/K *is a compact Lie group, then the moduli space* $\mathcal{M}^*_{-}(M,G/K)$ *is a smooth manifold of dimension equal to* $\dim H^1_{(A,\phi)}$.

(2) If G/K *is of compact type, then* $\{[A,\phi] \in \mathcal{M}^*_{-}(M,G/K) \mid A$ *is irreducible*$\}$ *is a smooth manifold of dimension equal to* $\dim H^1_{(A,\phi)}$.

By the result of Hitchin [7], we know that, if G/K is a compact Lie group, then the moduli space $\mathcal{M}_{-}$ has a hyperKähler structure.

The following are interesting questions:

What kinds of geometric structures do the moduli spaces $\mathcal{M}_{-}$ have for each compact symmetric space $N = G/K$? How does the geometry of a compact symmetric space $N = G/K$ reflect the geometry of the moduli space $\mathcal{M}_{-}$?

Remark. Very recently the second named author [12] gives some results for the above questions

6 Bochner-Weitzenböck formulas

In general the Bochner-Weitzenböck formula for the Laplace-Beltami operator $\Delta^A = d_A d^*_A + d^*_A d_A$ on $\Omega^1(\mathfrak{m}_P)$ is given as

$$\Delta^A \phi = \nabla^* \nabla \phi + \phi \circ \mathrm{Ric}^M + \mathcal{F}^A(\phi), \tag{6.1}$$

where $(\mathcal{F}^A(\phi))(X) = \sum_i [F(A)(e_i, X), \phi(e_i)]$, $\mathrm{Ric}^M(X) = \sum_i R^M(X, e_i)e_i$.

Let $(A,\phi) \in \mathcal{A}_P \times \Omega^1(\mathfrak{m}_P)$ be a solution to $(*)_{\pm}$. By using (6.1), we obtain

$$\Delta^M |\phi|^2 = -K_M |\phi|^2 \pm \sum_{i,j} |[\phi(e_i), \phi(e_j)]|^2 - |\nabla^A \phi|^2. \tag{6.2}$$

Remark.

(1) If M is a Riemann sphere and (A, ϕ) is a solution to $(*)_-$, then $\phi \equiv 0$, and hence A is a flat connection.

(2) If M is a torus and (A, ϕ) is a solution to $(*)_-$, then A is a flat connection, $\mathrm{Im}\phi$ is abelian and $\nabla^A \phi = 0$.

7 Weak compactness theorem

Let (A, ϕ) be a solution to $(*)_\pm$. Since $\Delta^A \phi = 0$, the Schauder estimates for an elliptic linear operator imply

$$(7.1) \qquad ||\phi||_{C^{\ell+2+\alpha}} \leq C_1 ||\Delta^A \phi||_{C^{\ell+\alpha}} + C_2 ||\phi||_{C^0} \leq C_2 ||\phi||_{C^0},$$

where C_1 and C_2 are positive constants independent of ϕ.

Define the *energy function* $\mathcal{E}$ on $\mathcal{M}_\pm$ by $\mathcal{E}([A, \phi]) = ||\phi||^2_{L^2}$. For each $c \geq 0$, we set

$$\mathcal{M}^c_\pm = \{[A, \phi] \in \mathcal{M}_\pm \mid \mathcal{E}([A, \phi]) \leq c\}.$$

Theorem 7.1 *For each $c \geq 0$, $\mathcal{M}^c_-$ is compact.*

Proof. In the case of $(*)_-$, by (6.2) we have

$$(7.2) \qquad \Delta^M |\phi|^2 \leq C_3 |\phi|^2.$$

By a maximum principle (cf. Schoen [14], Morrey [10], Gilberg and Trudinger [3]) it follows from (7.2) that

$$(7.3) \qquad \sup_M |\phi|^2 \leq C_4 ||\phi||^2_{L^2},$$

where $C_3, C_4 > 0$ does not depend on ϕ. Thus the $||\phi_i||^2_{C^0}$ are uniformly bounded. By the Schauder estimates (7.1), after choosing a suitable subsequence, ϕ_i converges in C^k to $\phi_\infty \in \Omega^1(\mathfrak{m}_P)$. Using the equation $F(A_i) = \frac{1}{2}[\phi_i \wedge \phi_i]$ and a result of Uhlenbeck [15], we can show that, after choosing gauge transformations k_i and a suitable subsequence of A_i, $k_j^* A_j$ converges in C^k to $B_\infty \in \mathcal{A}_P$, and by the same argument again we can show that, after choosing a suitable subsequence of ϕ_j, $k_j^* \phi_j$ converges in C^k to $\psi_\infty \in \Omega^1(\mathfrak{m}_P)$. □

Remark. This is a slight extension of Proposition (7.1)(1) of Hitchin [7].

In general, $\mathcal{M}_+^c$ is not compact. It is an interesting problem to study the behaviour of the limit of a sequence in $\mathcal{M}_+^c$ and to make a compactification of $\mathcal{M}_+^c$.

In the case of $(*)_+$, by (6.2) we have

$$\Delta^M |\phi|^2 \le C_5 |\phi|^2 + C_6 |\phi|^4.$$

Then we can prove the following a priori estimate in the same way as in Schoen[14] by using a local maximum principle Morrey[10],Gilberg and Trudinger[3].

Proposition 7.2 *There are constants $C > 0$ and $\varepsilon_0 > 0$ such that if a solution (A, ϕ) to $(*)_+$ on M satisfies $\|\phi|_{D(2r)}\|_{L^2}^2 \le \varepsilon_0$ on $D(2r) = \{|z| < 2r\} \subset M$, then*

$$\sup_{D(r)} |\phi|^2 \le \frac{C}{r^2} \|\phi|_{D(2r)}\|_{L^2}^2.$$

We use this to establish a weak compactness theorem for solutions to the gauge-theoretic equations in the case when $N = G/K$ is a compact Lie group.

Theorem 7.3 *Assume that $N = G/K$ is a compact Lie group. Let $(A_i, \phi_i) \in \widetilde{\mathcal{M}}_+$ be a sequence of solutions to $(*)_+$ on M with $\|\phi\|_{L^2}^2 \le c$. Then there exists a subsequence $(A_j, \phi_j) \in \widetilde{\mathcal{M}}_+$, gauge transformations $k_j \in \mathcal{K}_P$, and a solution (A_∞, ϕ_∞) to the equations $(*)_+$ on M such that $k_i^*(A_i, \phi_i)$ converges to (A_∞, ϕ_∞) in C^k on each compact subset X in $M \setminus \{p_1, \cdots, p_s\}$. Moreover, at each singular point p_α $(\alpha = 1, \cdots, s)$, a nontrivial solution $(\tilde{A}_\alpha, \tilde{\phi}_\alpha)$ to the equations $(*)_+$ appears in the trivial principal G-bundle on a Riemann sphere S^2 as the phenomenon of bubbling .*

Proof. Following the observation in Section 2, we may consider the corresponding principal bundle P_1 over M with the structure group G. We denote by $\mathcal{G}$ the group of gauge transformations of P_1. We can consider a sequence of pairs $(A_i - \phi_i, \phi_i)$ instead of a sequence of (A_i, ϕ_i). Since each $A_i - \phi_i$ is a flat connection, by the compactness of the moduli space of flat conncetions, after choosing a subsequence there are gauge transformations $g_j \in \mathcal{G}$ such that $g_j^*(A_j - \phi_j)$ converges in C^k to a flat connection C on M. Set $B_j = g_j^* A_j$ and $\psi_j = g_j^* \phi_j$. Then each (B_j, ψ_j) is also a solution to $(*)_+$ and $B_j - \psi_j$ converges in C^k to the flat connection C. In order to examine the convergence of $(B_j = (B_j - \psi_j) + \psi_j, \psi_j)$, we have to discuss the convergence of the Higgs fields ψ_j with $\|\psi_j\|_{L^2}^2 = \|\phi_j\|_{L^2}^2 \le c$ satisfying $d_{A_j} \psi_j = d_{A_j} * \psi_j = 0$. By virtue of (7.1) and Proposition 7.2, the covering argument similar to Sacks-Uhlenbeck[13] implies that, after choosing a suitable subsequence, ψ_j converges to ψ_∞ in C^k on each compact subset K in $M \setminus \{p_1, \cdots, p_s\}$, and thus B_j also converges in

C^k to $B_\infty = C + \psi_\infty$ on each compact subset K in $M \setminus \{p_1, \cdots, p_s\}$. We see that (B_∞, ψ_∞) is a solution to $(*)_+$ defined on $M \setminus \{p_1, \cdots, p_s\}$ with

$$\int_{M\setminus\{p_1,\cdots,p_s\}} |\psi_\infty|^2 dv_M \leq c < \infty$$

By the next theorem, (B_∞, ψ_∞) extends smoothly to a solution on M. As j goes to ∞, Higgs fields $|\psi_j|^2$ concentrates at each singular point p_α. By applying the renormalization procedure of Sacks-Uhlenbeck to (B_j, ψ_j) around each singular point p_α $(\alpha = 1, \cdots, s)$, a nontrivial solution $(\tilde{A}_\alpha, \tilde{\phi}_\alpha)$ to the equations $(*)_+$ appears in the trivial principal G-bundle on a Riemann sphere S^2. □

Finally we prove the removability theorem for an isolated singularity for a solution to the gauge-theoretic equations $(*)_G$ for compact G .

Theorem 7.4 *Let $P = D \times G$ be the trivial principal G-bundle over $D = \{z = x + \sqrt{-1}y \in \mathbf{C} \mid |z| < 1\}$ and $(A, \phi) \in \mathcal{A}_P \times \Omega^1(\mathfrak{g}_P)$ be a smooth solution to the equations $(*)_G$ defined on $D \setminus \{0\}$ such that $A - \phi$ is the trivial connection with respect to the trivialization $P = D \times G$. Assume that*

$$\int_{D\setminus\{0\}} |\phi|^2 dx \wedge dy < \infty.$$

Then (A, ϕ) extends smoothly to the whole disk D.

Proof. We denote by $\pi : P \longrightarrow D$ the projection map. The flat connection $A+\phi$ defines an integrable horizontal distribution on $P|_{D\setminus\{0\}}$. Suppose that the flat connection $A + \phi$ has non-trivial holonomy around 0. Let S be a maximal connected integral submanifold of the integrable horizontal distribution on $P|_{D\setminus\{0\}}$. The projection π restricts to $\pi : S \longrightarrow D \setminus \{0\}$. Let (r, θ) be the polar coordinate system on $D \setminus \{0\}$ with $z = re^{\sqrt{-1}\theta}$ $(0 < r < 1, \theta \in \mathbf{R})$. For each $0 < r < 1$, we define a horizontal lift

$$s = s(r, \theta) = ((r, \theta), \varphi(r, \theta)) \quad (\theta \in \mathbf{R})$$

of a circle $z = re^{\sqrt{-1}\theta}$ $(\theta \in \mathbf{R})$ relative to $\pi : S \longrightarrow D \setminus \{0\}$. Since the flat connection $A + \phi$ has non-trivial holonomy around 0, we can write

$$s(r, \theta + 2\pi) = s(r, \theta)a \qquad (0 < r < 1, \theta \in \mathbf{R})$$

for some $a \in G$ which is not the identity element e. We denote by $\ell > 0$ the minimum for energy values of piecewise smooth paths joining e with a in G with respect to a bi-invariant Riemannian metric of G. Then we see that for each $0 < r < 1$ the minimum for energy values of piecewise smooth paths joining $\varphi(r, 0)$ with $\varphi(r, 2\pi) = \varphi(r, 0)a$ is also equal to ℓ. Thus we have

$$\frac{1}{2}\int_0^{2\pi} |d\varphi_{(r,\theta)}(\frac{\partial}{\partial\theta})|^2 d\theta \geq \ell > 0$$

for each $0 < r < 1$. We can estimate the energy of ϕ from below;

$$\begin{aligned}
\int_{D\setminus\{0\}} |\phi|^2 dx \wedge dy &\geq \frac{1}{4}\int_{D\setminus\{0\}} \left(|d\varphi(\frac{\partial}{\partial r})|^2 + |d\varphi(\frac{1}{r}\frac{\partial}{\partial\theta})|^2\right) r dr \wedge d\theta \\
&\geq \frac{1}{4}\int_{D\setminus\{0\}} \frac{1}{r^2}|d\varphi(\frac{\partial}{\partial\theta})|^2 r dr \wedge d\theta \\
&= \frac{1}{4}\int_0^1 \frac{1}{r}\left(\int_0^{2\pi} |d\varphi(\frac{\partial}{\partial\theta})|^2 d\theta\right) dr \\
&\geq \ell \int_0^1 \frac{1}{r} dr = \infty.
\end{aligned}$$

This is a contradiction to the finiteness of energy for ϕ. We conclude that the flat connection $A + \phi$ has trivial holonomy around 0. Thus we can find a smooth section s of the trivial principal bundle $P = D \times G$ defined on $D \setminus \{0\}$ which is horizontal with respect to the flat connection $A + \phi$, and hence by setting $s(z) = (z, \varphi(z))$ for each $z \in D \setminus \{0\}$ we obtain a harmonic map $\varphi : D \setminus \{0\} \longrightarrow G$ with finite energy. By Sacks-Uhlenbeck's theorem [13] on the removability of isolated singularities of harmonic maps, φ extends to a smooth harmonic map from D to G. Therefore the Higgs field ϕ and the connection A extend smoothly to the whole disk D. □

In a later paper, we will discuss the compactification of the moduli spaces.

Acknowledgements

The authors would like to thank M.A. Guest for his interest in this work and valuable comments.

References

1. J. Dorfmeister, F. Pedit and H. Wu, Weierstrass type representation of harmonic maps into symmetric spaces, to appear in *Comm. Anal. Geom.*
2. S.K. Donaldson, Twisted harmonic maps and the self-duality equations, *Proc. London Math. Soc.(3)* **55**, 127–131 (1987).
3. D. Gilberg and N.S. Trudinger, *Elliptic Partial Differential Equations of Second Order, 2nd edition*, Springer, Berlin, 1983.

4. M.A. Guest and Y. Ohnita, Actions of loop groups, deformations of harmonic maps and their applications, *Sugaku* **46**, 228-242 (1994)(in Japanese). In *Selected Papers on Harmonic Analysis, Groups and Invariants*, Amer. Math. Soc. Translations Ser.2, Vol 183, (1998) (K.Nomizu, editor), 33-50.
5. S. Helgason, *Differential Geometry, Lie Groups and Symmetric Spaces*, Academic Press, Inc., New York, San Diego, London, Tokyo, 1978.
6. N.J. Hitchin, *Monopoles, Minimal Surfaces and Algebraic Curves*, Seminaire de mathematiques supérieures, Nato ASI, 105, Les Presses de l'Université de Montréal, Montréal, Canada, 1987.
7. N.J. Hitchin, The self-duality equations on a Riemann surface, *Proc. London Math. Soc. (3)* **55**, 59–126 (1987).
8. N.J. Hitchin, Harmonic maps from a 2-torus to the 3-sphere, *J. Differential Geom.* **31**, 627–710 (1990).
9. S. Kobayashi, *Differential Geometry of Complex Vector Bundles*, Publ. Math. Soc. Japan, Vol 15, Iwanami Shoten Publ. and Princeton University Press, 1987.
10. C.B. Morrey, *Multiple Integrals in the Calculus of Variations*, Springer, Berlin, 1966.
11. M. Mukai, Moduli spaces of solutions to the gauge-theoretic equations for harmonic maps, preprint.
12. Y. Ohnita, Gauge-theoretic equations for symmetric spaces and certain minimal submanifolds in moduli spaces, preprint.
13. J. Sacks and K. Uhlenbeck, The existence of minimal immersions of 2-spheres, *Ann. of Math.* **113**, 1–24 (1981).
14. R. Schoen, Analytic aspects of the harmonic map problem, *Lectures on Partial Differential Equations*(S.S. Chern, ed.), Math. Sci. Res. Inst. Publ. Series, Springer, Berlin, 1984, 321–358.
15. K. Uhlenbeck, Connections with L^p bounds on curvature *Commun. Math. Phys.* **83**, 31–42 (1982).
16. K. Uhlenbeck, Harmonic maps into Lie groups (Classical solutions of the chiral model), *J. Differential Geom.* **30**, 1–50 (1989).
17. G. Valli, Some remarks on geodesics in gauge groups and harmonic maps, *J. Geom. Phys.* **4**, 335–359 (1987).
18. G. Valli, Harmonic gauges on Riemann surfaces and stable bundles, *Ann. Inst. H. Poincaré Anal. Nonlinéare* **6**, 233–245 (1989).

GRASSMANN GEOMETRIES ON COMPACT SYMMETRIC SPACES

HIROO NAITOH
Department of Mathematics, Yamaguchi University,
Yamaguchi 753-8512, JAPAN
E-mail: naitoh@po.yb.cc.yamaguchi-u.ac.jp

A Grassmann geometry produces a class of submanifolds. Using this, R. Harvey and H.B. Lawson introduced a calibrated geometry. In this paper we consider another Grassmann geometry on a compact simply connected riemannian symmetric space, which is closely related to a class of submanifolds in the symmetric space, with congruent tangent spaces.

1 Introduction

Let M be an m-dimensional smooth manifold, and fix a positive integer s such that $1 \leq s \leq m$. Let $G^s(T_pM)$ be the Grassmann manifold of all the s-dimensional linear subspaces of the tangent space T_pM at p and let $G^s(TM)$ be the Grassmann bundle over M with the fibres $G^s(T_pM)$ $(p \in M)$. Take an arbitrary subset Σ of $G^s(TM)$. Then an s-dimensional connected submanifold S of M is said to be a Σ-*submanifold* if the tangent spaces T_pS $(p \in S)$ of S belong to Σ. The collection of Σ-submanifolds constitutes a Σ-*geometry*. A Grassmann geometry is a collected name for such a Σ-geometry, which has been introduced by R. Harvey-H. B. Lawson [3]. They constructed a subset Σ from a suitable differential form on a riemannian manifold M and using it, they studied homologically volume-minimizing submanifolds of M. This Σ-geometry is called a calibrated geometry. We now consider a Grassmann geometry of another type, namely *orbital type*. Some geometries of orbital type are closely related to calibrated geometries. In fact, many of our considerable Σ-geometries are also typical examples of the calibrated geometries.

Let M be a compact simply connected riemannian symmetric space and denote by G the identity component of the isometry group. Then G acts isometrically and transitively on M and at the same time acts on the Grassmann bundle $G^s(TM)$ via the differentials of isometries. As a subset Σ, we take a G-orbit on $G^s(TM)$ by this action. In the following we use a notation O in place of Σ. Obviously an O-geometry gives a class of submanifolds of M with congruent tangent spaces and typical examples of O-submanifolds are *equivariant submanifolds*, namely, a connected submanifold S is said to be equivariant if for any two points p and q in S there exists an isometry g in G such that $g(S) = S$ and $g(p) = q$. Particularly a complete totally geodesic submanifold of a symmetric space is an equivariant submanifold. The aim of our study is summarized into the following problems.

Problem I. Study whether an arbitrary O-geometry is empty or not.

Problem II. Moreover study details of non-empty O-geometries. For example, among the O-geometries with totally geodesic O-submanifolds, determine the ones that admit non-totally geodesic O-submanifolds.

In connection with Problem I, an arbitrary O-geometry of curves, i.e., a case that $s = 1$, is nonempty, since a complete geodesic in M with every direction can be realized as an orbit by a one-parameter subgroup of G. In this paper we will prove that *an arbitrary O-geometry of real hypersurfaces, i.e., a case that $s = m - 1$, is also nonempty.*

In connection with Problem 2, a large class of O-geometries has been studied by Naitoh [10,11,12,13]. Such an O-geometry is called *of SC-type* and has a totally geodesic O-submanifold defined by a connected component of the fixed-point-set of an involutive isometry of M. The above papers have determined all the O-geometries of SC-type which admit non-totally geodesic O-submanifolds.

We first review O-geometries of SC-type and summarize the results in the above papers. Next we give the proof of the above claim on an O-geometry of real hypersurfaces. Last we give a necessary and sufficient condition that a general O-geometry is nonempty, in terms of an involutive distribution on the orbit O.

2 O-geometries of SC-type

Let V be an s-dimensional linear subspace of T_pM. Then V is called *strongly curvature-invariant* if it satisfies the following condition (SC):

$$R_p(V,V)V \subset V \qquad \text{and} \qquad R_p(V^\perp, V^\perp)V^\perp \subset V^\perp$$

where R denotes the curvature tensor on M and $V^\perp$ denotes the orthogonal complement of V in T_pM. We take such a subspace V and consider a G-orbit O in $G^s(TM)$ through V. Since an element of G is an isometry of M, any subspace in O also satisfies the condition (SC). Such a G-orbit O and the associated O-geometry are called *of SC-type.*

We first explain a relation between the G-orbits of SC-type and the involutive isometries of M.

Let O be a G-orbit of SC-type. Fix a point p in M and take a subspace V of T_pM such that $V \in$ O. Moreover let f be an involutive linear isometry of T_pM such that V and $V^\perp$ are respectively the (± 1)-eigenspaces of f. Then the condition (SC) implies that f preserves the curvarure tensor R_p. Hence by a result of the symmetric space theory, there exists an involutive isometry F of M such that $F(p) = p$ and $(dF)_p = f$. Let S be the connected component of the fixed-point-set of F through p. Then a mapping $\mathsf{O} \to S$ induces a bijective correspondence from the collection of the G-orbits of SC-type over M to the G-congruence classes of the connected components of the fixed-point-sets of the involutive isometries of M. Here G-congruence implies that two connected components are congruent by some isometry in G. Moreover from our assumption $s \geq 1$, we don't consider the case that a connected component consists of one point. Under this correspondence, S gives a compact totally geodesic O-submanifold. Particularly the associated O-geometry is nonempty.

Remark. The collection of all involutive isometries generates the full isometry group of M, and every involutive isometry gives a reflection with respect to any connected component of the fixed-point-set. Therefore these facts may suggest that on the submanifold theory of symmetric spaces, it is natural to consider geometries of SC-type.

We next explain the classification of the G-orbits of SC-type, thus, the O-geometries of SC-type.

A G-orbit O of SC-type over M is said to be *not substantial* if M is a product of symmetric spaces M_1 and M_2, and moreover $V \subset TM_1$ for any element V in O, where TM_1 denotes the tangent subbundle of TM along M_1. Otherwise it is said to be *substantial*. Also, the associated O-geometry is called substantial or not substantial according as the orbit O is substantial or not. If an O-geometry of SC-type over M is not substantial, it can be regarded as a geometry of SC-type over M_1, under the above situation. So we may consider only substantial ones.

Next let $\mathfrak{g}$ be a compact semisimple Lie algebra and σ be an involution on $\mathfrak{g}$. Then a pair $(\mathfrak{g},\sigma)$ is called a *symmetric Lie algebra* if the $(+1)$-eigenspace $\mathfrak{k}$ of σ acts faithfully on the (-1)-eigenspace $\mathfrak{p}$ of σ by the adjoint action. Moreover an orderd triple $(\mathfrak{g},\sigma,\tau)$ of symmetric Lie algebras $(\mathfrak{g},\sigma)$ and $(\mathfrak{g},\tau)$ is called a *pairwise symmetric Lie algebra*, abbreviated to PSLA, if the involutions σ and τ are commutative. The classification of the substantial G-orbits of SC-type is closely related to that of the PSLA's. Now let's see this relation.

Let M be a simply connected commpact riemannian symmetric space and O be a substantial G-orbit of SC-type over M. Fix a point p in M and take a subspace V in O such that $V \subset T_pM$. Moreover let s_p be the (geodesic) symmetry of the symmetric space M at p, and t_p be an involutive isometry of M such that

$$t_p(p) = p, \quad \text{and} \quad (dt_p)(v) = \begin{cases} -v & \text{if } v \in V, \\ v & \text{if } v \in V^{\perp}. \end{cases}$$

The existance of the isometry t_p is also assured by the condition (SC). Now let $\mathfrak{g}$ be the Lie algebra of G, and let σ and τ be the differentials of inner automorphisms of G induced from s_p and t_p. Then the triple $(\mathfrak{g},\sigma,\tau)$ is a PSLA. The substantiality of geometry gives a necessary and sufficient condition that the pair $(\mathfrak{g},\tau)$ is a symmetric Lie algebra. This construction $\mathrm{O} \rightarrow (\mathfrak{g},\sigma,\tau)$ gives a bijective correspondence between the collection of the substantial geometries of SC-type over M and the equivalence classes of PSLA's with the underlying symmetric Lie algebra $(\mathfrak{g},\sigma)$. Here the equivalence of PSLA's is the usual one given by the inner automorphisms of $\mathfrak{g}$. We here remark the construction $M \rightarrow (\mathfrak{g},\sigma)$ gives a bijective correspondence between the isomorphism classes of the compact simply connected symmetric spaces and the equivalence classes of the symmetric Lie algebras.

Next a PSLA is said to be *irreducible* if it is not decomposed into a sum of more than two PSLA's. A PSLA with a simple Lie algebra is irreducible, but an irreducible PSLA don't always have a simple Lie algebra. If a PSLA $(\mathfrak{g},\sigma,\tau)$ is irreducible, the underlying Lie algebra $\mathfrak{g}$ is a compact simple Lie algebra, or a direct sum of two or four copies of a compact simple Lie algebra. On the other hand, a substantial G-orbit of SC-type over M is said to be *reducible* if M is a product of symmetric spaces M_1 and M_2, and moreover the orbit over M is a product of substantial orbits of SC-type over M_1 and M_2. Otherwise it is said to be *irreducible*. The irreducibility of PSLA's corresponds to that of substantial orbits of SC-type. To classify the irreducible PSLA's, we can use the classification by Berger [1] of the irreducible local affine symmetric spaces of semisimple type, but we also have a similar diagram method based on the Dynkin's diagrams to Murakami's method [7] for the irreducible symmetric Lie algebras. (See Naitoh [11,12,13] for this diagram method.) Anyway, there are more than 100 kinds of irreducible PSLA's.

Next we review four examples of substantial geometries of SC-type which admit non-totally geodesic O-submanifolds. These geometries correspond to irreducible PSLA's.

Example 1. Let M be the m-dimensional standard sphere S^m $(m \geq 2)$ and fix a positive integer s such that $1 \leq s \leq m$. Put $\mathrm{O} = G^s(TM)$. Then O is only the G-orbit in $G^s(TM)$ and defines a substantial geometry of SC-type. In the geometry an O-submanifold implies an s-dimensional connected submanifold, and a compact totally geodesic O-submanifold is the s-dimensional great sphere.

Example 2. Let M be the n-dimensional complex projective space CP^n. (In this case $m = 2n$.) Then the following two geometries are considerable:

(1) *Complex type*: Fix an integer t such that $1 \leq t \leq n$ and let O be the set

of the t-dimensional tangential complex subspaces of M. (In this case $s = 2t$.) Then O is a G-orbit in $G^{2t}(TM)$ and defines a substantial geometry of SC-type. In the geometry an O-submanifold implies a t-dimensional connected complex submanifold, and a compact totally geodesic O-submanifold is the t-dimensional complex subspace.

(2) *Totally real type*: Let O be the set of the n-dimensional tangential totally real subspaces of M. Then O is a G-orbit in $G^n(TM)$ and defines a substantial geometry of SC-type. In the geometry an O-submanifold implies a connected totally real submanifold with the half dimension n, and a compact totally geodesic O-submanifold is the naturally imbedded n-dimensional real projective space.

Example 3. Let M be the n-dimensional quaternion projective space HP^n. (In this case $m = 4n$.) Then the following geometry is considerable: Let O be the set of the $2n$-dimensional tangential totally complex subspaces of M. Then O is a G-orbit in $G^{2n}(TM)$ and defines a substantial geometry of SC-type. In the geometry an O-submanifold implies a connected totally complex submanifold with the half dimension $2n$, and a compact totally geodesic O-submanifold is the naturally imbedded n-dimensional complex projective space.

Each geometry of Examples 1 through 3 has been greatly studied as a submanifold theory of rank one symmetric spaces, and from the study we know that it has non-totally geodesic O-submanifolds. (For example, see Naitoh-Takeuchi [14].) On the other hand, the following example is a new geometry whose ambient space is rather a higher rank symmetric space.

Example 4. In this case a geometry is closely related to an irreducible symmetric R-space. Such the spaces constitute a class of (not necessarily simply connected) compact symmetric spaces which contains the irreducible compact hermitian symmetric spaces.

Case (1): Let $\hat{M}$ be a compact irreducible hermitian symmetric space and $\hat{N}$ be a real form of $\hat{M}$. Namely, $\hat{N}$ is a half-dimensional compact totally real totally geodesic submanifold of $\hat{M}$. Such the real forms are exhausted by the canonical imbeddings of all the irreducible symmetric R-spaces of not hermitian type. (See Takeuchi [16].)

Take such a real form $\hat{N}$ and consider an orbit $\hat{\text{O}}$ over $\hat{M}$ which contains the tangent spaces of $\hat{N}$. Since a tangent space of $\hat{N}$ is strongly curvature-invariant, then $\hat{\text{O}}$ induces a substantial geometry of SC-type. Obviously, $\hat{N}$ is an $\hat{\text{O}}$-submanifold. Now denote by $\hat{G}$ the connected component of the isometry group of $\hat{M}$ and let M be the set of all the real forms of $\hat{M}$ which are $\hat{G}$-congruent to $\hat{N}$. Then M has a structure of compact riemannian symmetric space, naturally induced from that of $\hat{M}$. At each real form $\hat{R}$ in M, the geodesic symmetry is induced from the reflection with respect to $\hat{R}$. Next let $\hat{N} \to M$ be a mapping, so-called Gauss map, assigning to a point $\hat{p}$ in $\hat{N}$ the real form $\hat{R}$ through $\hat{p}$ and complementary orthogonal to $\hat{N}$. We here note that this $\hat{R}$ is also $\hat{G}$-congruent to $\hat{N}$. Then we can see that the image N of this mapping is a compact totally geodesic submanifold of M locally isometric to $\hat{N}$ and moreover that a G-orbit O over M which contains the tangent spaces of N also induces a substantial geometry of SC-type. ¿From a demand on our situation, we may here take a universal covering of M and a lift of N in place of M and N. This O-geometry corresponds to an irreducible PSLA obtained by an exchange of two involutions in the PSLA corresponding to the Ô-geometry, namely, if the PSLA for Ô is $(\mathfrak{g},\tau,\sigma)$, the PSLA for O is $(\mathfrak{g},\sigma,\tau)$.

Now we explain that this O-geometry admits non-totally geodesic O-submanifolds. Fix a point p in N and take a compact totally geodesic submanifold $N^\perp$ complementally orthogonal to N at p. Then $N^\perp$ is a symmetric space with 1-dimensional euclidean factor. Moreover take a complete geodesic $c(t)$ in $N^\perp$ starting from p with the direction of the euclidean factor. Then it is a simply closed curve with a minimum period ℓ, i.e., $c(0) = c(\ell) = p$. Take a quarter point m $(= c(\ell/4))$ and let K_m be the isotropy subgroup in G at m. Then a K_m-orbit $K_m(c(t))$ is an O-submanifold for each t except the values $\ell/4$ and $3\ell/4$, while for the exceptional values it is a one-point set. Moreover the orbits at $c(0)$ and $c(\ell/2)$ coincide with N, and the other orbits give non-totally geodesic O-submanifolds, which are imbedded into M as pseudo-umbilical submanifolds.

A typical example of this case is the hypersurface geometry of S^n, which also appeared in Example 1. Let $\hat{M}$ be the $(n-1)$-dimensional complex quadric CQ^{n-1} and let $\hat{N}$ be a sphere S^{n-1} canonically imbedded into CQ^{n-1} as a real form. Then M is the sphere S^n and N is a great hypersphere S^{n-1}. Moreover the orbits $K_m(c(t))$ where $t \neq 0,\ \ell/4,\ \ell/2,\ 3\ell/4$ are small hyperspheres.

Case (2): Next let L be a compact irreducible hermitian symmetric space and put $\hat{M} = L \times L$, provided that the complex structure on the second L is the minus multiple of the complex structure on the first L. Moreover let $\hat{N}$ be the diagonal real form L in the product space $L \times L$. Then, the same situation holds as the case (1), and we again obtain a substantial geometry of SC-type which admits non-totally geodesic submanifolds. In this case, M is a compact Lie group, precisely, the universal covering of the identity component of the isometry group of L. Moreover N is a lift of the image of the so-called Cartan imbedding from L to the isometry group of L. For the details of Example 4, we refer to Naitoh[8,11,12,13].

Now the results by Naitoh[10,11,12,13] are summarized as follows. Let M be a compact simply connected riemannian symmetric space and O be a substantial

G-orbit of SC-type over M. Decompose O into irreducible ones;

$$M = M_1 \times \cdots \times M_r \quad \text{and} \quad \mathsf{O} = \mathsf{O}_1 \times \cdots \times \mathsf{O}_r$$

where each O_i is an irreducible substantial orbit of SC-type over M_i. Then we have the following.

Theorem 1. (1) *An arbitrary O-submanifold S of M is locally decomposed into a product of O_i-submanifolds S_i of M_i, inclusively of the way of imbedding it. Moreover, if S is complete, the global version holds; S_i's are complete and the imbedding $S \to M$ is the global product of the imbeddings $S_i \to M_i$. Next,*

(2) *if O is an irreducible orbit except Examples 1 through 4, an O-submanifold is always totally geodesic in M.*

Next we explain an application of Theorem 1. Let M be a compact simply connected symmetric space. Then a connected submanifold S of M is called a *symmetric submanifold* if for each point p in S there exists an extrinsic symmetry t_p of S at p, namely, t_p is an involutive isometry of M such that

$$t_p(p) = p, \quad t_p(S) = S, \quad \text{and} \quad (dt_p)(v) = \begin{cases} -v & \text{if } v \in T_pS, \\ v & \text{if } v \in N_pS. \end{cases}$$

Obviously a symmetric submanifold is a riemannian symmetric space, and the definition is analogous to that of a riemannian symmetric space.

Now Theorem 1 completes the classification of symmetric submanifolds, together with the following facts:

(a) A symmetric submanifold, by the definition, belongs to some geometry of SC-type. Moreover in the case that M is a product of symmetric spaces M_1 and M_2, if S is a symmetric submanifold of M contained in M_1, then it is induced from a symmetric submanifold of M_1. So we consider a symmetric submanifold in a substantial geometry of SC-type.

(b) Next, under the situation of Theorem 1,(1), if S is a symmetric submanifold of M, then all S_i's are also symmetric submanifolds of M_i. Hence a symmetric submanifold can be decomposed into irreducible ones. Moreover, a compact totally geodesic O-submanifold is obviously a symmetric submanifold. So, by virtue of Theorem 1,(2), it is sufficient to classify the non-totally geodesic symmetric submanifolds in the geometries of Examples 1 through 4.

(c) But the symmetric submanifolds in the geometries of Examples 1 through 4 have been already classified inclusively of non-totally geodesic ones. Generally, in our geometry of SC-type, a symmetric submanifold is equivalently a complete parallel submanifold. Such the submanifolds of Example 1 have been classified by Ferus, the ones of Example 2 by Takagi-Nakagawa (complex type) and the author-Takeuchi (totally real type), the ones of Example 3 by Tsukada, and the ones of Example 4 by the author. For the classifications see Naitoh-Takeuchi [14] (Examples 1 through 3) and Naitoh [8] (Example 4). Particularly in Example 4, the symmetric submanifolds are exhausted by the isotropy orbits $K_m(c(t))$ ($t \neq \ell/4,\ 3\ell/4$), except the hypersurface geometry of S^n. In the exceptional case there exist the generalized Clifford torus besides great or small hyperspheres.

3 O-geometries of real hypersurfaces

In this section we consider an O-geometry of real hypersurfaces over a compact simply connected riemannian symmetric space M, i.e., O is a G-orbit in $G^{m-1}(TM)$ where m is the dimension of M. As described in Introduction, we claim the following theorem in connection with Problem I. We here remark that generally M has no totally geodesic real hypersurface, except the case $M = S^m$.

Theorem 2. *Let O be a G-orbit of real hypersurfaces over M. Then the O-geometry is nonempty.*

First suppose that rank $M = 1$. In this case, since G acts transitively on $G^{m-1}(TM)$, this $G^{m-1}(TM)$ is only the G-orbit and thus the O-geometry consists of all the connected real hypersurfaces of M. Particularly geodesic hyperspheres of M with small radius give compact O-submanifolds.

Next suppose that rank $M \geq 2$. Then there exists infinitely many G-orbits of real hypersurfaces, and for each G-orbit O, O-submanifolds are constructed as follows.

Fix a point p in M and let $\mathfrak{g}$ be the Lie algebra of G. Denote by σ the involutive automorphism of $\mathfrak{g}$ induced from the geodesic symmetry of M at p. Moreover decompose $\mathfrak{g}$ into the sum of the $(+1)$-eigenspace $\mathfrak{k}$ and the (-1)-eigenspace $\mathfrak{p}$ of σ. Then the vector space $\mathfrak{p}$ is identified with the tangent space T_pM. Take a maximal abelian subspace $\mathfrak{a}$ of $\mathfrak{p}$ and let $\mathfrak{h}$ be the maximal abelian subspace of $\mathfrak{g}$ which contains $\mathfrak{a}$. For α in $\mathfrak{h}$ put

$$\mathfrak{g}_\alpha = \{\ X \in \mathfrak{g}^C;\ [H, X] = \sqrt{-1} < \alpha, H > X \text{ for any } H \in\ \mathfrak{h}\ \}$$

where $\mathfrak{g}^C$ denotes the complexification of $\mathfrak{g}$ and $<,>$ denotes an inner product on $\mathfrak{g}$ such that (i) it is σ-invariant, (ii) the endomorphisms $\mathrm{ad}(X)$ ($X \in \mathfrak{g}$) are skew symmetric with respect to it, and (iii) on $\mathfrak{p}$ it is equal to the riemannian metric on T_pM. Moreover put as follows:

$$\Sigma(\mathfrak{g}) = \{\ \alpha \neq 0 \in \mathfrak{h}\ ;\ \mathfrak{g}_\alpha \neq \{0\}\ \}, \quad \Sigma_0(\mathfrak{g}) = \{\ \alpha \in \Sigma(\mathfrak{g});\ \bar{\alpha} = 0\ \}$$

where $\bar{\ }$ is the canonical projection of $\mathfrak{h}$ onto $\mathfrak{a}$. Then the image of $\Sigma(\mathfrak{g}) - \Sigma_0(\mathfrak{g})$ by the projection, denoted by $\Sigma(\mathfrak{g},\mathfrak{k})$, is a root system in $\mathfrak{a}$.

Let K be the isotropy subgroup of G at p and let K_0 be the centralizer of $\mathfrak{a}$ in K. We now consider a smooth mapping $\Phi : K/K_0 \times \mathfrak{a} \ni (kK_0, H) \rightarrow$ Exp Ad $(k)H \in M$. Putting

$$D(\mathfrak{g},\mathfrak{k}) = \{\ H \in \mathfrak{a};\ <\gamma, H> \in \pi\mathbf{Z} \text{ for some } \gamma \in \Sigma(\mathfrak{g},\mathfrak{k})\ \},$$

then we can see that Φ is regular at (kK_0, H) if and only if $H \notin D(\mathfrak{g},\mathfrak{k})$. (See Helgason[4].)

Next take an order in $\mathfrak{a}$ and let $\{\gamma_1, \cdots, \gamma_r\}$ be the system of simple roots with respect to this order. Moreover take the highest roots of the irreducible components of the root system $\Sigma(\mathfrak{g},\mathfrak{k})$ and let δ be the sum of them. Putting

$$W = \{\ H \in \mathfrak{a};\ < H, \gamma_i > > \ 0\ (1 \leq i \leq r),\ < H, \delta > < \ \pi\ \},$$

then we can see that W is an open connected component of $\mathfrak{a} - D(\mathfrak{g},\mathfrak{k})$. Moreover from the simple connectedness of M, it follows $\Phi : K/K_0 \times W \rightarrow M$ is an imbedding with open image. (See Takeuchi[15].)

Now let O be a G-orbit of real hypersurfaces. Noting that $\mathfrak{p} = \cup_{k \in K} k(\mathfrak{a})$, we can take a following subspace V in O: $\mathfrak{a}^{\perp} \subset V \subset T_pM$ where $\mathfrak{a}^{\perp}$ denotes the orthogonal complement of $\mathfrak{a}$ in T_pM. Take a hyperplane L in $\mathfrak{a}$ which is parallel to $V \cap \mathfrak{a}$ and satisfies $L_W = L \cap W \neq \emptyset$. Then $\Phi(K/K_0 \times L_W)$ is a real hypersurface in M. It follows by the following lemma that this real hypersurface is an O-submanifold.

Lemma (cf. Tasaki[17]). *For a point (kK_0, H) in $K/K_0 \times L_W$, it holds*

$$(d\Phi)_{(kK_0,H)}(T_{(kK_0,H)}(K/K_0 \times L_W)) = (d\tau_{k exp\ H})_p(V)$$

where τ_g $(g \in G)$ denotes the mapping $M \ni q \rightarrow g(q) \in M$.

Remark. In the case rank $M = 1$, if we take a one-point subset of W in place of the hyperplane L, the above argument is usable. Then the real hypersurface $\Phi(K/K_0 \times L_W)$ is a geodesic hypersphere, thus, a compact hypersurface. But, in the case that M is irreducible and rank $M \geq 2$, we can ascertain that this hypersurface is not complete. Hence we have the following question: *Let M be an irreducible compact simply connected symmetric space. Then does M have rank one if and only if there exists a G-orbit O of real hypersurfaces with a complete O-submanifold?* If this is true, the following holds: If an irreducible compact simply connected symmetric space has an equivariant hypersurface, it has rank one. This is closely related to the following result by B.Y.Chen-T.Nagano[2]: If an irreducible compact simply connected symmetric space has a totally geodesic hypersurface, it has rank one.

4 O-geometries of general type

In this section we consider an orbital geometry of general type over a compact simply connected riemannian symmetric space M and, in connection with Problem I, give a necessary and sufficient condition that the geometry is nonempty.

Let O be a G-orbit in $G^s(TM)$. Then the fibre bundle $\mathsf{O} \to M$ is a homogeneous G-bundle. Denote by π the natural projection. Fixing a subspace V in O where $\pi(V) = p$, and putting

$$K = \{ g \in G;\ g(p) = p \}, \qquad K_+ = \{ g \in K;\ g(V) = V \},$$

then we can identify this bundle with the canonical G-bundle $G/K_+ \to G/K$.

We first give a bundle decomposition of the tangent bundle $T\mathsf{O}$. Let K^- and H be the vertical and the horizontal subbundles of $T\mathsf{O}$ associated with the bundle $\mathsf{O} \to M$. Since $d\pi_V : H_V \to T_pM$ is a linear isomorphism, then we can take the horizontal lifts P_V^- and P_V^+ in H_V of the subspaces V and $V^\perp$, respectively. Moreover, putting $P^- = \cup_{V \in \mathsf{O}} P_V^-$ and $P^+ = \cup_{V \in \mathsf{O}} P_V^+$, we have the following bundle decomposition:

$$T\mathsf{O} = K^- \oplus H = K^- \oplus P^- \oplus P^+.$$

We here remark that these bundles are all G-invariant. Next, under the identification of the bundle $\mathsf{O} \to M$ with the canonical bundle $G/K_+ \to G/K$, we moreover consider the canonical bundle $G \to G/K_+$. Let ${}^GK^+$ and ${}^GH$ be the vertical and the horizontal subbundles in TG associated with this bundle, and let ${}^GK^-$ and ${}^GP^\pm$ be the horizontal lifts in ${}^GH$ of K^- and $P^\pm$, respectively. Then we also have the following bundle decomposition:

$$TG =^G K^+ \oplus^G H =^G K^+ \oplus^G K^- \oplus^G P^+ \oplus^G P^-.$$

By the above remark, these subbundles ${}^GK^\pm$ and ${}^GP^\pm$ are left G-invariant and right K_+-invariant.

Now, for a local section λ of the homomorphism bundle $\mathrm{Hom}(P^-, K^-)$ over O, we define an s-dimensional local distribution D^λ on $T\mathsf{O}$ as follows:

$$\mathsf{D}^\lambda = \{\mathsf{D}_V^\lambda\}, \quad \mathsf{D}_V^\lambda = \{ \lambda_V(x) + x \in K_V^- \oplus P_V^-;\ x \in P_V^- \}.$$

Then, if there exists an s-dimensional integral submanifold $\bar{S}$ of D^λ, the projection $\pi(\bar{S})$ $(= S)$ is a local O-submanifold. Conversely, if S is an O-submanifold, we can locally construct a desired involutive distribution D^λ by the following way. Set $\bar{S} = \{ T_pS \in \mathsf{O};\ p \in S \}$. Then, since $\pi(\bar{S}) = S$, it follows that for each V in $\bar{S}$, $T_V\bar{S} \subset K_V^- \oplus P_V^-$ and the projection of $T_V\bar{S}$ to P_V^- is a linear isomorphism. Fix a subspace V in $\bar{S}$ and recall the canonical bundle $G \ni g \to g(V) \in \mathsf{O}$. Then by the above fact, we can take a small open set W of ${}^G K_e^- \oplus {}^G P_e^+$ such that $(\exp W)(\bar{S})$ contains some open neighbourhood of V in O. Hence the slice submanifolds $(\exp X)(\bar{S})$ $(X \in W)$ locally define an involutive distribution D on $T\mathsf{O}$. This distribution D again satisfies that $\mathsf{D} \subset K^- \oplus P^-$ and the projection $\mathsf{D} \to P^-$ is an isomorphism on the domain of D. Hence $\mathsf{D} = \mathsf{D}^\lambda$ for some local section λ of $\mathrm{Hom}(P^-, K^-)$. We can summarize the above arguments as follows.

Theorem 3. *Let O be a G-orbit over a compact simply connected riemannian symmetric space M. Then the O-geometry is nonempty if and only if there exists a local section λ of Hom(P^-, K^-) such that D^λ is involutive.*

The integrability condition of D^λ can be written as follows: For X, Y in P^-,

$$[\lambda(X) + X, \lambda(Y) + Y] \in K^- \oplus P^-, \tag{1}$$

and

$$[\lambda(X) + X, \lambda(Y) + Y]_{K^-} = \lambda([\lambda(X) + X, \lambda(Y) + Y]_{P^-}). \tag{2}$$

Noting that $[K^-, K^-] \subset K^-$ and $[P^-, P^-] \subset K^- \oplus P^-$, we can moreover rewrite the conditions (1), (2) as follows:

$$[\lambda(X), Y]_{P^+} + [X, \lambda(Y)]_{P^+} = 0, \tag{3}$$

and

$$\begin{aligned}&[\lambda(X), \lambda(Y)]_{K^-} + [X, Y]_{K^-} + [\lambda(X), Y]_{K^-} + [X, \lambda(Y)]_{K^-} \\ &= \lambda([X, Y]_{P^-} + [\lambda(X), Y]_{P^-} + [X, \lambda(Y)]_{P^-}).\end{aligned} \tag{4}$$

Here the condition (3) is a linear condition with respect to X, Y and λ, and the condition (4) is a linear condition with respect to X, Y and a 1st order PDE with respect to λ.

We moreover rewrite these conditions by ones on TG. Again, we fix a subspace V in O and recall the canonical bundle $G \to \mathsf{O}$. Put ${}^G K_e^- = \mathfrak{k}_-$ and ${}^G P_e^\pm = \mathfrak{p}_\pm$. Moreover put

$$\mathsf{S}_e = \{\ \alpha \in \mathrm{Hom}\ (\mathfrak{p}_-, \mathfrak{k}_-);\ [\alpha(x), y]_{\mathfrak{p}_+} + [x, \alpha(y)]_{\mathfrak{p}_+} = 0 \text{ for } x, y \in \mathfrak{p}_-\ \}$$

and let S be the left G-invariant subbundle in $\mathrm{Hom}({}^G P^-, {}^G K^-)$ induced by S_e. Then the equations (3) and (4) can be respectively rewritten by the following conditions with respect to the horizontal lift ${}^G\lambda$ of λ, i.e., a right K_+-invariant local section of $\mathrm{Hom}({}^G P^-, {}^G K^-)$: (3) implies that ${}^G\lambda$ is a local section of S, and moreover (4) can be done as follows.

$$\begin{aligned}&[{}^G\lambda(U), {}^G\lambda(V)]_{{}^G K^-} + [U, V]_{{}^G K^-} + [{}^G\lambda(U), V]_{{}^G K^-} \\ &+ [U, {}^G\lambda(V)]_{{}^G K^-} = {}^G\lambda([{}^G\lambda(U), V]_{{}^G P^-} + [U, {}^G\lambda(V)]_{{}^G P^-})\end{aligned} \tag{5}$$

for U, V in ${}^{G}P^{-}$.

Remark. (a) To prove Theorem 1,(2), Naitoh [10,11,12,13] show that $\mathsf{S}_e = \{0\}$ for the irreducible substantial geometries of SC-type except Examples 1 through 4. This condition induces that the second fundamental form of an O-submanifold vanishes.

(b) Next, in the case of an O-geometry of real hypersurfaces, we can see that S_e is linearly isomorphic to the space of all the symmetric endomorphisms of $\mathfrak{k}_{-}$.

Acknowledgments

This work is partly supported by the Grant-in-aid for Scientic Research (B) (No. 09440035), The Ministry of Education, Science, Sports and Culture, Japan.

References

1. M.Berger, *Ann. Sci. Ecole Norm. Sup.* **74**, 85 (1957).
2. B.Y. Chen-T. Nagano, *Duke Math. J.* **45**, 405 (1978).
3. R. Harvey-H.B. Lawson, *Acta Math.* **148**, 47 (1982).
4. S.Helgason, *Differential Geometry, Lie groups and Symmetric spaces* (Academic Press, New York, 1978)
5. S.Kobayashi–K.Nomizu, *Foundations of differential geometry I,II* (Wiley, New York, 1963, 1969)
6. J.D. Moore, *J. Diff. Geom.* **5**, 159 (1971)
7. S. Murakami, *Osaka J. Math.* **2**, 291 (1965)
8. H. Naitoh, *Tsukuba J. Math.* **10**, 215 (1986)
9. H.Naitoh, *Adv. Stu. in Pure Math.* **22**, 197 (1993)
10. H.Naitoh, *Osaka J. Math.* **30**, 653, 691 (1993)
11. H.Naitoh, *Grassmann geometries on compact symmetric spaces of classical type*, (to appear)
12. H.Naitoh, *Grassmann geometries on compact symmetric spaces of exceptional type*, (to appear)
13. H.Naitoh, *Grassmann geometries on compact symmetric spaces of general type*, (to appear in *Math. Soc. of Japan* **50** (1998))
14. H.Naitoh-M.Takeuchi, *Sugaku Exp.* **2**, 157 (1989)
15. M.Takeuchi, *Tsukuba J. Math.* **2**, 35 (1977)
16. M.Takeuchi, *Tohoku J. Math.* **36**, 293 (1984)
17. H.Tasaki, *Tsukuba J. Math.* **9**, 117 (1985)

COMBINATORIAL FORMULAE FOR FUTAKI CHARACTERS AND GENERALIZED KILLING FORMS OF TORIC FANO ORBIFOLDS

Dedicated to Professor Shingo Murakami on his seventieth birthday

YASUHIRO NAKAGAWA
Mathematical Institute
Tôhoku University
Sendai 980-8578
Japan
E-mail: nakagawa@math.tohoku.ac.jp

The main purpose of this paper is to establish combinatorial formulae for Futaki characters and generalized Killing forms of toric Fano orbifolds. As applications, we can prove Mabuchi's conjectures for two-dimensional toric Fano orbifolds and non-singular toric Fano manifolds of dimension $r \leqq 4$.

1 Introduction

Let X be an r-dimensional compact complex orbifold with positive first Chern class. Then one can naturally ask whether X admits an Einstein-Kähler metric. As to such existence of Einstein-Kähler metrics, two obstructions are known, one of which is Matsushima's obstruction (see Fact 2.4) and the other is the Futaki character (see Definition 2.5 and Fact 2.6). These obstructions are defined on the Lie algebra of holomorphic vector fields. We begin with the following conjecture:

Conjecture 1.1 *If an r-dimensional compact complex orbifold with positive first Chern class has no non-trivial holomorphic vector fields, then it admits an Einstein-Kähler metric.*

However, Tian recently constructed counter-examples for Conjecture 1.1 in [32], which are non-singular Fano manifolds. We therefore consider the following conjectures for r-dimensional toric Fano orbifolds (see Definition 3.1 for the definition of toric Fano orbifolds).

Conjecture 1.2 *If an r-dimensional toric Fano orbifold has vanishing Futaki character, then its automorphism group is a reductive algebraic group.*

Conjecture 1.3 *If an r-dimensional toric Fano orbifold has vanishing Futaki character, then it admits an Einstein-Kähler metric.*

Note that if Conjecture 1.3 is true, then Conjecture 1.2 is also true (see Fact 2.4). Whenever X is non-singular and $r \leqq 4$, Conjecture 1.3 for X was solved affirmatively except the four-dimensional toric Fano manifold of type X-1 in Table 3 (see Fact 5.3).

In this paper, we shall establish a combinatorial formula for the Futaki characters of a toric Fano orbifold (see Corollary 4.3). Using this formula and classifications of toric Fano orbifolds (cf. [2], [3], [4], [19], [30], [34]), we can prove Conjecture 1.2 for two-dimensional toric Fano orbifolds and non-singular toric Fano manifolds of dimension $r \leqq 4$ (see Theorems 5.2 and 5.4). Moreover, we can also prove Conjecture 1.3 for two-dimensional toric Fano orbifolds (see Theorem 5.1).

In [6], Calabi introduced the notion of extremal Kähler metrics (see Definition 2.9). Moreover, Calabi proved the structure theorem (see Fact 2.10) for the automorphism group of a compact Kähler manifold admitting an extremal Kähler metric, in [7]. Inspired by this Calabi's structure theorem, Mabuchi posed two conjectures (see Conjectures 2.11 and 2.12) on the automorphism group of a compact complex orbifold with positive first Chern class. We shall also consider these conjectures for toric Fano orbifolds.

2 Futaki characters and generalized Killing forms

In this section, we review some known facts concerning Einstein-Kähler orbifold metrics (see [13] or [14] for more details).

Let X be an r-dimensional compact connected complex orbifold with positive first Chern class (see [29] for the definition of orbifolds). We denote by $G_X := \mathrm{Aut}^\circ(X)$ the identity component of the group of holomorphic automorphisms of X, and by U_X the unipotent radical of G_X. The Chevalley decomposition allows us to write G_X as a semidirect product

$$G_X = K_X^{\mathbb{C}} \ltimes U_X, \tag{1}$$

where K_X is a compact subgroup of G_X and $K_X^{\mathbb{C}} \subseteqq G_X$ is its complexification. We now take a K_X-invariant Kähler orbifold metric g on X such that $[\omega_g] = 2\pi c_1(X)$, where ω_g is the Kähler form of g and written as

$$\omega_g := \sqrt{-1} \sum_{\alpha,\beta=1}^{r} g_{\alpha\overline{\beta}} dz^\alpha \wedge \overline{dz^\beta}$$

in terms of holomorphic local coordinates $(z^1, z^2, \ldots, z^r)$ on X, and $[\omega_g]$ is its de Rham cohomology class. Let $\mathfrak{g}_X$, $\mathfrak{k}_X$, $\mathfrak{k}_X^{\mathbb{C}}$, $\mathfrak{u}_X$ be the Lie algebras of G_X, K_X, $K_X^{\mathbb{C}}$, U_X, respectively. Note that we regard $\mathfrak{g}_X$ as the Lie algebra of holomorphic vector fields on X, and $\mathfrak{k}_X^{\mathbb{C}}$ and $\mathfrak{u}_X$ are complex Lie subalgebras of $\mathfrak{g}_X$. Moreover, $\mathfrak{k}_X$ is a Lie subalgebra of the Lie algebra of Killing vector fields on (X, g), which is identified with a real Lie subalgebra of $\mathfrak{g}_X$ by an injective Lie algebra homomorphism

$$\{\text{Killing vector fields on } (X, g)\} \ni V \mapsto \frac{1}{2}\left(V - \sqrt{-1}JV\right) \in \mathfrak{g}_X,$$

where J is the complex structure on X. Therefore, we can regard $\mathfrak{k}_X$ as a real Lie subalgebra of $\mathfrak{g}_X$. From the decomposition (1), we obtain

$$\mathfrak{g}_X = \mathfrak{k}_X^{\mathbb{C}} + \mathfrak{u}_X. \tag{2}$$

Let $C^\infty(X)_{\mathbb{C}}$, $C^\infty(X)_{\mathbb{R}}$ be respectively the set of complex, real valued C^∞ functions on X. For $\psi \in C^\infty(X)_{\mathbb{C}}$, we define the *gradient vector field* $\mathrm{grad}_g\psi$ of ψ with respect to g by

$$\mathrm{grad}_g\psi := \frac{1}{\sqrt{-1}} \sum_{\alpha,\beta=1}^{r} g^{\overline{\beta}\alpha} \frac{\partial \psi}{\partial z^{\overline{\beta}}} \frac{\partial}{\partial z^\alpha},$$

where $\left(g^{\overline{\beta}\alpha}\right)$ is the inverse matrix of $\left(g_{\alpha\overline{\beta}}\right)$. Now, we put

$$\mathcal{H}_g^{\mathbb{C}} := \left\{\psi \in C^\infty(X)_{\mathbb{C}}\,;\, \mathrm{grad}_g\psi \in \mathfrak{k}_X^{\mathbb{C}} \text{ and } \int_X \psi \left(\frac{\omega_g}{2\pi}\right)^r = 0\right\},$$

and $\mathcal{H}_g^{\mathbb{R}} := \mathcal{H}_g^{\mathbb{C}} \cap C^\infty(X)_{\mathbb{R}}$. Then we have an isomorphism

$$\Psi_g\colon \mathcal{H}_g^{\mathbb{C}} \ni \psi \mapsto \mathrm{grad}_g\psi \in \mathfrak{k}_X^{\mathbb{C}},$$

between $\mathcal{H}_g^{\mathbb{C}}$ and $\mathfrak{k}_X^{\mathbb{C}}$ (cf. [14, Lemma 1.4]), which induces

$$\Psi_g|_{\mathcal{H}_g^{\mathbb{R}}}\colon \mathcal{H}_g^{\mathbb{R}} \xrightarrow{\cong} \mathfrak{k}_X,$$

(cf. [18, Theorem 4.4]). For a holomorphic vector field $V \in \mathfrak{k}_X^{\mathbb{C}}$, we put $\psi_V := \Psi_g^{-1}(V)$. Recall the following definition due to Futaki and Mabuchi:

Definition 2.1 (Futaki and Mabuchi [13]) *The $\mathbb{C}$-bilinear form $B_X\colon \mathfrak{k}_X^{\mathbb{C}} \times \mathfrak{k}_X^{\mathbb{C}} \longrightarrow \mathbb{C}$ defined by*

$$B_X(V_1, V_2) := \int_X \psi_{V_1}\psi_{V_2} \left(\frac{\omega_g}{2\pi}\right)^r, \qquad V_1, V_2 \in \mathfrak{k}_X^{\mathbb{C}},$$

is called the generalized Killing form of X.

Concerning the generalized Killing form B_X, Futaki and Mabuchi proved the following fact:

Fact 2.2 (Futaki and Mabuchi [13]) *For X and B_X as above, the following hold:*

(i) *B_X does not depend on the choice of g satisfying $[\omega_g] = 2\pi c_1(X)$;*

(ii) *B_X is a non-degenerate symmetric $\mathbb{C}$-bilinear form on $\mathfrak{k}_X^{\mathbb{C}}$.*

By Ric_g, we denote the Ricci form of g. Since ω_g and Ric_g are in the same de Rham cohomology class $2\pi c_1(X)$, there exists a real-valued C^∞-function f_g on X such that $\mathrm{Ric}_g - \omega_g = \sqrt{-1}\partial\bar{\partial} f_g$, where f_g is unique up to additive constants.

Definition 2.3 *g is called an Einstein-Kähler orbifold metric if* $\mathrm{Ric}_g = \omega_g$.

The following fact on automorphism groups is known as Matsushima's obstruction to the existence of Einstein-Kähler orbifold metrics.

Fact 2.4 (Matsushima [22]) *Let X be as above. If X admits an Einstein-Kähler orbifold metric, then $G_X = \mathrm{Aut}^\circ(X)$ is a reductive algebraic group (defined over $\mathbb{C}$).*

In [12], Futaki introduced an obstruction to the existence of Einstein-Kähler orbifold metrics as follows:

Definition 2.5 (Futaki [12]) *The $\mathbb{C}$-linear map $F_X: \mathfrak{g}_X \longrightarrow \mathbb{C}$ defined by*

$$F_X(V) := \frac{1}{\sqrt{-1}} \int_X (V f_g) \left(\frac{\omega_g}{2\pi}\right)^r, \qquad V \in \mathfrak{g}_X,$$

is called the Futaki character of X.

We now quote the following fundamental facts about Futaki characters:

Fact 2.6 (Futaki [12]) *For X and F_X as above, the following hold:*

(i) *F_X does not depend on the choice of g satisfying $[\omega_g] = 2\pi c_1(X)$;*

(ii) *If X admits an Einstein-Kähler orbifold metric, then F_X vanishes;*

(iii) *F_X vanishes on the commutator subalgebra $[\mathfrak{g}_X, \mathfrak{g}_X]$ of $\mathfrak{g}_X$.*

Fact 2.7 (Mabuchi [21]) *Let X and F_X be as above. Then, F_X vanishes on $\mathfrak{u}_X$.*

We denote by $\sigma_g \in C^\infty(X)_{\mathbb{R}}$ the scalar curvature of g, and put $V_g := \mathrm{grad}_g \sigma_g$. This vector field V_g may not be a holomorphic vector field in general, and so we define a holomorphic vector field V_g' by

$$V_g' := \mathrm{grad}_g \sigma_g' \in \mathfrak{k}_X,$$

where σ'_g is the image of σ_g under the L^2-orthogonal projection $C^\infty(X)_\mathbb{R} \longrightarrow \mathcal{H}_g^\mathbb{R}$.

In [13], Futaki and Mabuchi proved the following relationship between F_X and V'_g, which means that F_X coincides with V'_g via the non-degenerate bilinear form B_X.

Fact 2.8 (Futaki and Mabuchi [13]) *Let X, F_X, B_X, and V'_g be as above. Then, we have*

$$F_X(V) = B_X(V, V'_g),$$

for every $V \in \mathfrak{k}_X^\mathbb{C}$.

In [6], Calabi introduced the notion of *extremal Kähler* metrics as a generalization of Einstein-Kähler metrics or constant scalar curvature Kähler metrics.

Definition 2.9 (Calabi [6]) *We say that g is an extremal Kähler orbifold metric, if V_g is a holomorphic vector fields on X.*

We recall Calabi's structure theorem on automorphism groups of compact complex orbifolds admitting an extremal Kähler orbifold metric, which is a generalization of Fact 2.4:

Fact 2.10 (Calabi [7]) *Let X be as above. If X admits a K_X-invariant extremal Kähler orbifold metric h such that $[\omega_h] = 2\pi c_1(X)$, then the following hold:*

(i) $\mathfrak{k}_X^\mathbb{C} = Z_{\mathfrak{g}_X}\left(\sqrt{-1}V_h\right) := \left\{V \in \mathfrak{g}_X\,;\, \left[\sqrt{-1}V_h, V\right] = 0\right\}$;

(ii) $\mathfrak{u}_X = \bigoplus_{\lambda>0} \mathfrak{h}_\lambda$, *where* $\mathfrak{h}_\lambda := \left\{V \in \mathfrak{g}_X\,;\, \left[\sqrt{-1}V_h, V\right] = \lambda V\right\}$.

Based on this Calabi's structure theorem, Mabuchi posed the following two conjectures:

Conjecture 2.11 *Let (X, g) be as above. Then,*

$$\mathfrak{k}_X^\mathbb{C} = Z_{\mathfrak{g}_X}\left(\sqrt{-1}V'_g\right) := \left\{V \in \mathfrak{g}_X\,;\, \left[\sqrt{-1}V'_g, V\right] = 0\right\}.$$

Conjecture 2.12 *Let (X, g) be as above. Then*

$$\mathfrak{u}_X = \bigoplus_{\lambda>0} \mathfrak{h}'_\lambda,$$

where $\mathfrak{h}'_\lambda := \left\{V \in \mathfrak{g}_X\,;\, \left[\sqrt{-1}V'_g, V\right] = \lambda V\right\}$.

REMARK. If Conjecture 2.11 holds for a toric Fano orbifold X (see Definition 3.1 for the definition of toric Fano orbifolds), then Conjecture 1.2 also holds for X.

For (X, g) as above, Yau's theorem in [35] says that there exists a unique Kähler orbifold metric g' on X such that $\mathrm{Ric}_{g'} = \omega_g$. For each holomorphic vector field $V \in \mathfrak{g}_X$ on X, we define a smooth section $\mathcal{L}_V^{g'}$ of the endomorphism bundle $\mathrm{End}\left(T^{1,0}X\right)$ of $T^{1,0}X$ over X, by

$$\mathcal{L}_V^{g'} := \nabla_V^{g'} - L_V \in C^\infty\left(X, \mathrm{End}\left(T^{1,0}X\right)\right),$$

where $T^{1,0}X$ denotes the vector bundle of tangent vectors of type $(1,0)$ of X, $\nabla^{g'}$ the Levi-Civita connection corresponding to g', and L_V the Lie differentiation for vector fields with respect to V. For any $V \in \mathfrak{g}_X$, we can prove

$$\sqrt{-1}\mathrm{grad}_g \mathrm{tr}\mathcal{L}_V^{g'} = -V. \tag{3}$$

For a $GL(r, \mathbb{C})$-invariant symmetric polynomial ϕ of degree $(r+l)$ on $\mathfrak{gl}(r, \mathbb{C})$, we define a map $\mathcal{F}_\phi \colon \mathfrak{g}_X \longrightarrow \mathbb{C}$ by

$$\mathcal{F}_\phi(V) := \int_X \phi\left(\mathcal{L}_V^{g'} + \Theta_{g'}\right), \qquad V \in \mathfrak{g}_X,$$

where $\Theta_{g'}$ is the curvature form of g'. At a zero point $p \in X$ of V, $\mathcal{L}_V^{g'}$ induces the linear map

$$\mathcal{L}_{V,p}^{g'} = -\left(L_V\right)_p : T_p^{1,0}X \longrightarrow T_p^{1,0}X. \tag{4}$$

A holomorphic vector field $V \in \mathfrak{g}_X$ on X is called *non-degenerate*, if the following two conditions hold:

(i) The zero set $\mathrm{Zero}(V)$ of V is finite;

(ii) For every zero point $p \in \mathrm{Zero}(V)$ of V, the linear map $\mathcal{L}_{V,p}^{g'}$ is non-singular.

The following localization formula for $\mathcal{F}_\phi$ is fundamental in the construction of our combinatorial formulae in Section 4.

Fact 2.13 (cf. [5], [11] and [15]) *Let X be as above, V a non-degenerate holomorphic vector field on X, and ϕ a $GL(r, \mathbb{C})$-invariant symmetric polynomial of degree $(r+l)$ on $\mathfrak{gl}(r, \mathbb{C})$. Then, we have*

$$\mathcal{F}_\phi(V) = \sum_{p \in \mathrm{Zero}(V)} \frac{\phi\left(\mathcal{L}_{V,p}^{g'}\right)}{|\Gamma_p| \det \frac{\sqrt{-1}}{2\pi}\mathcal{L}_{V,p}^{g'}},$$

where Γ_p is the isotropy group of p (cf. Satake [29]), and $|\Gamma_p|$ denotes the order of Γ_p.

EXAMPLE. The polynomial $c_1^{r+l} := \left(\frac{\sqrt{-1}}{2\pi}\mathrm{tr}\right)^{r+l}$ is a $GL(r,\mathbb{C})$-invariant symmetric polynomial of degree $(r+l)$ on $\mathfrak{gl}(r,\mathbb{C})$. First of all, we consider the case $l=0$. For an arbitrary holomorphic vector field V on X, we have

$$\begin{aligned}\mathcal{F}_{c_1^r}(V) &= \int_X \left(\frac{\sqrt{-1}}{2\pi}\mathrm{tr}\left(\mathcal{L}_V^{g'}+\Theta_{g'}\right)\right)^r \\ &= \int_X \left(\frac{\omega_g}{2\pi}\right)^r = c_1(X)^r[X]. \end{aligned} \tag{5}$$

Note that $\mathcal{F}_{c_1^r}(V)$ does not depend on the choice of a holomorphic vector field V on X. Next, we consider the case $l=1$. For an arbitrary holomorphic vector field V on X, we have

$$\begin{aligned}\mathcal{F}_{c_1^{r+1}}(V) &= \int_X \left(\frac{\sqrt{-1}}{2\pi}\mathrm{tr}\left(\mathcal{L}_V^{g'}+\Theta_{g'}\right)\right)^{r+1} \\ &= \frac{\sqrt{-1}(r+1)}{2\pi}\int_X \mathrm{tr}\mathcal{L}_V^{g'}\left(\frac{\omega_g}{2\pi}\right)^r \\ &= \frac{\sqrt{-1}(r+1)}{2\pi}\int_X (Vf_g)\left(\frac{\omega_g}{2\pi}\right)^r \\ &= -\frac{r+1}{2\pi}F_X(V) \end{aligned} \tag{6}$$

(cf. [14, pp. 22–24]). Moreover, in view of (3), we have

$$\begin{aligned}B_X(V_1,V_2) &= -\int_X \mathrm{tr}\mathcal{L}_{V_1}^{g'}\mathrm{tr}\mathcal{L}_{V_2}^{g'}\left(\frac{\omega_g}{2\pi}\right)^r \\ &\quad -\frac{4\pi^2\mathcal{F}_{c_1^{r+1}}(V_1)\mathcal{F}_{c_1^{r+1}}(V_2)}{(r+1)^2c_1(X)^r[X]} \\ &= \frac{\pi^2(4\mathcal{F}_{c_1^{r+2}}(V_1+V_2)-\mathcal{F}_{c_1^{r+2}}(2V_1)-\mathcal{F}_{c_1^{r+2}}(2V_2))}{(r+2)(r+1)} \\ &\quad -\frac{4\pi^2\mathcal{F}_{c_1^{r+1}}(V_1)\mathcal{F}_{c_1^{r+1}}(V_2)}{(r+1)^2\mathcal{F}_{c_1^r}(V_1)}, \end{aligned} \tag{7}$$

for all $V_1, V_2 \in \mathfrak{g}_X$.

3 Toric Fano orbifolds

In this section, we recall some basic notations and facts concerning toric Fano orbifolds (see [1], [9], [10], [16], [17], or [26] for more detail).

Let $\mathbb{R}_{\geqq 0}$ be the set of non-negative real numbers, r a positive integer, and $T_r := (\mathbb{C}^*)^r$ an r-dimensional algebraic torus. We put $N := \mathbb{Z}^r$ and $M := \mathrm{Hom}_{\mathbb{Z}}(N, \mathbb{Z}) (\cong \mathbb{Z}^r)$, where we regard an element of N and M as an r-dimensional column vector and row vector, respectively. The natural pairing $\langle\, , \rangle : M \times N \to \mathbb{Z}$ is extended to the bilinear form $\langle\, , \rangle : M_{\mathbb{R}} \times N_{\mathbb{R}} \to \mathbb{R}$ where $M_{\mathbb{R}} := M \otimes_{\mathbb{Z}} \mathbb{R}(\cong \mathbb{R}^r)$ and $N_{\mathbb{R}} := N \otimes_{\mathbb{Z}} \mathbb{R}(\cong \mathbb{R}^r)$.

Definition 3.1 (a) *An r-dimensional irreducible complete normal algebraic variety X over $\mathbb{C}$ with at most quotient singularities is called an r-dimensional compact toric orbifold, if T_r acts algebraically on X with an open dense orbit isomorphic to T_r.*

(b) *An r-dimensional compact toric orbifold is called an r-dimensional toric Fano orbifold, if its anti-canonical divisor is an ample Cartier divisor.*

Definition 3.2 *An r-dimensional convex polytope P in $N_{\mathbb{R}}$ is called a Fano polytope if the following three conditions are satisfied:*

(i) *P is an integral polytope, namely, the set $\mathcal{V}(P)$ of vertices of P is contained in $N = \mathbb{Z}^r$;*

(ii) $\mathrm{Int}\,(P) \cap N = \{0\}$, *where* $\mathrm{Int}\,(P)$ *is the interior of P;*

(iii) *P is a simplicial polytope, that is, each face (which is always assumed to be closed) of P is a simplex.*

An r-dimensional Fano polytope P is called non-singular, if P satisfies an additional condition:

(iv) *For each facet F of P, i.e., F is a codimension one face of P, let $a_1(F), \ldots, a_r(F)$ be its vertices. Then $\{a_1(F), \ldots, a_r(F)\}$ forms a $\mathbb{Z}$-basis of N.*

Let P be an r-dimensional Fano polytope in $N_{\mathbb{R}}$. For each $(k-1)$-dimensional face E of P, let $b_1, b_2, \ldots, b_k$ be its vertices, and we put

$$\sigma(E) := \mathbb{R}_{\geqq 0} b_1 + \mathbb{R}_{\geqq 0} b_2 + \cdots + \mathbb{R}_{\geqq 0} b_k,$$
$$\Delta_P(k) := \{\sigma(E)\, ;\, E \text{ is a } (k-1)\text{-dimensional face of } P\},$$
$$\Delta_P(0) := \{0\},$$
$$\Delta_P := \bigcup_{k=0}^{r} \Delta_P(k).$$

Then $\sigma(E)$ is a *strongly convex rational polyhedral cone* in $N_{\mathbb{R}}$ (see [26, p. 1]) and Δ_P is a *fan* of N (see [26, p. 2]). The following fact is fundamental in the study of toric Fano orbifolds.

Fact 3.3 (cf. [26]) (a) *For each r-dimensional Fano polytope P in $N_{\mathbb{R}}$, there exists a unique r-dimensional toric Fano orbifold X_P satisfying the following*:

(i) *To each $(k-1)$-dimensional face σ of P, $0 \leqq k \leqq r$, there corresponds a unique $(r-k)$-dimensional T_r-orbit, denoted by $\mathfrak{O}(\sigma)$, such that* $X_P = \bigcup_{\sigma \in \Delta_P} \mathfrak{O}(\sigma)$;

(ii) *For each $\sigma \in \Delta_P(k)$, $0 \leqq k \leqq r$, the closure $V(\sigma)$ of $\mathfrak{O}(\sigma)$ in X_P is an irreducible normal $(r-k)$-dimensional T_r-invariant subvariety of X_P of the form $V(\sigma) = \bigcup_{\sigma \prec \tau} \mathfrak{O}(\tau)$, where $\sigma \prec \tau$ means that σ is a face of τ (see* [26, p. 2]).

(b) *Every r-dimensional toric Fano orbifold X is T_r-equivariantly isomorphic to X_P for some r-dimensional Fano polytope P in $N_{\mathbb{R}}$.*

(c) *X_P is non-singular if and only if P is non-singular.*

We now explain the outline of the construction of X_P from P in Fact 3.3. Let P be an r-dimensional Fano polytope in $N_{\mathbb{R}}$. For each facet F of P, let

$$a_1(F) = \begin{pmatrix} a_1^1(F) \\ a_1^2(F) \\ \vdots \\ a_1^r(F) \end{pmatrix}, a_2(F) = \begin{pmatrix} a_2^1(F) \\ a_2^2(F) \\ \vdots \\ a_2^r(F) \end{pmatrix}, \ldots, a_r(F) = \begin{pmatrix} a_r^1(F) \\ a_r^2(F) \\ \vdots \\ a_r^r(F) \end{pmatrix}$$

be its vertices, and we put

$$A(F) := \left(a_i^j(F)\right) = (a_1(F), a_2(F), \ldots, a_r(F)) \in GL(r, \mathbb{R}).$$

Then X_P is expressible as

$$X_P = \bigcup_{F \in \Xi(P)} \widetilde{U}_F / \Lambda_F, \qquad \widetilde{U}_F = \mathbb{C}^r,$$

where Λ_F is a finite group of order $\gamma(F) := |\det A(F)|$ acting on $\widetilde{U}_F = \mathbb{C}^r$ (cf. [26, Corollary 1.16]), and $\Xi(P)$ is the set of facets of P. Let $\left(t^1, t^2, \ldots, t^r\right)$ and $\left(z^1(F), z^2(F), \ldots, z^r(F)\right)$ be the standard coordinate system on $\mathfrak{O}(\{0\}) = T_r = (\mathbb{C}^*)^r$ and $\widetilde{U}_F = \mathbb{C}^r$, respectively. The following system of identities is the coordinate transformation between these coordinates:

$$t^i = \prod_{j=1}^{r} z^j(F)^{a_j^i(F)}, \qquad i = 1, 2, \ldots, r. \tag{8}$$

REMARK. In [9], Delzant proved that every toric Fano orbifold can be constructed as a symplectic quotient of some symplectic manifold by a compact torus action (see also [1] and [16]). Therefore, every toric Fano orbifold has positive first Chern class.

In order to consider the Conjectures 1.2, 1.3, 2.11 and 2.12 for toric Fano orbifolds, we prepare some notations. We define a basis

$$\left\{\tau_i := t^i \frac{\partial}{\partial t^i}\,;\, i = 1, 2, \ldots, r\right\}, \tag{9}$$

of the Lie algebra $\mathfrak{t}_r$ of T_r, where we regard $\mathfrak{t}_r$ as a complex Lie subalgebra of the Lie algebra $\mathfrak{g}_P$ of holomorphic vector fields on X_P. For $n = {}^t\left(a^1, a^2, \ldots, a^r\right) \in N_{\mathbb{R}}$, we define a holomorphic vector field $\varepsilon_n \in \mathfrak{t}_r$ by

$$\varepsilon_n := \sum_{k=1}^{r} a^k \tau_k.$$

Let P be an r-dimensional Fano polytope in $N_{\mathbb{R}}$, and we now put

$$R(P) := \left\{\alpha \in M;\ \begin{array}{l}\langle \alpha, v(\alpha)\rangle = 1 \text{ for some } v(\alpha) \in \mathcal{V}(P) \text{ and}\\ \langle \alpha, v\rangle \leqq 0 \text{ for all } v \in \mathcal{V}(P) \setminus \{v(\alpha)\}\end{array}\right\}, \tag{10}$$

$$R_s(P) := R(P) \cap (-R(P)), \tag{11}$$

$$R_u(P) := R(P) \setminus R_s(P), \tag{12}$$

where $\mathcal{V}(P)$ is the set of vertices of P. For

$$\alpha = (b_1(\alpha), b_2(\alpha), \ldots, b_r(\alpha)) \in R(P),$$

we write $v(\alpha)$ in (10) as

$$v(\alpha) = {}^t\left(a^1(\alpha), a^2(\alpha), \ldots, a^r(\alpha)\right) \in N,$$

and we define a holomorphic vector field δ_α on X_P by

$$\delta_\alpha := \left(\prod_{k=1}^{r} \left(t^k\right)^{-b_k(\alpha)}\right) \varepsilon_{v(\alpha)} = \left(\prod_{k=1}^{r} \left(t^k\right)^{-b_k(\alpha)}\right) \sum_{i=1}^{r} a^i(\alpha)\tau_i,$$

in terms of the standard coordinate system $\left(t^1, t^2, \ldots, t^r\right)$ on $T_r = \mathfrak{O}(\{0\})$, and the basis $\{\tau_1, \tau_2, \ldots, \tau_r\}$ defined by (9) (cf. [26, p. 137]). For every $\alpha \in R(P)$ and $n \in N_{\mathbb{R}}$, we can prove

$$[\varepsilon_n, \delta_\alpha] = -\langle \alpha, n\rangle \delta_\alpha. \tag{13}$$

The following result was proved by Demazure for non-singular toric Fano manifolds, and by Cox for toric Fano orbifolds.

Fact 3.4 (Demazure [10] and Cox [8]) *Let P be an r-dimensional Fano polytope, X_P the corresponding r-dimensional toric Fano orbifold, $\mathfrak{g}_P$ the Lie algebra of holomorphic vector fields on X_P, and $\mathfrak{g}_P = \mathfrak{k}_P^{\mathbb{C}} + \mathfrak{u}_P$ the corresponding Chevalley decomposition. Then $\mathfrak{k}_P^{\mathbb{C}}$ is a reductive Lie algebra with $\mathfrak{t}_r$ as a maximal torus and with $R_s(P)$ as the root system, and can be expressed as*

$$\mathfrak{k}_P^{\mathbb{C}} = \mathfrak{t}_r \oplus \left(\bigoplus_{\alpha \in R_s(P)} \mathbb{C}\delta_\alpha \right).$$

On the other hand, $\mathfrak{u}_P$ can be written as

$$\mathfrak{u}_P = \bigoplus_{\alpha \in R_u(P)} \mathbb{C}\delta_\alpha .$$

From Facts 2.2, 2.6, 2.7, and 2.8, we obtain $\sqrt{-1}V_g' = \varepsilon_{e(P)}$, where $e(P) \in N_{\mathbb{R}}$ is defined by

$$e(P) := \begin{pmatrix} B_{X_P}(\tau_1, \tau_1) & \dots & B_{X_P}(\tau_1, \tau_r) \\ \vdots & \dots & \vdots \\ B_{X_P}(\tau_r, \tau_1) & \dots & B_{X_P}(\tau_r, \tau_r) \end{pmatrix}^{-1} \begin{pmatrix} \sqrt{-1}F_{X_P}(\tau_1) \\ \vdots \\ \sqrt{-1}F_{X_P}(\tau_r) \end{pmatrix}. \tag{14}$$

In view of (13) and Fact 3.4, we can interpret Conjectures 2.11 and 2.12 for toric Fano orbifolds as the following two conjectures.

Conjecture 3.5 *Let P be an r-dimensional Fano polytope. Then*

$$R_s(P) = \{\alpha \in R(P)\,;\, \langle \alpha, e(P) \rangle = 0\}.$$

Conjecture 3.6 *Let P be an r-dimensional Fano polytope. Then*

$$R_u(P) = \{\alpha \in R(P)\,;\, \langle \alpha, e(P) \rangle < 0\}.$$

4 Combinatorial formulae for Futaki characters and generalized Killing forms

In this section, we shall establish combinatorial formulae for Futaki characters and generalized Killing forms of toric Fano orbifolds. Throughout this section, we shall use the notations introduced in Sections 2 and 3.

We here fix, once for all, an r-dimensional Fano polytope P in $N_{\mathbb{R}}$. For each facet $F \in \Xi(P)$ of P, and $i = 1, 2, \dots, r$, we define a holomorphic vector field $\zeta_i(F)$ on $\widetilde{U}_F$ by

$$\zeta_i(F) := z^i(F) \frac{\partial}{\partial z^i(F)},$$

in terms of the standard coordinate system $\left(z^1(F), \ldots, z^r(F)\right)$ on $\widetilde{U}_F = \mathbb{C}^r$. From (8), we obtain

$$(\tau_1, \tau_2, \ldots, \tau_r) = (\zeta_1(F), \zeta_2(F), \ldots, \zeta_r(F))\, A(F)^{-1}. \tag{15}$$

For $S \in GL(r, \mathbb{C})$ and $F \in \Xi(P)$, we put

$$Q(S; F) = \left(q^i_j(S; F)\right) := A(F)^{-1} S \in GL(r, \mathbb{C}).$$

A non-singular matrix $S \in GL(r, \mathbb{C})$ is called *non-degenerate*, if S satisfies

$(*)$ $\quad q^i_j(S; F) \neq 0, \quad$ for all $i, j = 1, 2, \ldots, r$, and $F \in \Xi(P)$.

On the other hand, if S satisfies

$(**)$ $\quad q^i_j(S; F) + q^i_k(S; F) \neq 0, \quad$ for all $i, j, k = 1, 2, \ldots, r$, and $F \in \Xi(P)$,

then S is called *strictly non-degenerate*,

REMARK. In the condition $(**)$ above, put $k = j$. Then, for all $i, j = 1, 2, \ldots, r$, we obtain $2q^i_j(S; F) \neq 0$. Therefore, if a non-singular matrix $S \in GL(r, \mathbb{C})$ is strictly non-degenerate, then S is also non-degenerate.

EXAMPLE. For example, a non-singular matrix

$$S_0 := \begin{pmatrix} 1 & 1 & \ldots & 1 \\ \pi & \pi^2 & \ldots & \pi^r \\ \pi^2 & \pi^4 & \ldots & \pi^{2r} \\ \vdots & \vdots & \ldots & \vdots \\ \pi^{r-1} & \pi^{2(r-1)} & \ldots & \pi^{r(r-1)} \end{pmatrix}$$

satisfies the strictly non-degenerate condition $(**)$.

For a non-degenerate matrix $S = \left(s^i_j\right) \in GL(r, \mathbb{C})$, and $i = 1, 2 \ldots, r$, we define a holomorphic vector field $V_i(S)$ by

$$V_i(S) := \sum_{k=1}^{r} s^k_i \tau_k.$$

Then, from (15), we obtain

$$\begin{aligned} (V_1(S), V_2(S), \ldots, V_r(S)) &= (\tau_1, \tau_2, \ldots, \tau_r)\, S \\ &= (\zeta_1(F), \zeta_2(F), \ldots, \zeta_r(F))\, Q(S; F), \end{aligned}$$

for every facet $F \in \Xi(P)$ of P, that is,

$$V_i(S) = \sum_{k=1}^{r} q_i^k(S;F) z^k(F) \frac{\partial}{\partial z^k(F)}, \qquad i = 1,2,\ldots,r. \tag{16}$$

From (16) and the non-degenerate condition $(*)$, we obtain

$$\mathrm{Zero}\,(V_i(S)) = \left\{ \text{The origin } o(F) \text{ of } \widetilde{U}_F = \mathbb{C}^r;\ F \in \Xi(P) \right\}, \tag{17}$$

for $i = 1,2,\ldots,r$. Moreover, we have $\Gamma_{o(F)} = \Lambda_F$ for any $F \in \Xi(P)$, where $\Gamma_{o(F)}$ is the isotropy group of $o(F)$ (cf. Fact 2.13), and we also have

$$\left[V_i(S), \frac{\partial}{\partial z^j(F)} \right]_{o(F)} = -q_i^j(S;F) \left(\frac{\partial}{\partial z^j(F)} \right)_{o(F)}, \tag{18}$$

for all $i,j = 1,2,\ldots,r$, and $F \in \Xi(P)$. Hence, for $i = 1,2,\ldots,r$, (18) allows us to express the linear map $\mathcal{L}^{g'}_{V_i(S),o(F)} : T^{1,0}_{o(F)} X_P \longrightarrow T^{1,0}_{o(F)} X_P$ defined by (4) as

$$\mathcal{L}^{g'}_{V_i(S),o(F)} = \begin{pmatrix} q_i^1(S;F) & & & 0 \\ & q_i^2(S;F) & & \\ & & \ddots & \\ 0 & & & q_i^r(S;F) \end{pmatrix}, \tag{19}$$

with respect to a basis $\left\{ \left(\frac{\partial}{\partial z^1(F)} \right)_{o(F)}, \ldots, \left(\frac{\partial}{\partial z^r(F)} \right)_{o(F)} \right\}$ of $T^{1,0}_{o(F)} X_P$.

REMARK. From the expression (19) above, the non-degenerate condition $(*)$ means that $V_i(S)$ is a non-degenerate holomorphic vector field on X_P for all $i = 1,2,\ldots,r$, and the strictly non-degenerate condition $(**)$ means that $V_i(S) + V_j(S)$ is a non-degenerate holomorphic vector field on X_P for all $i,j = 1,2,\ldots,r$.

Therefore, from Fact 2.13, (17), and (19), we obtain the following combinatorial formula for $\mathcal{F}_{c_1^{r+l}}$, on a toric Fano orbifold X_P.

Theorem 4.1 *Let P be an r-dimensional Fano polytope, and $S \in GL(r,\mathbb{C})$ a non-degenerate matrix. Then, for $i = 1,2,\ldots,r$, and $l = 0,1,2,\ldots$,*

$$\mathcal{F}_{c_1^{r+l}}(V_i(S)) = \left(\frac{\sqrt{-1}}{2\pi} \right)^l \sum_{F \in \Xi(P)} \frac{\left(\sum_{k=1}^{r} q_i^k(S;F) \right)^{r+l}}{\gamma(F) \prod_{k=1}^{r} q_i^k(S;F)}.$$

In view of (5), (6), and (7), we obtain the following three corollaries:

Corollary 4.2 *Let P be an r-dimensional Fano polytope, and $S \in GL(r, \mathbb{C})$ a non-degenerate matrix. Then, for $i = 1, 2, \ldots, r$,*

$$c_1(X_P)^r[X_P] = \sum_{F \in \Xi(P)} \frac{\left(\sum_{k=1}^{r} q_i^k(S;F)\right)^r}{\gamma(F) \prod_{k=1}^{r} q_i^k(S;F)}.$$

Corollary 4.3 *Let P be an r-dimensional Fano polytope, and $S \in GL(r, \mathbb{C})$ a non-degenerate matrix. Then, for $i = 1, 2, \ldots, r$,*

$$\sqrt{-1}F_{X_P}(V_i(S)) = \frac{1}{r+1} \sum_{F \in \Xi(P)} \frac{\left(\sum_{k=1}^{r} q_i^k(S;F)\right)^{r+1}}{\gamma(F) \prod_{k=1}^{r} q_i^k(S;F)}.$$

Corollary 4.4 *Let P be an r-dimensional Fano polytope, and $S \in GL(r, \mathbb{C})$ a strictly non-degenerate matrix. Then,*

$$\begin{aligned} &B_{X_P}(V_i(S), V_j(S)) = \\ &-\frac{1}{(r+1)(r+2)} \left(\sum_{F \in \Xi(P)} \frac{\left(\sum_{k=1}^{r} (q_i^k(S;F) + q_j^k(S;F))\right)^{r+2}}{\gamma(F) \prod_{k=1}^{r} (q_i^k(S;F) + q_j^k(S;F))} \right) \\ &+\frac{1}{(r+1)(r+2)} \left(\sum_{F \in \Xi(P)} \frac{\left(\sum_{k=1}^{r} q_i^k(S;F)\right)^{r+2}}{\gamma(F) \prod_{k=1}^{r} (q_i^k(S;F))} \right) \\ &+\frac{1}{(r+1)(r+2)} \left(\sum_{F \in \Xi(P)} \frac{\left(\sum_{k=1}^{r} q_j^k(S;F)\right)^{r+2}}{\gamma(F) \prod_{k=1}^{r} (q_j^k(S;F))} \right) \end{aligned}$$

$$
+\frac{\left(\sum_{F\in\Xi(P)}\frac{\left(\sum_{k=1}^{r}q_i^k(S;F)\right)^{r+1}}{\gamma(F)\prod_{k=1}^{r}q_i^k(S;F)}\right)\left(\sum_{F\in\Xi(P)}\frac{\left(\sum_{k=1}^{r}q_j^k(S;F)\right)^{r+1}}{\gamma(F)\prod_{k=1}^{r}q_j^k(S;F)}\right)}{(r+1)^2\sum_{F\in\Xi(P)}\frac{\left(\sum_{k=1}^{r}q_i^k(S;F)\right)^{r}}{\gamma(F)\prod_{k=1}^{r}q_i^k(S;F)}},
$$

for $i, j = 1, 2, \ldots, r$.

REMARK. Let P be an r-dimensional Fano polytope, and $S \in GL(r, \mathbb{C})$ a strictly non-degenerate matrix. Then we have the following relations:

$$
\begin{aligned}
&(F_{X_P}(\tau_1), F_{X_P}(\tau_2), \ldots, F_{X_P}(\tau_r)) = \\
&\qquad (F_{X_P}(V_1(S)), F_{X_P}(V_2(S)), \ldots, F_{X_P}(V_r(S)))\, S^{-1}, \\
&\begin{pmatrix} B_{X_P}(\tau_1,\tau_1) & \ldots & B_{X_P}(\tau_1,\tau_r) \\ \vdots & \ldots & \vdots \\ B_{X_P}(\tau_r,\tau_1) & \ldots & B_{X_P}(\tau_r,\tau_r) \end{pmatrix} = \\
&\qquad {}^tS^{-1} \begin{pmatrix} B_{X_P}(V_1(S),V_1(S)) & \ldots & B_{X_P}(V_1(S),V_r(S)) \\ \vdots & \ldots & \vdots \\ B_{X_P}(V_r(S),V_1(S)) & \ldots & B_{X_P}(V_r(S),V_r(S)) \end{pmatrix} S^{-1}.
\end{aligned}
$$

In view of (14), Corollaries 4.3, 4.4, and the relations above, $e(P)$ can be explicitly calculated.

5 Applications

In this section, we shall consider Conjectures 1.3, 3.5, and 3.6 for low dimensional cases, that is, two, three and four dimensional cases.

First of all, we consider the two-dimensional case. Two-dimensional toric Fano orbifolds were classified by Batyrev [3] and Koelman [19], independently, i.e., there exist exactly 16 mutually non-isomorphic two-dimensional toric Fano orbifolds. According to this classification, we can calculate $R_s(P)$, $R_u(P)$, and $e(P)$, which are defined by (11), (12), and (14), respectively, for all two-dimensional Fano polytopes. We carried out our calculations by means of

Risa/Asir[a] on the Linux system. Therefore, we obtain Table 1 in Section 6. From Table 1, we can prove Conjecture 1.3 for two-dimensional case affirmatively.

Theorem 5.1 *For a two-dimensional toric Fano orbifold X, the following are equivalent:*

(i) *The Futaki character F_X of X vanishes;*

(ii) *X admits an Einstein-Kähler metric.*

Proof. From Table 1, we obtain exactly five two-dimensional Fano polytopes, which are associated with two-dimensional toric Fano orbifolds with vanishing Futaki characters, that is, two-dimensional Fano polytopes of type A-3, A-5, B-1, B-2, and D-1 in Table 1. The polytopes of type A-3, A-5, B-1, and B-2 correspond to $\mathbb{P}^2(\mathbb{C})$, $\mathbb{P}^2(\mathbb{C})/\mathbb{Z}_3$, $\mathbb{P}^1(\mathbb{C}) \times \mathbb{P}^1(\mathbb{C})$, and $(\mathbb{P}^1(\mathbb{C}) \times \mathbb{P}^1(\mathbb{C}))/\mathbb{Z}_2$, respectively. Here the action of $\mathbb{Z}_3$ on $\mathbb{P}^2(\mathbb{C})$ is generated by the automorphism

$$\mathbb{P}^2(\mathbb{C}) \ni [z_0 : z_1 : z_2] \mapsto \left[z_0 : e^{\frac{2\pi\sqrt{-1}}{3}} z_1 : e^{\frac{4\pi\sqrt{-1}}{3}} z_2\right] \in \mathbb{P}^2(\mathbb{C})$$

of $\mathbb{P}^2(\mathbb{C})$, and that of $\mathbb{Z}_2$ on $\mathbb{P}^1(\mathbb{C}) \times \mathbb{P}^1(\mathbb{C})$ is generated by

$$\begin{aligned}\mathbb{P}^1(\mathbb{C}) \times \mathbb{P}^1(\mathbb{C}) \ni ([z_0 : z_1], [w_0 : w_1]) \mapsto \\ ([z_0 : -z_1], [w_0 : -w_1]) \in \mathbb{P}^1(\mathbb{C}) \times \mathbb{P}^1(\mathbb{C}).\end{aligned}$$

All of these two-dimensional toric Fano orbifolds have an Einstein-Kähler orbifold metric defined from Fubini-Study metrics. Therefore, the remaining case is the polytope of type D-1. The two-dimensional Fano polytope of type D-1 corresponds to the non-singular projective algebraic surface obtained from $\mathbb{P}^2(\mathbb{C})$ by blowing up three points, on which Siu [31], Tian and Yau [33], and Nadel [23] proved the existence of an Einstein-Kähler metric. This completes the proof of Theorem 5.1. □

Moreover, we can also prove Conjectures 3.5 and 3.6 for two-dimensional Fano polytopes.

Theorem 5.2 *Let P be a two-dimensional Fano polytope. Then*

$$\begin{aligned} R_s(P) &= \{\alpha \in R(P) ; \langle \alpha, e(P) \rangle = 0\}, \\ R_u(P) &= \{\alpha \in R(P) ; \langle \alpha, e(P) \rangle < 0\}. \end{aligned}$$

[a] Risa/Asir is an experimental computer algebra system developed at Fujitsu Laboratories Limited.

Proof. The theorem is straightforward from Table 1. □

Non-singular three-dimensional toric Fano manifolds were classified by Batyrev [2], and K. Watanabe and M. Watanabe [34], independently. Furthermore, Batyrev also classified non-singular four-dimensional toric Fano manifolds in [4]. Hence, we have exactly 18 and 124 isomorphic classes of non-singular three-dimensional and four-dimensional toric Fano manifolds, respectively (see also [30]). In view of these classifications, we obtain Tables 2 and 3 in Section 6, by calculating $R_s(P)$, $R_u(P)$, and $e(P)$. In these calculations, we also used Risa/Asir. As for Conjecture 1.3 for non-singular toric Fano manifolds of dimension $r \leqq 4$, the author proved the following fact in [25], by using results of Sakane [27], Mabuchi [20], Siu [31], Tian and Yau [33], and Nadel [23].

Fact 5.3 ([25]) *For a non-singular toric Fano manifold X of dimension $r \leqq 4$ except the four-dimensional one of type* X-1 *in Table* 3,[b] *the following are equivalent*:

(i) *The Futaki character F_X of X vanishes*;

(ii) *X admits an Einstein-Kähler metric.*

Finally, we consider Conjectures 3.5 and 3.6 for non-singular Fano polytopes of dimension $r \leqq 4$. The following theorem follows from Tables 1, 2, and 3, immediately.

Theorem 5.4 *Let P be a non-singular Fano polytope of dimension $r \leqq 4$. Then*

$$\begin{aligned} R_s(P) &= \{\alpha \in R(P)\,;\, \langle \alpha, e(P)\rangle = 0\}\,, \\ R_u(P) &= \{\alpha \in R(P)\,;\, \langle \alpha, e(P)\rangle < 0\}\,. \end{aligned}$$

REMARK. Conjecture 3.5 for all non-singular three-dimensional Fano polytopes was originally proved by Sakuma in [28].

6 Tables

[b]In [25], the author classified Einstein-Kähler toric Fano fourfolds by using Batyrev's classification of toric Fano fourfolds ([4]). However, Sato informed the author in June, 1998 that he found a new toric Fano fourfold which is not contained in Batyrev's list (see [30]). This new toric Fano fourfold is the one of type X-1 in Table 3 and has vanishing Futaki character. The author does not know whether it admits an Einstein-Kähler metric.

Table 1: Two-dimensional Fano polytopes

P	$\mathcal{V}(P)$	$R_s(P)$	$R_u(P)$	$e(P)$
A-1	$\begin{pmatrix}1\\-1\end{pmatrix}, \begin{pmatrix}0\\1\end{pmatrix}, \begin{pmatrix}-1\\-1\end{pmatrix}$	$\pm(1, 0)$	$(1, 1)$, $(0, 1)$, $(-1, 1)$	$\begin{pmatrix}0\\-\frac{3}{2}\end{pmatrix}$
A-2	$\begin{pmatrix}2\\-1\end{pmatrix}, \begin{pmatrix}-1\\1\end{pmatrix}, \begin{pmatrix}-1\\-1\end{pmatrix}$		$(1, 2)$, $(1, 1)$, $(0, 1)$	$\begin{pmatrix}\frac{2}{3}\\-\frac{4}{3}\end{pmatrix}$
A-3	$\begin{pmatrix}1\\0\end{pmatrix}, \begin{pmatrix}0\\1\end{pmatrix}, \begin{pmatrix}-1\\-1\end{pmatrix}$	$\pm(1, 0)$, $\pm(0, 1)$, $\pm(1, -1)$		$\begin{pmatrix}0\\0\end{pmatrix}$
A-4	$\begin{pmatrix}3\\-1\end{pmatrix}, \begin{pmatrix}-1\\1\end{pmatrix}, \begin{pmatrix}-1\\-1\end{pmatrix}$		$(0, 1)$	$\begin{pmatrix}\frac{3}{2}\\-\frac{3}{2}\end{pmatrix}$
A-5	$\begin{pmatrix}2\\-1\end{pmatrix}, \begin{pmatrix}-1\\2\end{pmatrix}, \begin{pmatrix}-1\\-1\end{pmatrix}$			$\begin{pmatrix}0\\0\end{pmatrix}$
B-1	$\begin{pmatrix}1\\0\end{pmatrix}, \begin{pmatrix}-1\\1\end{pmatrix}, \begin{pmatrix}-1\\0\end{pmatrix}, \begin{pmatrix}1\\-1\end{pmatrix}$	$\pm(1, 1)$, $\pm(0, 1)$		$\begin{pmatrix}0\\0\end{pmatrix}$
B-2	$\begin{pmatrix}1\\1\end{pmatrix}, \begin{pmatrix}-1\\1\end{pmatrix}, \begin{pmatrix}-1\\-1\end{pmatrix}, \begin{pmatrix}1\\-1\end{pmatrix}$			$\begin{pmatrix}0\\0\end{pmatrix}$
B-3	$\begin{pmatrix}1\\0\end{pmatrix}, \begin{pmatrix}0\\1\end{pmatrix}, \begin{pmatrix}-1\\-1\end{pmatrix}, \begin{pmatrix}1\\-1\end{pmatrix}$		$(0, 1)$, $(-1, 0)$, $(-1, 1)$	$\begin{pmatrix}\frac{456}{2113}\\-\frac{2496}{2113}\end{pmatrix}$
B-4	$\begin{pmatrix}1\\1\end{pmatrix}, \begin{pmatrix}0\\1\end{pmatrix}, \begin{pmatrix}-1\\-1\end{pmatrix}, \begin{pmatrix}1\\-1\end{pmatrix}$		$(-1, 1)$, $(-1, 0)$	$\begin{pmatrix}\frac{42}{65}\\-\frac{36}{65}\end{pmatrix}$
B-5	$\begin{pmatrix}1\\2\end{pmatrix}, \begin{pmatrix}0\\1\end{pmatrix}, \begin{pmatrix}-1\\-1\end{pmatrix}, \begin{pmatrix}1\\-1\end{pmatrix}$		$(-1, 0)$	$\begin{pmatrix}\frac{648}{529}\\\frac{264}{529}\end{pmatrix}$
B-6	$\begin{pmatrix}1\\2\end{pmatrix}, \begin{pmatrix}-1\\0\end{pmatrix}, \begin{pmatrix}-1\\-1\end{pmatrix}, \begin{pmatrix}1\\-1\end{pmatrix}$			$\begin{pmatrix}\frac{12}{13}\\\frac{6}{13}\end{pmatrix}$
B-7	$\begin{pmatrix}1\\1\end{pmatrix}, \begin{pmatrix}-1\\0\end{pmatrix}, \begin{pmatrix}-1\\-1\end{pmatrix}, \begin{pmatrix}0\\-1\end{pmatrix}$	$\pm(1. -1)$	$(1, 0)$, $(0, 1)$	$\begin{pmatrix}-\frac{6}{11}\\-\frac{6}{11}\end{pmatrix}$
C-1	$\begin{pmatrix}1\\0\end{pmatrix}, \begin{pmatrix}0\\1\end{pmatrix}, \begin{pmatrix}-1\\1\end{pmatrix}, \begin{pmatrix}-1\\0\end{pmatrix}, \begin{pmatrix}0\\-1\end{pmatrix}$		$(0, -1)$, $(1, 0)$	$\begin{pmatrix}-\frac{168}{409}\\\frac{168}{409}\end{pmatrix}$
C-2	$\begin{pmatrix}1\\1\end{pmatrix}, \begin{pmatrix}0\\1\end{pmatrix}, \begin{pmatrix}-1\\0\end{pmatrix}, \begin{pmatrix}-1\\-1\end{pmatrix}, \begin{pmatrix}1\\-1\end{pmatrix}$			$\begin{pmatrix}\frac{120}{253}\\-\frac{120}{253}\end{pmatrix}$
C-3	$\begin{pmatrix}1\\0\end{pmatrix}, \begin{pmatrix}0\\1\end{pmatrix}, \begin{pmatrix}-1\\0\end{pmatrix}, \begin{pmatrix}-1\\-1\end{pmatrix}, \begin{pmatrix}1\\-1\end{pmatrix}$		$(0, 1)$	$\begin{pmatrix}0\\-\frac{36}{37}\end{pmatrix}$
D-1	$\begin{pmatrix}1\\0\end{pmatrix}, \begin{pmatrix}1\\1\end{pmatrix}, \begin{pmatrix}0\\1\end{pmatrix}, \begin{pmatrix}-1\\0\end{pmatrix}, \begin{pmatrix}-1\\-1\end{pmatrix}, \begin{pmatrix}0\\-1\end{pmatrix}$			$\begin{pmatrix}0\\0\end{pmatrix}$

Table 2: Non-singular three-dimensional Fano polytopes

P	$\mathcal{V}(P)$	$R_s(P)$	$R_u(P)$	$e(P)$
A-1	$\begin{pmatrix}1\\0\\0\end{pmatrix}, \begin{pmatrix}0\\1\\0\end{pmatrix}, \begin{pmatrix}0\\0\\1\end{pmatrix}, \begin{pmatrix}-1\\-1\\-1\end{pmatrix}$	$\pm(1,0,0)$, $\pm(1,0,-1)$, $\pm(1,-1,0)$, $\pm(0,1,0)$, $\pm(0,1,-1)$, $\pm(0,0,1)$		$\begin{pmatrix}0\\0\\0\end{pmatrix}$
B-1	$\begin{pmatrix}1\\0\\0\end{pmatrix}, \begin{pmatrix}0\\1\\0\end{pmatrix}, \pm\begin{pmatrix}0\\0\\1\end{pmatrix}, \begin{pmatrix}-1\\-1\\0\end{pmatrix}$	$\pm(1,0,0)$, $\pm(0,1,0)$, $\pm(0,0,1)$, $\pm(1,-1,0)$		$\begin{pmatrix}0\\0\\0\end{pmatrix}$
B-2	$\begin{pmatrix}1\\0\\0\end{pmatrix}, \begin{pmatrix}0\\1\\0\end{pmatrix}, \pm\begin{pmatrix}0\\0\\1\end{pmatrix}, \begin{pmatrix}-1\\-1\\-1\end{pmatrix}$	$\pm(1,0,0)$, $\pm(0,1,0)$, $\pm(1,-1,0)$	$(0,0,1)$, $(0,-1,1)$, $(-1,0,1)$	$\begin{pmatrix}0\\0\\-\frac{70}{97}\end{pmatrix}$
B-3	$\begin{pmatrix}1\\0\\0\end{pmatrix}, \begin{pmatrix}0\\1\\0\end{pmatrix}, \pm\begin{pmatrix}0\\0\\1\end{pmatrix}, \begin{pmatrix}-1\\-1\\-2\end{pmatrix}$	$\pm(1,0,0)$, $\pm(0,1,0)$, $\pm(1,-1,0)$	$(0,0,1)$, $(-1,-1,1)$, $(0,-2,1)$, $(-1,0,1)$, $(-2,0,1)$, $(0,-1,1)$	$\begin{pmatrix}0\\0\\-\frac{620}{349}\end{pmatrix}$
B-4	$\begin{pmatrix}1\\0\\0\end{pmatrix}, \begin{pmatrix}0\\1\\0\end{pmatrix}, \begin{pmatrix}0\\0\\1\end{pmatrix}, \begin{pmatrix}-1\\-1\\0\end{pmatrix}, \begin{pmatrix}-1\\-1\\-1\end{pmatrix}$	$\pm(1,-1,0)$, $\pm(0,0,1)$	$(1,0,0)$, $(1,0,-1)$, $(0,1,0)$, $(0,1,-1)$	$\begin{pmatrix}-\frac{20}{43}\\-\frac{20}{43}\\0\end{pmatrix}$
C-1	$\pm\begin{pmatrix}1\\0\\0\end{pmatrix}, \pm\begin{pmatrix}0\\1\\0\end{pmatrix}, \pm\begin{pmatrix}0\\0\\1\end{pmatrix}$	$\pm(1,0,0)$, $\pm(0,1,0)$, $\pm(0,0,1)$		$\begin{pmatrix}0\\0\\0\end{pmatrix}$
C-2	$\begin{pmatrix}1\\0\\0\end{pmatrix}, \begin{pmatrix}0\\1\\0\end{pmatrix}, \pm\begin{pmatrix}0\\0\\1\end{pmatrix}, \begin{pmatrix}-1\\0\\-1\end{pmatrix}, \begin{pmatrix}0\\-1\\-1\end{pmatrix}$	$\pm(1,0,0)$, $\pm(0,1,0)$	$(0,0,1)$, $(-1,-1,1)$ $(0,-1,1)$, $(-1,0,1)$	$\begin{pmatrix}0\\0\\-\frac{260}{219}\end{pmatrix}$
C-3	$\begin{pmatrix}1\\0\\0\end{pmatrix}, \begin{pmatrix}0\\1\\0\end{pmatrix}, \pm\begin{pmatrix}0\\0\\1\end{pmatrix}, \begin{pmatrix}-1\\0\\-1\end{pmatrix}, \begin{pmatrix}0\\-1\\1\end{pmatrix}$	$\pm(1,0,0)$, $\pm(0,1,0)$		$\begin{pmatrix}0\\0\\0\end{pmatrix}$
C-4	$\begin{pmatrix}1\\0\\0\end{pmatrix}, \pm\begin{pmatrix}0\\1\\0\end{pmatrix}, \pm\begin{pmatrix}0\\0\\1\end{pmatrix}, \begin{pmatrix}-1\\-1\\0\end{pmatrix}$	$\pm(1,0,0)$, $\pm(0,0,1)$	$(-1,1,0)$, $(0,1,0)$	$\begin{pmatrix}0\\-\frac{6}{11}\\0\end{pmatrix}$
C-5	$\begin{pmatrix}1\\0\\0\end{pmatrix}, \begin{pmatrix}0\\1\\0\end{pmatrix}, \pm\begin{pmatrix}0\\0\\1\end{pmatrix}, \begin{pmatrix}-1\\0\\-1\end{pmatrix}, \begin{pmatrix}-1\\-1\\-1\end{pmatrix}$	$\pm(0,1,0)$	$(-1,0,1)$, $(0,-1,1)$, $(0,0,1)$, $(1,-1,0)$, $(1,0,0)$	$\begin{pmatrix}-\frac{7600}{17787}\\0\\-\frac{17750}{17787}\end{pmatrix}$
D-1	$\begin{pmatrix}1\\0\\0\end{pmatrix}, \pm\begin{pmatrix}0\\1\\0\end{pmatrix}, \begin{pmatrix}0\\0\\1\end{pmatrix}, \begin{pmatrix}-1\\-1\\-1\end{pmatrix}, \begin{pmatrix}1\\-1\\0\end{pmatrix}$	$\pm(0,0,1)$	$(-1,0,0)$, $(-1,0,1)$, $(-1,1,0)$, $(0,1,-1)$, $(0,1,0)$	$\begin{pmatrix}\frac{99600}{467581}\\-\frac{627000}{467581}\\0\end{pmatrix}$
D-2	$\begin{pmatrix}1\\0\\0\end{pmatrix}, \pm\begin{pmatrix}0\\1\\0\end{pmatrix}, \begin{pmatrix}0\\0\\1\end{pmatrix}, \begin{pmatrix}-1\\0\\-1\end{pmatrix}, \begin{pmatrix}1\\-1\\0\end{pmatrix}$	$\pm(0,0,1)$	$(0,1,0)$, $(-1,0,0)$, $(-1,0,1)$	$\begin{pmatrix}\frac{219420}{650251}\\-\frac{318320}{650251}\\0\end{pmatrix}$
E-1	$\pm\begin{pmatrix}1\\0\\0\end{pmatrix}, \pm\begin{pmatrix}0\\1\\0\end{pmatrix}, \pm\begin{pmatrix}0\\0\\1\end{pmatrix}, \begin{pmatrix}-1\\-1\\0\end{pmatrix}$	$\pm(0,0,1)$	$(1,0,0)$, $(0,1,0)$	$\begin{pmatrix}-\frac{168}{409}\\-\frac{168}{409}\\0\end{pmatrix}$
E-2	$\pm\begin{pmatrix}1\\0\\0\end{pmatrix}, \pm\begin{pmatrix}0\\1\\0\end{pmatrix}, \begin{pmatrix}0\\0\\1\end{pmatrix}, \begin{pmatrix}-1\\-1\\0\end{pmatrix}, \begin{pmatrix}-1\\0\\-1\end{pmatrix}$	$\pm(0,0,1)$	$(1,0,0)$, $(0,1,0)$, $(1,0,-1)$	$\begin{pmatrix}-\frac{2646160}{2735927}\\-\frac{982960}{2735927}\\0\end{pmatrix}$

P	$\mathcal{V}(P)$	$R_s(P)$	$R_u(P)$	$e(P)$
E-3	$\pm\begin{pmatrix}1\\0\\0\end{pmatrix}, \pm\begin{pmatrix}0\\1\\0\end{pmatrix}, \begin{pmatrix}0\\0\\1\end{pmatrix}, \begin{pmatrix}-1\\-1\\0\end{pmatrix}, \begin{pmatrix}0\\1\\-1\end{pmatrix}$	$\pm(0,0,1)$	$(1,0,0)$	$\begin{pmatrix}-\frac{34208}{78995}\\ \frac{7936}{78995}\\ 0\end{pmatrix}$
E-4	$\pm\begin{pmatrix}1\\0\\0\end{pmatrix}, \pm\begin{pmatrix}0\\1\\0\end{pmatrix}, \begin{pmatrix}0\\0\\1\end{pmatrix}, \begin{pmatrix}-1\\-1\\0\end{pmatrix}, \begin{pmatrix}-1\\-1\\-1\end{pmatrix}$	$\pm(0,0,1)$	$(1,0,0)$, $(1,0,-1)$, $(0,1,-1)$, $(0,1,0)$	$\begin{pmatrix}-\frac{17020}{19651}\\ -\frac{17020}{19651}\\ 0\end{pmatrix}$
F-1	$\pm\begin{pmatrix}1\\0\\0\end{pmatrix}, \pm\begin{pmatrix}0\\1\\0\end{pmatrix}, \pm\begin{pmatrix}0\\0\\1\end{pmatrix}, \pm\begin{pmatrix}-1\\1\\0\end{pmatrix}$	$\pm(0,0,1)$		$\begin{pmatrix}0\\0\\0\end{pmatrix}$
F-2	$\pm\begin{pmatrix}1\\0\\0\end{pmatrix}, \pm\begin{pmatrix}0\\1\\0\end{pmatrix}, \begin{pmatrix}0\\0\\1\end{pmatrix}, \pm\begin{pmatrix}-1\\1\\0\end{pmatrix}, \begin{pmatrix}0\\1\\-1\end{pmatrix}$	$\pm(0,0,1)$		$\begin{pmatrix}0\\ \frac{36}{67}\\ 0\end{pmatrix}$

Table 3: Non-singular four-dimensional Fano polytopes

P	$\mathcal{V}(P)$	$R_s(P)$	$R_u(P)$	$e(P)$
A-1	$\begin{pmatrix}1\\0\\0\\0\end{pmatrix}, \begin{pmatrix}0\\1\\0\\0\end{pmatrix}, \begin{pmatrix}0\\0\\1\\0\end{pmatrix}, \begin{pmatrix}0\\0\\0\\1\end{pmatrix}, \begin{pmatrix}-1\\-1\\-1\\-1\end{pmatrix}$	$\pm(1,0,0,0)$, $\pm(1,0,0,-1)$, $\pm(1,0,-1,0)$, $\pm(1,-1,0,0)$, $\pm(0,1,0,0)$, $\pm(0,1,0,-1)$, $\pm(0,1,-1,0)$, $\pm(0,0,1,0)$, $\pm(0,0,1,-1)$, $\pm(0,0,0,1)$		$\begin{pmatrix}0\\0\\0\\0\end{pmatrix}$
B-1	$\begin{pmatrix}1\\1\\1\\0\end{pmatrix}, \begin{pmatrix}1\\1\\0\\1\end{pmatrix}, \begin{pmatrix}1\\0\\1\\1\end{pmatrix}, \begin{pmatrix}0\\1\\1\\1\end{pmatrix}, \pm\begin{pmatrix}1\\1\\1\\1\end{pmatrix}$	$\pm(1,0,0,-1)$, $\pm(1,0,-1,0)$, $\pm(1,-1,0,0)$, $\pm(0,1,0,-1)$, $\pm(0,1,-1,0)$, $\pm(0,0,1,-1)$	$(-1,-1,-1,2)$, $(-1,-1,0,1)$, $(-1,-1,1,0)$, $(-1,-1,2,-1)$, $(-1,0,-1,1)$, $(-1,0,0,0)$, $(-1,0,1,-1)$, $(-1,1,-1,0)$, $(-1,1,0,-1)$, $(-1,2,-1,-1)$, $(0,-1,-1,1)$, $(0,-1,0,0)$, $(0,-1,1,-1)$, $(0,0,-1,0)$, $(0,0,0,-1)$, $(0,1,-1,-1)$, $(1,-1,-1,0)$, $(1,-1,0,-1)$, $(1,0,-1,-1)$, $(2,-1,-1,-1)$	$\begin{pmatrix}\frac{400500}{108133}\\\frac{400500}{108133}\\\frac{400500}{108133}\\\frac{400500}{108133}\end{pmatrix}$
B-2	$\begin{pmatrix}1\\0\\0\\0\end{pmatrix}, \begin{pmatrix}0\\1\\1\\0\end{pmatrix}, \begin{pmatrix}0\\1\\0\\1\end{pmatrix}, \begin{pmatrix}1\\0\\1\\1\end{pmatrix}, \pm\begin{pmatrix}1\\1\\1\\1\end{pmatrix}$	$\pm(1,1,-1,-1)$, $\pm(1,0,0,-1)$, $\pm(1,0,-1,0)$, $\pm(0,1,0,-1)$, $\pm(0,1,-1,0)$, $\pm(0,0,1,-1)$	$(-2,-1,1,1)$, $(-1,-1,0,1)$, $(-1,-1,1,0)$, $(-1,0,0,0)$, $(0,-1,-1,1)$, $(0,-1,0,0)$, $(0,-1,1,-1)$, $(0,0,-1,0)$, $(0,0,0,-1)$, $(0,1,-1,-1)$	$\begin{pmatrix}\frac{3150}{1577}\\\frac{3150}{1577}\\\frac{3150}{1577}\\\frac{3150}{1577}\end{pmatrix}$
B-3	$\begin{pmatrix}1\\0\\0\\0\end{pmatrix}, \begin{pmatrix}0\\1\\0\\0\end{pmatrix}, \begin{pmatrix}0\\0\\1\\0\end{pmatrix}, \begin{pmatrix}0\\0\\0\\1\end{pmatrix}, \pm\begin{pmatrix}1\\1\\1\\1\end{pmatrix}$	$\pm(1,0,0,-1)$, $\pm(1,0,-1,0)$, $\pm(1,-1,0,0)$, $\pm(0,1,0,-1)$, $\pm(0,1,-1,0)$, $\pm(0,0,1,-1)$	$(-1,0,0,0)$, $(0,-1,0,0)$, $(0,0,-1,0)$, $(0,0,0,-1)$	$\begin{pmatrix}\frac{82620}{101357}\\\frac{82620}{101357}\\\frac{82620}{101357}\\\frac{82620}{101357}\end{pmatrix}$
B-4	$\begin{pmatrix}1\\0\\0\\0\end{pmatrix}, \begin{pmatrix}0\\1\\0\\0\end{pmatrix}, \begin{pmatrix}0\\0\\1\\0\end{pmatrix}, \pm\begin{pmatrix}0\\0\\0\\1\end{pmatrix}, \begin{pmatrix}-1\\-1\\-1\\0\end{pmatrix}$	$\pm(1,0,0,0)$, $\pm(1,0,-1,0)$, $\pm(1,-1,0,0)$, $\pm(0,1,0,0)$, $\pm(0,1,-1,0)$, $\pm(0,0,1,0)$, $\pm(0,0,0,1)$		$\begin{pmatrix}0\\0\\0\\0\end{pmatrix}$
B-5	$\begin{pmatrix}1\\0\\0\\0\end{pmatrix}, \begin{pmatrix}0\\1\\0\\0\end{pmatrix}, \begin{pmatrix}0\\0\\1\\0\end{pmatrix}, \begin{pmatrix}0\\0\\0\\1\end{pmatrix}, \begin{pmatrix}0\\0\\1\\1\end{pmatrix}, \begin{pmatrix}-1\\-1\\-1\\-1\end{pmatrix}$	$\pm(1,0,0,0)$, $\pm(1,-1,0,0)$, $\pm(0,1,0,0)$, $\pm(0,0,1,-1)$	$(0,0,-1,0)$, $(0,0,0,-1)$, $(0,1,-1,0)$, $(0,1,0,-1)$, $(1,0,-1,0)$, $(1,0,0,-1)$	$\begin{pmatrix}0\\0\\\frac{30}{71}\\\frac{30}{71}\end{pmatrix}$

P	$\mathcal{V}(P)$	$R_s(P)$	$R_u(P)$	$e(P)$
C-1	$\begin{pmatrix}1\\0\\0\\0\end{pmatrix}, \begin{pmatrix}0\\1\\0\\0\end{pmatrix}, \begin{pmatrix}0\\0\\1\\0\end{pmatrix}, \begin{pmatrix}0\\0\\0\\1\end{pmatrix}, \begin{pmatrix}-1\\-1\\0\\0\end{pmatrix}, \begin{pmatrix}2\\0\\-1\\-1\end{pmatrix}$	$\pm(0,1,0,0)$, $\pm(0,0,1,0)$, $\pm(0,0,1,-1)$, $\pm(0,0,0,1)$	$(-1,0,-2,0)$, $(-1,0,-1,-1)$, $(-1,0,-1,0)$, $(-1,0,0,-2)$, $(-1,0,0,-1)$, $(-1,0,0,0)$, $(-1,1,-2,0)$, $(-1,1,-1,-1)$, $(-1,1,-1,0)$, $(-1,1,0,-2)$, $(-1,1,0,-1)$, $(-1,1,0,0)$	$\begin{pmatrix}\frac{3740}{2803}\\0\\0\\0\end{pmatrix}$
C-2	$\begin{pmatrix}1\\0\\0\\0\end{pmatrix}, \begin{pmatrix}0\\1\\0\\0\end{pmatrix}, \begin{pmatrix}0\\0\\1\\0\end{pmatrix}, \begin{pmatrix}0\\0\\0\\1\end{pmatrix}, \begin{pmatrix}-1\\-1\\0\\0\end{pmatrix}, \begin{pmatrix}1\\0\\-1\\-1\end{pmatrix}$	$\pm(0,1,0,0)$, $\pm(0,0,1,0)$, $\pm(0,0,1,-1)$, $\pm(0,0,0,1)$	$(-1,0,-1,0)$, $(-1,0,0,-1)$, $(-1,0,0,0)$, $(-1,1,-1,0)$, $(-1,1,0,-1)$, $(-1,1,0,0)$	$\begin{pmatrix}\frac{3040}{4961}\\0\\0\\0\end{pmatrix}$
C-3	$\begin{pmatrix}1\\0\\0\\0\end{pmatrix}, \begin{pmatrix}0\\1\\0\\0\end{pmatrix}, \begin{pmatrix}0\\0\\1\\0\end{pmatrix}, \begin{pmatrix}0\\0\\0\\1\end{pmatrix}, \begin{pmatrix}-1\\-1\\0\\0\end{pmatrix}, \begin{pmatrix}1\\1\\-1\\-1\end{pmatrix}$	$\pm(1,-1,0,0)$, $\pm(0,0,1,0)$, $\pm(0,0,1,-1)$, $\pm(0,0,0,1)$	$(-1,0,-1,0)$, $(-1,0,0,-1)$, $(-1,0,0,0)$, $(0,-1,-1,0)$, $(0,-1,0,-1)$, $(0,-1,0,0)$	$\begin{pmatrix}\frac{2660}{2921}\\\frac{2660}{2921}\\0\\0\end{pmatrix}$
C-4	$\begin{pmatrix}1\\0\\0\\0\end{pmatrix}, \begin{pmatrix}0\\1\\0\\0\end{pmatrix}, \begin{pmatrix}0\\0\\1\\0\end{pmatrix}, \begin{pmatrix}0\\0\\0\\1\end{pmatrix}, \begin{pmatrix}-1\\-1\\0\\0\end{pmatrix}, \begin{pmatrix}0\\0\\-1\\-1\end{pmatrix}$	$\pm(1,0,0,0)$, $\pm(1,-1,0,0)$, $\pm(0,1,0,0)$, $\pm(0,0,1,0)$, $\pm(0,0,1,-1)$, $\pm(0,0,0,1)$		$\begin{pmatrix}0\\0\\0\\0\end{pmatrix}$
D-1	$\begin{pmatrix}1\\0\\0\\0\end{pmatrix}, \begin{pmatrix}0\\1\\0\\0\end{pmatrix}, \begin{pmatrix}0\\0\\1\\0\end{pmatrix}, \pm\begin{pmatrix}0\\0\\0\\1\end{pmatrix}, \begin{pmatrix}-1\\-1\\0\\2\end{pmatrix}, \begin{pmatrix}0\\0\\-1\\1\end{pmatrix}$	$\pm(1,0,0,0)$, $\pm(1,-1,0,0)$, $\pm(0,1,0,0)$, $\pm(0,0,1,0)$	$(-2,0,-1,-1)$, $(-2,0,0,-1)$, $(-1,-1,-1,-1)$, $(-1,-1,0,-1)$, $(-1,0,-1,-1)$, $(-1,0,0,-1)$, $(0,-2,-1,-1)$, $(0,-2,0,-1)$, $(0,-1,-1,-1)$, $(0,-1,0,-1)$, $(0,0,-1,-1)$, $(0,0,0,-1)$	$\begin{pmatrix}0\\0\\0\\\frac{21830}{8057}\end{pmatrix}$
D-2	$\begin{pmatrix}1\\0\\0\\0\end{pmatrix}, \begin{pmatrix}0\\1\\0\\0\end{pmatrix}, \begin{pmatrix}0\\0\\1\\0\end{pmatrix}, \pm\begin{pmatrix}0\\0\\0\\1\end{pmatrix}, \begin{pmatrix}-1\\-1\\2\\0\end{pmatrix}, \begin{pmatrix}0\\0\\-1\\1\end{pmatrix}$	$\pm(1,0,0,0)$, $\pm(0,1,0,0)$, $\pm(1,-1,0,0)$	$(-2,0,-1,-1)$, $(-2,0,-1,0)$, $(-1,-1,-1,-1)$, $(-1,-1,-1,0)$, $(-1,0,-1,-1)$, $(-1,0,-1,0)$, $(0,-2,-1,-1)$, $(0,-2,-1,0)$, $(0,-1,-1,-1)$, $(0,-1,-1,0)$, $(0,0,-1,-1)$, $(0,0,-1,0)$, $(0,0,0,-1)$	$\begin{pmatrix}0\\0\\\frac{487160}{376369}\\\frac{196020}{376369}\end{pmatrix}$
D-3	$\begin{pmatrix}1\\0\\0\\0\end{pmatrix}, \begin{pmatrix}0\\1\\0\\0\end{pmatrix}, \begin{pmatrix}0\\0\\1\\0\end{pmatrix}, \pm\begin{pmatrix}0\\0\\0\\1\end{pmatrix}, \begin{pmatrix}-1\\-1\\1\\1\end{pmatrix}, \begin{pmatrix}0\\0\\-1\\1\end{pmatrix}$	$\pm(1,0,0,0)$, $\pm(0,1,0,0)$, $\pm(1,-1,0,0)$	$(-2,0,-1,-1)$, $(-1,-1,-1,-1)$, $(-1,0,-1,-1)$, $(-1,0,-1,0)$, $(-1,0,0,-1)$, $(0,-2,-1,-1)$, $(0,-1,-1,-1)$, $(0,-1,-1,0)$, $(0,-1,0,-1)$, $(0,0,-1,-1)$, $(0,0,-1,0)$, $(0,0,0,-1)$	$\begin{pmatrix}0\\0\\\frac{600}{1261}\\\frac{23783700}{13920179}\end{pmatrix}$

P	$\mathcal{V}(P)$	$R_s(P)$	$R_u(P)$	$e(P)$
D-4	$\begin{pmatrix}1\\0\\0\\0\end{pmatrix}, \begin{pmatrix}0\\1\\0\\0\end{pmatrix}, \begin{pmatrix}0\\0\\1\\0\end{pmatrix}, \pm\begin{pmatrix}0\\0\\0\\1\end{pmatrix}, \begin{pmatrix}-1\\-1\\0\\2\end{pmatrix}, \begin{pmatrix}1\\0\\-1\\0\end{pmatrix}$	$\pm(0,1,0,0)$, $\pm(0,0,1,0)$	$(-2,0,-2,-1)$, $(-2,0,-1,-1)$, $(-2,0,0,-1)$, $(-1,-1,-1,-1)$, $(-1,-1,0,-1)$, $(-1,0,-1,-1)$, $(-1,0,-1,0)$, $(-1,0,0,-1)$, $(-1,0,0,0)$, $(-1,1,-1,0)$, $(-1,1,0,0)$, $(0,-2,0,-1)$, $(0,-1,0,-1)$, $(0,0,0,-1)$	$\begin{pmatrix}\frac{1389500}{5141177}\\0\\0\\\frac{10897600}{5141177}\end{pmatrix}$
D-5	$\begin{pmatrix}1\\0\\0\\0\end{pmatrix}, \begin{pmatrix}0\\1\\0\\0\end{pmatrix}, \pm\begin{pmatrix}0\\0\\1\\0\end{pmatrix}, \pm\begin{pmatrix}0\\0\\0\\1\end{pmatrix}, \begin{pmatrix}-1\\-1\\1\\0\end{pmatrix}$	$\pm(1,0,0,0)$, $\pm(1,-1,0,0)$, $\pm(0,1,0,0)$, $\pm(0,0,0,1)$	$(-1,0,-1,0)$, $(0,-1,-1,0)$, $(0,0,-1,0)$	$\begin{pmatrix}0\\0\\\frac{70}{97}\\0\end{pmatrix}$
D-6	$\begin{pmatrix}1\\0\\0\\0\end{pmatrix}, \begin{pmatrix}0\\1\\0\\0\end{pmatrix}, \begin{pmatrix}0\\0\\1\\0\end{pmatrix}, \pm\begin{pmatrix}0\\0\\0\\1\end{pmatrix}, \begin{pmatrix}-1\\-1\\0\\1\end{pmatrix}, \begin{pmatrix}0\\0\\-1\\1\end{pmatrix}$	$\pm(1,0,0,0)$, $\pm(1,-1,0,0)$, $\pm(0,1,0,0)$, $\pm(0,0,1,0)$	$(-1,0,-1,-1)$, $(-1,0,0,-1)$, $(0,-1,-1,-1)$, $(0,-1,0,-1)$, $(0,0,-1,-1)$, $(0,0,0,-1)$	$\begin{pmatrix}0\\0\\0\\\frac{2790}{1973}\end{pmatrix}$
D-7	$\begin{pmatrix}1\\0\\0\\0\end{pmatrix}, \begin{pmatrix}0\\1\\0\\0\end{pmatrix}, \begin{pmatrix}0\\0\\1\\0\end{pmatrix}, \begin{pmatrix}0\\0\\0\\1\end{pmatrix}, \begin{pmatrix}1\\0\\0\\-1\end{pmatrix}, \begin{pmatrix}-1\\-1\\0\\0\end{pmatrix}, \begin{pmatrix}1\\0\\-1\\0\end{pmatrix}$	$\pm(0,1,0,0)$, $\pm(0,0,1,0)$, $\pm(0,0,0,1)$	$(-1,0,-1,-1)$, $(-1,0,-1,0)$, $(-1,0,0,-1)$, $(-1,0,0,0)$, $(-1,1,-1,-1)$, $(-1,1,-1,0)$, $(-1,1,0,-1)$, $(-1,1,0,0)$	$\begin{pmatrix}\frac{495}{523}\\0\\0\\0\end{pmatrix}$
D-8	$\begin{pmatrix}1\\0\\0\\0\end{pmatrix}, \begin{pmatrix}0\\1\\0\\0\end{pmatrix}, \begin{pmatrix}0\\0\\1\\0\end{pmatrix}, \pm\begin{pmatrix}0\\0\\0\\1\end{pmatrix}, \begin{pmatrix}-1\\-1\\1\\0\end{pmatrix}, \begin{pmatrix}0\\0\\-1\\1\end{pmatrix}$	$\pm(1,0,0,0)$, $\pm(1,-1,0,0)$, $\pm(0,1,0,0)$	$(-1,0,-1,-1)$, $(-1,0,-1,0)$, $(0,-1,-1,-1)$, $(0,-1,-1,0)$, $(0,0,-1,-1)$, $(0,0,-1,0)$, $(0,0,0,-1)$	$\begin{pmatrix}0\\0\\\frac{6612000}{11697163}\\\frac{6884100}{11697163}\end{pmatrix}$
D-9	$\begin{pmatrix}1\\0\\0\\0\end{pmatrix}, \begin{pmatrix}0\\1\\0\\0\end{pmatrix}, \pm\begin{pmatrix}0\\0\\1\\0\end{pmatrix}, \pm\begin{pmatrix}0\\0\\0\\1\end{pmatrix}, \begin{pmatrix}-1\\-1\\1\\1\end{pmatrix}$	$\pm(1,0,0,0)$, $\pm(1,-1,0,0)$, $\pm(0,1,0,0)$	$(-1,0,-1,0)$, $(-1,0,0,-1)$, $(0,-1,-1,0)$, $(0,-1,0,-1)$, $(0,0,-1,0)$, $(0,0,0,-1)$	$\begin{pmatrix}0\\0\\\frac{870}{1177}\\\frac{870}{1177}\end{pmatrix}$
D-10	$\begin{pmatrix}1\\0\\0\\0\end{pmatrix}, \begin{pmatrix}0\\1\\0\\0\end{pmatrix}, \begin{pmatrix}0\\0\\1\\0\end{pmatrix}, \pm\begin{pmatrix}0\\0\\0\\1\end{pmatrix}, \begin{pmatrix}-1\\-1\\0\\1\end{pmatrix}, \begin{pmatrix}1\\0\\-1\\0\end{pmatrix}$	$\pm(0,1,0,0)$, $\pm(0,0,1,0)$	$(-1,0,-1,-1)$, $(-1,0,-1,0)$, $(-1,0,0,-1)$, $(-1,0,0,0)$, $(-1,1,-1,0)$, $(-1,1,0,0)$, $(0,-1,0,-1)$, $(0,0,0,-1)$	$\begin{pmatrix}\frac{18125}{46732}\\0\\0\\\frac{34945}{46732}\end{pmatrix}$
D-11	$\begin{pmatrix}1\\0\\0\\0\end{pmatrix}, \begin{pmatrix}0\\1\\0\\0\end{pmatrix}, \begin{pmatrix}0\\0\\1\\0\end{pmatrix}, \begin{pmatrix}0\\0\\0\\1\end{pmatrix}, \begin{pmatrix}-1\\-1\\0\\0\end{pmatrix}, \begin{pmatrix}1\\0\\-1\\0\end{pmatrix}, \begin{pmatrix}0\\0\\1\\-1\end{pmatrix}$	$\pm(0,1,0,0)$, $\pm(0,0,0,1)$	$(-1,0,-1,-1)$, $(-1,0,-1,0)$, $(-1,0,0,0)$, $(-1,1,-1,-1)$, $(-1,1,-1,0)$, $(-1,1,0,0)$, $(0,0,-1,-1)$, $(0,0,-1,0)$	$\begin{pmatrix}\frac{465}{1007}\\0\\\frac{375}{1007}\\0\end{pmatrix}$

P	$\mathcal{V}(P)$	$R_s(P)$	$R_u(P)$	$e(P)$
D-12	$\begin{pmatrix}1\\0\\0\\0\end{pmatrix}, \begin{pmatrix}0\\1\\0\\0\end{pmatrix}, \pm\begin{pmatrix}0\\0\\1\\0\end{pmatrix}, \pm\begin{pmatrix}0\\0\\0\\1\end{pmatrix}, \begin{pmatrix}-1\\-1\\2\\0\end{pmatrix}$	$\pm(1,0,0,0)$, $\pm(1,-1,0,0)$, $\pm(0,1,0,0)$, $\pm(0,0,0,1)$	$(-2,0,-1,0)$, $(-1,-1,-1,0)$, $(-1,0,-1,0)$, $(0,-2,-1,0)$, $(0,-1,-1,0)$, $(0,0,-1,0)$	$\begin{pmatrix}0\\0\\\frac{620}{349}\\0\end{pmatrix}$
D-13	$\begin{pmatrix}1\\0\\0\\0\end{pmatrix}, \begin{pmatrix}0\\1\\0\\0\end{pmatrix}, \pm\begin{pmatrix}0\\0\\1\\0\end{pmatrix}, \pm\begin{pmatrix}0\\0\\0\\1\end{pmatrix}, \begin{pmatrix}-1\\-1\\0\\0\end{pmatrix}$	$\pm(1,0,0,0)$, $\pm(1,-1,0,0)$, $\pm(0,1,0,0)$, $\pm(0,0,1,0)$, $\pm(0,0,0,1)$		$\begin{pmatrix}0\\0\\0\\0\end{pmatrix}$
D-14	$\begin{pmatrix}1\\0\\0\\0\end{pmatrix}, \begin{pmatrix}0\\1\\0\\0\end{pmatrix}, \pm\begin{pmatrix}0\\0\\1\\0\end{pmatrix}, \begin{pmatrix}0\\0\\0\\1\end{pmatrix}, \begin{pmatrix}-1\\-1\\0\\0\end{pmatrix}, \begin{pmatrix}1\\0\\0\\-1\end{pmatrix}$	$\pm(0,1,0,0)$, $\pm(0,0,1,0)$, $\pm(0,0,0,1)$	$(-1,0,0,-1)$, $(-1,0,0,0)$, $(-1,1,0,-1)$, $(-1,1,0,0)$	$\begin{pmatrix}\frac{20}{43}\\0\\0\\0\end{pmatrix}$
D-15	$\begin{pmatrix}1\\0\\0\\0\end{pmatrix}, \begin{pmatrix}0\\1\\0\\0\end{pmatrix}, \pm\begin{pmatrix}0\\0\\1\\0\end{pmatrix}, \begin{pmatrix}0\\0\\0\\1\end{pmatrix}, \begin{pmatrix}-1\\-1\\0\\0\end{pmatrix}, \begin{pmatrix}0\\0\\1\\-1\end{pmatrix}$	$\pm(1,0,0,0)$, $\pm(1,-1,0,0)$, $\pm(0,1,0,0)$, $\pm(0,0,0,1)$	$(0,0,-1,-1)$, $(0,0,-1,0)$	$\begin{pmatrix}0\\0\\\frac{6}{11}\\0\end{pmatrix}$
D-16	$\begin{pmatrix}1\\0\\0\\0\end{pmatrix}, \begin{pmatrix}0\\1\\0\\0\end{pmatrix}, \begin{pmatrix}0\\0\\1\\0\end{pmatrix}, \pm\begin{pmatrix}0\\0\\0\\1\end{pmatrix}, \begin{pmatrix}-1\\-1\\1\\-1\end{pmatrix}, \begin{pmatrix}0\\0\\-1\\1\end{pmatrix}$	$\pm(1,0,0,0)$, $\pm(1,-1,0,0)$, $\pm(0,1,0,0)$	$(-1,0,-1,0)$, $(0,-1,-1,0)$, $(0,0,-1,-1)$, $(0,0,-1,0)$	$\begin{pmatrix}0\\0\\\frac{796536}{1061921}\\-\frac{335484}{1061921}\end{pmatrix}$
D-17	$\begin{pmatrix}1\\0\\0\\0\end{pmatrix}, \begin{pmatrix}0\\1\\0\\0\end{pmatrix}, \begin{pmatrix}0\\0\\1\\0\end{pmatrix}, \begin{pmatrix}0\\0\\0\\1\end{pmatrix}, \begin{pmatrix}-1\\-1\\0\\0\end{pmatrix}, \begin{pmatrix}1\\0\\-1\\0\end{pmatrix}, \begin{pmatrix}0\\1\\0\\-1\end{pmatrix}$	$\pm(0,0,1,0)$, $\pm(0,0,0,1)$	$(-1,0,-1,0)$, $(-1,0,0,0)$, $(0,-1,0,-1)$, $(0,-1,0,0)$	$\begin{pmatrix}\frac{75}{134}\\\frac{75}{134}\\0\\0\end{pmatrix}$
D-18	$\begin{pmatrix}1\\0\\0\\0\end{pmatrix}, \begin{pmatrix}0\\1\\0\\0\end{pmatrix}, \begin{pmatrix}0\\0\\1\\0\end{pmatrix}, \pm\begin{pmatrix}0\\0\\0\\1\end{pmatrix}, \begin{pmatrix}-1\\-1\\0\\-2\end{pmatrix}, \begin{pmatrix}0\\0\\-1\\1\end{pmatrix}$	$\pm(1,0,0,0)$, $\pm(1,-1,0,0)$, $\pm(0,1,0,0)$, $\pm(0,0,1,0)$		$\begin{pmatrix}0\\0\\0\\-\frac{5250}{5177}\end{pmatrix}$
D-19	$\begin{pmatrix}1\\0\\0\\0\end{pmatrix}, \begin{pmatrix}0\\1\\0\\0\end{pmatrix}, \begin{pmatrix}0\\0\\1\\0\end{pmatrix}, \pm\begin{pmatrix}0\\0\\0\\1\end{pmatrix}, \begin{pmatrix}-1\\-1\\0\\-1\end{pmatrix}, \begin{pmatrix}0\\0\\-1\\1\end{pmatrix}$	$\pm(1,0,0,0)$, $\pm(1,-1,0,0)$, $\pm(0,1,0,0)$, $\pm(0,0,1,0)$		$\begin{pmatrix}0\\0\\0\\-\frac{250}{1613}\end{pmatrix}$
E-1	$\pm\begin{pmatrix}1\\0\\0\\0\end{pmatrix}, \begin{pmatrix}0\\1\\0\\0\end{pmatrix}, \begin{pmatrix}0\\0\\1\\0\end{pmatrix}, \begin{pmatrix}0\\0\\0\\1\end{pmatrix}, \begin{pmatrix}2\\-1\\-1\\-1\end{pmatrix}, \begin{pmatrix}1\\1\\0\\0\end{pmatrix}$	$\pm(0,0,1,0)$, $\pm(0,0,1,-1)$, $\pm(0,0,0,1)$	$(-1,-2,0,0)$, $(-1,-1,-1,0)$, $(-1,-1,0,-1)$, $(-1,-1,0,0)$, $(-1,0,-2,0)$, $(-1,0,-1,-1)$, $(-1,0,-1,0)$, $(-1,0,0,-2)$, $(-1,0,0,-1)$, $(-1,0,0,0)$, $(0,-1,0,0)$, $(0,-1,0,1)$, $(0,-1,1,0)$	$\begin{pmatrix}\frac{4447297800}{1528096589}\\\frac{166293600}{1528096589}\\0\\0\end{pmatrix}$

P	$\mathcal{V}(P)$	$R_s(P)$	$R_u(P)$	$e(P)$
E-2	$\pm\begin{pmatrix}1\\0\\0\\0\end{pmatrix}, \begin{pmatrix}0\\1\\0\\0\end{pmatrix}, \begin{pmatrix}0\\0\\1\\0\end{pmatrix}, \begin{pmatrix}0\\0\\0\\1\end{pmatrix}, \begin{pmatrix}1\\-1\\-1\\-1\end{pmatrix}, \begin{pmatrix}1\\1\\0\\0\end{pmatrix}$	$\pm(0,0,1,0)$, $\pm(0,0,1,-1)$, $\pm(0,0,0,1)$	$(-1,-1,0,0)$, $(-1,0,-1,0)$, $(-1,0,0,-1)$, $(-1,0,0,0)$, $(0,-1,0,0)$, $(0,-1,0,1)$, $(0,-1,1,0)$	$\begin{pmatrix}\frac{1517634320}{1038057499}\\ \frac{215186080}{1038057499}\\ 0\\ 0\end{pmatrix}$
E-3	$\pm\begin{pmatrix}1\\0\\0\\0\end{pmatrix}, \begin{pmatrix}0\\1\\0\\0\end{pmatrix}, \begin{pmatrix}0\\0\\1\\0\end{pmatrix}, \begin{pmatrix}0\\0\\0\\1\end{pmatrix}, \begin{pmatrix}0\\-1\\-1\\-1\end{pmatrix}, \begin{pmatrix}1\\1\\0\\0\end{pmatrix}$	$\pm(0,0,1,0)$, $\pm(0,0,1,-1)$, $\pm(0,0,0,1)$	$(-1,0,0,0)$, $(0,-1,0,0)$, $(0,-1,0,1)$, $(0,-1,1,0)$	$\begin{pmatrix}\frac{3423372660}{6541476757}\\ \frac{1994323200}{6541476757}\\ 0\\ 0\end{pmatrix}$
F-1	$\pm\begin{pmatrix}1\\0\\0\\0\end{pmatrix}, \begin{pmatrix}0\\1\\0\\0\end{pmatrix}, \begin{pmatrix}0\\0\\1\\0\end{pmatrix}, \begin{pmatrix}0\\0\\0\\1\end{pmatrix}, \begin{pmatrix}1\\-1\\-1\\0\end{pmatrix}, \begin{pmatrix}2\\0\\-1\\-1\end{pmatrix}$	$\pm(0,1,0,0)$, $\pm(0,0,0,1)$	$(-1,-1,0,-2)$, $(-1,-1,0,-1)$, $(-1,-1,0,0)$, $(-1,0,-1,-1)$, $(-1,0,-1,0)$, $(-1,0,0,-2)$, $(-1,0,0,-1)$, $(-1,0,0,0)$, $(0,-1,1,-1)$, $(0,-1,1,0)$, $(0,0,1,-1)$, $(0,0,1,0)$	$\begin{pmatrix}\frac{37598378760}{16641653953}\\ 0\\ -\frac{8882851620}{16641653953}\\ 0\end{pmatrix}$
F-2	$\begin{pmatrix}1\\0\\0\\0\end{pmatrix}, \begin{pmatrix}0\\1\\0\\0\end{pmatrix}, \begin{pmatrix}0\\0\\1\\0\end{pmatrix}, \begin{pmatrix}0\\0\\0\\1\end{pmatrix}, \begin{pmatrix}-1\\-1\\0\\0\end{pmatrix}, \begin{pmatrix}2\\0\\-1\\-1\end{pmatrix}, \begin{pmatrix}-1\\0\\1\\0\end{pmatrix}$	$\pm(0,1,0,0)$, $\pm(0,0,0,1)$	$(-1,0,-2,0)$, $(-1,0,-1,-1)$, $(-1,0,-1,0)$, $(-1,1,-2,0)$, $(-1,1,-1,-1)$, $(-1,1,-1,0)$, $(0,0,-1,0)$, $(0,0,-1,1)$	$\begin{pmatrix}\frac{470033500}{600967411}\\ 0\\ \frac{122316000}{600967411}\\ 0\end{pmatrix}$
F-3	$\pm\begin{pmatrix}1\\0\\0\\0\end{pmatrix}, \begin{pmatrix}0\\1\\0\\0\end{pmatrix}, \begin{pmatrix}0\\0\\1\\0\end{pmatrix}, \begin{pmatrix}0\\0\\0\\1\end{pmatrix}, \begin{pmatrix}0\\-1\\-1\\0\end{pmatrix}, \begin{pmatrix}1\\0\\-1\\-1\end{pmatrix}$	$\pm(0,1,0,0)$, $\pm(0,0,0,1)$	$(-1,0,0,-1)$, $(-1,0,0,0)$, $(0,-1,1,-1)$, $(0,-1,1,0)$, $(0,0,1,-1)$, $(0,0,1,0)$	$\begin{pmatrix}\frac{1203559872}{1907713037}\\ 0\\ -\frac{1490831124}{1907713037}\\ 0\end{pmatrix}$
F-4	$\begin{pmatrix}1\\0\\0\\0\end{pmatrix}, \begin{pmatrix}0\\1\\0\\0\end{pmatrix}, \begin{pmatrix}0\\0\\1\\0\end{pmatrix}, \begin{pmatrix}0\\0\\0\\1\end{pmatrix}, \begin{pmatrix}-1\\-1\\0\\0\end{pmatrix}, \begin{pmatrix}1\\1\\-1\\-1\end{pmatrix}, \begin{pmatrix}-1\\0\\1\\0\end{pmatrix}$	$\pm(0,0,0,1)$	$(-1,0,-1,0)$, $(0,-1,-1,0)$, $(0,-1,0,-1)$, $(0,-1,0,0)$, $(0,0,-1,0)$, $(0,0,-1,1)$, $(1,-1,0,0)$	$\begin{pmatrix}\frac{10356297206700}{48057952407691}\\ \frac{33083176976940}{48057952407691}\\ \frac{16155069457920}{48057952407691}\\ 0\end{pmatrix}$
F-5	$\begin{pmatrix}1\\0\\0\\0\end{pmatrix}, \begin{pmatrix}0\\1\\0\\0\end{pmatrix}, \begin{pmatrix}0\\0\\1\\0\end{pmatrix}, \begin{pmatrix}0\\0\\0\\1\end{pmatrix}, \begin{pmatrix}-1\\-1\\0\\0\end{pmatrix}, \begin{pmatrix}0\\0\\-1\\-1\end{pmatrix}, \begin{pmatrix}-1\\0\\1\\0\end{pmatrix}$	$\pm(0,1,0,0)$, $\pm(0,0,0,1)$	$(0,0,-1,0)$, $(0,0,-1,1)$, $(1,-1,0,0)$, $(1,0,0,0)$	$\begin{pmatrix}-\frac{4015697280}{5976434111}\\ 0\\ \frac{2205554400}{5976434111}\\ 0\end{pmatrix}$
F-6	$\begin{pmatrix}1\\0\\0\\0\end{pmatrix}, \begin{pmatrix}0\\1\\0\\0\end{pmatrix}, \begin{pmatrix}0\\0\\1\\0\end{pmatrix}, \begin{pmatrix}0\\0\\0\\1\end{pmatrix}, \begin{pmatrix}-1\\-1\\0\\0\end{pmatrix}, \begin{pmatrix}1\\0\\-1\\-1\end{pmatrix}, \begin{pmatrix}-1\\0\\1\\0\end{pmatrix}$	$\pm(0,1,0,0)$, $\pm(0,0,0,1)$	$(-1,0,-1,0)$, $(-1,1,-1,0)$, $(0,0,-1,0)$, $(0,0,-1,1)$	$\begin{pmatrix}0\\ 0\\ \frac{1732320}{4388521}\\ 0\end{pmatrix}$

P	$\mathcal{V}(P)$	$R_s(P)$	$R_u(P)$	$e(P)$
G-1	$\begin{pmatrix}1\\0\\0\\0\end{pmatrix}, \begin{pmatrix}0\\1\\0\\0\end{pmatrix}, \begin{pmatrix}0\\0\\1\\0\end{pmatrix}, \begin{pmatrix}0\\0\\0\\1\end{pmatrix}, \begin{pmatrix}0\\-1\\-1\\-1\end{pmatrix}, \begin{pmatrix}-1\\-1\\0\\0\end{pmatrix}, \begin{pmatrix}1\\1\\0\\1\end{pmatrix}, \begin{pmatrix}-1\\0\\1\\0\end{pmatrix}$		$(-1, 1, -1, 0)$, $(0, 0, -1, 0)$, $(0, 0, 0, -1)$, $(0, 1, 0, -1)$	$\begin{pmatrix}0\\0\\\frac{598920}{1289443}\\\frac{598920}{1289443}\end{pmatrix}$
G-2	$\begin{pmatrix}1\\0\\0\\0\end{pmatrix}, \begin{pmatrix}0\\1\\0\\0\end{pmatrix}, \begin{pmatrix}0\\0\\1\\0\end{pmatrix}, \begin{pmatrix}0\\0\\0\\1\end{pmatrix}, \begin{pmatrix}0\\0\\-1\\-1\end{pmatrix}, \begin{pmatrix}-1\\-1\\0\\0\end{pmatrix}, \begin{pmatrix}1\\1\\0\\1\end{pmatrix}, \begin{pmatrix}-1\\0\\1\\0\end{pmatrix}$		$(0, -1, 0, 0)$, $(0, 0, -1, 0)$, $(0, 0, 0, -1)$, $(1, -1, 0, 0)$	$\begin{pmatrix}0\\\frac{1164120}{2180567}\\\frac{1120256220}{2592694163}\\\frac{1120256220}{2592694163}\end{pmatrix}$
H-1	$\pm\begin{pmatrix}1\\0\\0\\0\end{pmatrix}, \begin{pmatrix}0\\1\\0\\0\end{pmatrix}, \begin{pmatrix}0\\0\\1\\0\end{pmatrix}, \begin{pmatrix}0\\0\\0\\1\end{pmatrix}, \pm\begin{pmatrix}1\\-1\\0\\0\end{pmatrix}, \begin{pmatrix}0\\2\\-1\\-1\end{pmatrix}$	$\pm(0, 0, 1, 0)$, $\pm(0, 0, 1, -1)$, $\pm(0, 0, 0, 1)$	$(-1, -1, -2, 0)$, $(-1, -1, -1, -1)$, $(-1, -1, -1, 0)$, $(-1, -1, 0, -2)$, $(-1, -1, 0, -1)$, $(-1, -1, 0, 0)$, $(0, -1, -2, 0)$, $(0, -1, -1, -1)$, $(0, -1, -1, 0)$, $(0, -1, 0, -2)$, $(0, -1, 0, -1)$, $(0, -1, 0, 0)$	$\begin{pmatrix}0\\\frac{99820}{57409}\\0\\0\end{pmatrix}$
H-2	$\pm\begin{pmatrix}1\\0\\0\\0\end{pmatrix}, \begin{pmatrix}0\\1\\0\\0\end{pmatrix}, \begin{pmatrix}0\\0\\1\\0\end{pmatrix}, \begin{pmatrix}0\\0\\0\\1\end{pmatrix}, \pm\begin{pmatrix}1\\-1\\0\\0\end{pmatrix}, \begin{pmatrix}1\\1\\-1\\-1\end{pmatrix}$	$\pm(0, 0, 1, 0)$, $\pm(0, 0, 1, -1)$, $\pm(0, 0, 0, 1)$	$(-1, -1, -2, 0)$, $(-1, -1, -1, -1)$, $(-1, -1, -1, 0)$, $(-1, -1, 0, -2)$, $(-1, -1, 0, -1)$, $(-1, -1, 0, 0)$, $(0, -1, -1, 0)$, $(0, -1, 0, -1)$, $(0, -1, 0, 0)$	$\begin{pmatrix}\frac{12590952900}{12261327193}\\\frac{11321625300}{12261327193}\\0\\0\end{pmatrix}$
H-3	$\pm\begin{pmatrix}1\\0\\0\\0\end{pmatrix}, \begin{pmatrix}0\\1\\0\\0\end{pmatrix}, \begin{pmatrix}0\\0\\1\\0\end{pmatrix}, \begin{pmatrix}0\\0\\0\\1\end{pmatrix}, \pm\begin{pmatrix}1\\-1\\0\\0\end{pmatrix}, \begin{pmatrix}2\\0\\-1\\-1\end{pmatrix}$	$\pm(0, 0, 1, 0)$, $\pm(0, 0, 1, -1)$, $\pm(0, 0, 0, 1)$	$(-1, -1, -2, 0)$, $(-1, -1, -1, -1)$, $(-1, -1, -1, 0)$, $(-1, -1, 0, -2)$, $(-1, -1, 0, -1)$, $(-1, -1, 0, 0)$, $(0, -1, 0, 0)$	$\begin{pmatrix}\frac{1983987756}{999563887}\\\frac{260718408}{999563887}\\0\\0\end{pmatrix}$
H-4	$\pm\begin{pmatrix}1\\0\\0\\0\end{pmatrix}, \begin{pmatrix}0\\1\\0\\0\end{pmatrix}, \begin{pmatrix}0\\0\\1\\0\end{pmatrix}, \begin{pmatrix}0\\0\\0\\1\end{pmatrix}, \pm\begin{pmatrix}1\\-1\\0\\0\end{pmatrix}, \begin{pmatrix}0\\1\\-1\\-1\end{pmatrix}$	$\pm(0, 0, 1, 0)$, $\pm(0, 0, 1, -1)$, $\pm(0, 0, 0, 1)$	$(-1, -1, -1, 0)$, $(-1, -1, 0, -1)$, $(-1, -1, 0, 0)$, $(0, -1, -1, 0)$, $(0, -1, 0, -1)$, $(0, -1, 0, 0)$	$\begin{pmatrix}0\\\frac{795660}{778267}\\0\\0\end{pmatrix}$
H-5	$\pm\begin{pmatrix}1\\0\\0\\0\end{pmatrix}, \begin{pmatrix}0\\1\\0\\0\end{pmatrix}, \begin{pmatrix}0\\0\\1\\0\end{pmatrix}, \begin{pmatrix}0\\0\\0\\1\end{pmatrix}, \pm\begin{pmatrix}1\\-1\\0\\0\end{pmatrix}, \begin{pmatrix}1\\0\\-1\\-1\end{pmatrix}$	$\pm(0, 0, 1, 0)$, $\pm(0, 0, 1, -1)$, $\pm(0, 0, 0, 1)$	$(-1, -1, -1, 0)$, $(-1, -1, 0, -1)$, $(-1, -1, 0, 0)$, $(0, -1, 0, 0)$	$\begin{pmatrix}\frac{6818815200}{8410595791}\\\frac{224772300}{646968907}\\0\\0\end{pmatrix}$
H-6	$\pm\begin{pmatrix}1\\0\\0\\0\end{pmatrix}, \begin{pmatrix}0\\1\\0\\0\end{pmatrix}, \begin{pmatrix}0\\0\\1\\0\end{pmatrix}, \begin{pmatrix}0\\0\\0\\1\end{pmatrix}, \pm\begin{pmatrix}1\\-1\\0\\0\end{pmatrix}, \begin{pmatrix}2\\-1\\-1\\-1\end{pmatrix}$	$\pm(0, 0, 1, 0)$, $\pm(0, 0, 1, -1)$, $\pm(0, 0, 0, 1)$	$(-1, -1, -1, 0)$, $(-1, -1, 0, -1)$, $(-1, -1, 0, 0)$	$\begin{pmatrix}\frac{848077860}{521298061}\\-\frac{2015224800}{6776874793}\\0\\0\end{pmatrix}$

P	$\mathcal{V}(P)$	$R_s(P)$	$R_u(P)$	$e(P)$
H-7	$\pm\begin{pmatrix}1\\0\\0\\0\end{pmatrix}, \begin{pmatrix}0\\1\\0\\0\end{pmatrix}, \begin{pmatrix}0\\0\\1\\0\end{pmatrix}, \begin{pmatrix}0\\0\\0\\1\end{pmatrix}, \pm\begin{pmatrix}1\\-1\\0\\0\end{pmatrix}, \begin{pmatrix}2\\-2\\-1\\-1\end{pmatrix}$	$\pm(0,0,1,0)$, $\pm(0,0,1,-1)$, $\pm(0,0,0,1)$	$(-1,-1,0,0)$	$\begin{pmatrix}\frac{3404937900}{2191715291}\\-\frac{2489673540}{2191715291}\\0\\0\end{pmatrix}$
H-8	$\pm\begin{pmatrix}1\\0\\0\\0\end{pmatrix}, \begin{pmatrix}0\\1\\0\\0\end{pmatrix}, \begin{pmatrix}0\\0\\1\\0\end{pmatrix}, \begin{pmatrix}0\\0\\0\\1\end{pmatrix}, \pm\begin{pmatrix}1\\-1\\0\\0\end{pmatrix}, \begin{pmatrix}0\\0\\-1\\-1\end{pmatrix}$	$\pm(0,0,1,0)$, $\pm(0,0,1,-1)$, $\pm(0,0,0,1)$	$(-1,-1,0,0)$, $(0,-1,0,0)$	$\begin{pmatrix}0\\\frac{168}{409}\\0\\0\end{pmatrix}$
H-9	$\pm\begin{pmatrix}1\\0\\0\\0\end{pmatrix}, \begin{pmatrix}0\\1\\0\\0\end{pmatrix}, \begin{pmatrix}0\\0\\1\\0\end{pmatrix}, \begin{pmatrix}0\\0\\0\\1\end{pmatrix}, \pm\begin{pmatrix}1\\-1\\0\\0\end{pmatrix}, \begin{pmatrix}1\\-1\\-1\\-1\end{pmatrix}$	$\pm(0,0,1,0)$, $\pm(0,0,1,-1)$, $\pm(0,0,0,1)$	$(-1,-1,0,0)$	$\begin{pmatrix}\frac{225991260}{324937871}\\-\frac{1395187200}{5523943807}\\0\\0\end{pmatrix}$
H-10	$\pm\begin{pmatrix}1\\0\\0\\0\end{pmatrix}, \begin{pmatrix}0\\1\\0\\0\end{pmatrix}, \begin{pmatrix}0\\0\\1\\0\end{pmatrix}, \begin{pmatrix}0\\0\\0\\1\end{pmatrix}, \pm\begin{pmatrix}1\\-1\\0\\0\end{pmatrix}, \begin{pmatrix}0\\-1\\-1\\-1\end{pmatrix}$	$\pm(0,0,1,0)$, $\pm(0,0,1,-1)$, $\pm(0,0,0,1)$		$\begin{pmatrix}0\\-\frac{37440}{116609}\\0\\0\end{pmatrix}$
I-1	$\pm\begin{pmatrix}1\\0\\0\\0\end{pmatrix}, \begin{pmatrix}0\\1\\0\\0\end{pmatrix}, \begin{pmatrix}0\\0\\1\\0\end{pmatrix}, \begin{pmatrix}0\\0\\0\\1\end{pmatrix}, \begin{pmatrix}0\\-1\\1\\0\end{pmatrix}, \begin{pmatrix}2\\0\\-1\\-1\end{pmatrix}, \begin{pmatrix}-1\\0\\1\\0\end{pmatrix}$	$\pm(0,1,0,0)$, $\pm(0,0,0,1)$	$(-1,-2,-2,0)$, $(-1,-1,-2,0)$, $(-1,-1,-1,-1)$, $(-1,-1,-1,0)$, $(-1,0,-2,0)$, $(-1,0,-1,-1)$, $(-1,0,-1,0)$, $(0,-1,-1,0)$, $(0,-1,-1,1)$, $(0,0,-1,0)$, $(0,0,-1,1)$	$\begin{pmatrix}\frac{455864610}{298795537}\\0\\\frac{140393265}{298795537}\\0\end{pmatrix}$
I-2	$\pm\begin{pmatrix}1\\0\\0\\0\end{pmatrix}, \begin{pmatrix}0\\1\\0\\0\end{pmatrix}, \begin{pmatrix}0\\0\\1\\0\end{pmatrix}, \begin{pmatrix}0\\0\\0\\1\end{pmatrix}, \begin{pmatrix}1\\-1\\0\\0\end{pmatrix}, \begin{pmatrix}2\\0\\-1\\-1\end{pmatrix}, \begin{pmatrix}-1\\0\\1\\0\end{pmatrix}$	$\pm(0,1,0,0)$, $\pm(0,0,0,1)$	$(-1,-1,-2,0)$, $(-1,-1,-1,-1)$, $(-1,-1,-1,0)$, $(-1,0,-2,0)$, $(-1,0,-1,-1)$, $(-1,0,-1,0)$, $(0,0,-1,0)$, $(0,0,-1,1)$	$\begin{pmatrix}\frac{69547230}{35772317}\\0\\\frac{302232510}{1752843533}\\0\end{pmatrix}$
I-3	$\pm\begin{pmatrix}1\\0\\0\\0\end{pmatrix}, \begin{pmatrix}0\\1\\0\\0\end{pmatrix}, \begin{pmatrix}0\\0\\1\\0\end{pmatrix}, \begin{pmatrix}0\\0\\0\\1\end{pmatrix}, \begin{pmatrix}1\\-1\\0\\0\end{pmatrix}, \begin{pmatrix}1\\1\\-1\\-1\end{pmatrix}, \begin{pmatrix}-1\\0\\1\\0\end{pmatrix}$	$\pm(0,0,0,1)$	$(-1,-1,-2,0)$, $(-1,-1,-1,-1)$, $(-1,-1,-1,0)$, $(-1,0,-1,0)$, $(0,-1,-1,0)$, $(0,-1,0,-1)$, $(0,-1,0,0)$, $(0,0,-1,0)$, $(0,0,-1,1)$	$\begin{pmatrix}\frac{559725599420}{557417025511}\\\frac{553671270165}{1114834051022}\\\frac{212169307415}{1114834051022}\\0\end{pmatrix}$
I-4	$\pm\begin{pmatrix}1\\0\\0\\0\end{pmatrix}, \begin{pmatrix}0\\1\\0\\0\end{pmatrix}, \begin{pmatrix}0\\0\\1\\0\end{pmatrix}, \begin{pmatrix}0\\0\\0\\1\end{pmatrix}, \begin{pmatrix}-1\\-1\\1\\0\end{pmatrix}, \begin{pmatrix}2\\0\\-1\\-1\end{pmatrix}, \begin{pmatrix}-1\\0\\1\\0\end{pmatrix}$	$\pm(0,1,0,0)$, $\pm(0,0,0,1)$	$(-1,-1,-2,0)$, $(-1,0,-2,0)$, $(-1,0,-1,-1)$, $(-1,0,-1,0)$, $(0,-1,-1,0)$, $(0,-1,-1,1)$, $(0,0,-1,0)$, $(0,0,-1,1)$	$\begin{pmatrix}\frac{29597715}{42931741}\\0\\\frac{23135190}{42931741}\\0\end{pmatrix}$

P	$\mathcal{V}(P)$	$R_s(P)$	$R_u(P)$	$e(P)$
I-5	$\pm\begin{pmatrix}1\\0\\0\\0\end{pmatrix},\begin{pmatrix}0\\1\\0\\0\end{pmatrix},\begin{pmatrix}0\\0\\1\\0\end{pmatrix},\begin{pmatrix}0\\0\\0\\1\end{pmatrix},\begin{pmatrix}0\\-1\\0\\1\end{pmatrix},\begin{pmatrix}2\\0\\-1\\-1\end{pmatrix},\begin{pmatrix}-1\\0\\1\\0\end{pmatrix}$	$\pm(0,1,0,0)$	$(-1,-1,-1,-1)$, $(-1,0,-2,0)$, $(-1,0,-1,-1)$, $(-1,0,-1,0)$, $(0,-1,0,-1)$, $(0,0,-1,0)$, $(0,0,0,-1)$	$\begin{pmatrix}\frac{16299777247200}{13032955449721}\\0\\\frac{3178785237975}{13032955449721}\\\frac{4952504268375}{13032955449721}\end{pmatrix}$
I-6	$\pm\begin{pmatrix}1\\0\\0\\0\end{pmatrix},\begin{pmatrix}0\\1\\0\\0\end{pmatrix},\begin{pmatrix}0\\0\\1\\0\end{pmatrix},\begin{pmatrix}0\\0\\0\\1\end{pmatrix},\begin{pmatrix}0\\-1\\1\\0\end{pmatrix},\begin{pmatrix}1\\0\\-1\\-1\end{pmatrix},\begin{pmatrix}-1\\0\\1\\0\end{pmatrix}$	$\pm(0,1,0,0)$, $\pm(0,0,0,1)$	$(-1,-1,-1,0)$, $(-1,0,-1,0)$, $(0,-1,-1,0)$, $(0,-1,-1,1)$, $(0,0,-1,0)$, $(0,0,-1,1)$	$\begin{pmatrix}\frac{48154815}{230998253}\\0\\\frac{164153400}{230998253}\\0\end{pmatrix}$
I-7	$\pm\begin{pmatrix}1\\0\\0\\0\end{pmatrix},\pm\begin{pmatrix}0\\1\\0\\0\end{pmatrix},\begin{pmatrix}0\\0\\1\\0\end{pmatrix},\begin{pmatrix}0\\0\\0\\1\end{pmatrix},\begin{pmatrix}2\\0\\-1\\-1\end{pmatrix},\begin{pmatrix}-1\\0\\1\\0\end{pmatrix}$	$\pm(0,1,0,0)$, $\pm(0,0,0,1)$	$(-1,0,-2,0)$, $(-1,0,-1,-1)$, $(-1,0,-1,0)$, $(0,0,-1,0)$, $(0,0,-1,1)$	$\begin{pmatrix}\frac{527400}{467581}\\0\\\frac{99600}{467581}\\0\end{pmatrix}$
I-8	$\pm\begin{pmatrix}1\\0\\0\\0\end{pmatrix},\begin{pmatrix}0\\1\\0\\0\end{pmatrix},\begin{pmatrix}0\\0\\1\\0\end{pmatrix},\begin{pmatrix}0\\0\\0\\1\end{pmatrix},\begin{pmatrix}-1\\-1\\0\\0\end{pmatrix},\begin{pmatrix}1\\1\\-1\\-1\end{pmatrix},\begin{pmatrix}-1\\0\\1\\0\end{pmatrix}$	$\pm(0,0,0,1)$	$(0,-1,-1,0)$, $(0,-1,0,-1)$, $(0,-1,0,0)$, $(0,0,-1,0)$, $(0,0,-1,1)$, $(1,-1,0,0)$	$\begin{pmatrix}-\frac{354416417280}{1577014856729}\\\frac{991006584000}{1577014856729}\\\frac{428321791680}{1577014856729}\\0\end{pmatrix}$
I-9	$\pm\begin{pmatrix}1\\0\\0\\0\end{pmatrix},\begin{pmatrix}0\\1\\0\\0\end{pmatrix},\begin{pmatrix}0\\0\\1\\0\end{pmatrix},\begin{pmatrix}0\\0\\0\\1\end{pmatrix},\begin{pmatrix}-1\\-1\\1\\0\end{pmatrix},\begin{pmatrix}1\\0\\-1\\-1\end{pmatrix},\begin{pmatrix}-1\\0\\1\\0\end{pmatrix}$	$\pm(0,1,0,0)$, $\pm(0,0,0,1)$	$(-1,0,-1,0)$, $(0,-1,-1,0)$, $(0,-1,-1,1)$, $(0,0,-1,0)$, $(0,0,-1,1)$	$\begin{pmatrix}-\frac{11160375}{31553027}\\0\\\frac{2180325}{2868457}\\0\end{pmatrix}$
I-10	$\pm\begin{pmatrix}1\\0\\0\\0\end{pmatrix},\begin{pmatrix}0\\1\\0\\0\end{pmatrix},\begin{pmatrix}0\\0\\1\\0\end{pmatrix},\begin{pmatrix}0\\0\\0\\1\end{pmatrix},\begin{pmatrix}1\\-1\\0\\0\end{pmatrix},\begin{pmatrix}1\\0\\-1\\-1\end{pmatrix},\begin{pmatrix}-1\\0\\1\\0\end{pmatrix}$	$\pm(0,1,0,0)$, $\pm(0,0,0,1)$	$(-1,-1,-1,0)$, $(-1,0,-1,0)$, $(0,0,-1,0)$, $(0,0,-1,1)$	$\begin{pmatrix}\frac{95308890}{126010781}\\0\\\frac{302223825}{1008086248}\\0\end{pmatrix}$
I-11	$\pm\begin{pmatrix}1\\0\\0\\0\end{pmatrix},\pm\begin{pmatrix}0\\1\\0\\0\end{pmatrix},\begin{pmatrix}0\\0\\1\\0\end{pmatrix},\begin{pmatrix}0\\0\\0\\1\end{pmatrix},\begin{pmatrix}1\\1\\-1\\-1\end{pmatrix},\begin{pmatrix}-1\\0\\1\\0\end{pmatrix}$	$\pm(0,0,0,1)$	$(-1,0,-1,0)$, $(0,-1,-1,0)$, $(0,-1,0,-1)$, $(0,-1,0,0)$, $(0,0,-1,0)$, $(0,0,-1,1)$	$\begin{pmatrix}\frac{1527360}{7524593}\\\frac{5306880}{7524593}\\\frac{2119680}{7524593}\\0\end{pmatrix}$
I-12	$\pm\begin{pmatrix}1\\0\\0\\0\end{pmatrix},\begin{pmatrix}0\\1\\0\\0\end{pmatrix},\begin{pmatrix}0\\0\\1\\0\end{pmatrix},\begin{pmatrix}0\\0\\0\\1\end{pmatrix},\begin{pmatrix}-1\\-1\\0\\0\end{pmatrix},\begin{pmatrix}1\\0\\-1\\-1\end{pmatrix},\begin{pmatrix}-1\\0\\1\\0\end{pmatrix}$	$\pm(0,1,0,0)$, $\pm(0,0,0,1)$	$(0,0,-1,0)$, $(0,0,-1,1)$	$\begin{pmatrix}-\frac{5471496}{14696135}\\0\\\frac{5124843}{14696135}\\0\end{pmatrix}$

P	$\mathcal{V}(P)$	$R_s(P)$	$R_u(P)$	$e(P)$
I-13	$\pm\begin{pmatrix}1\\0\\0\\0\end{pmatrix}, \pm\begin{pmatrix}0\\1\\0\\0\end{pmatrix}, \begin{pmatrix}0\\0\\1\\0\end{pmatrix}, \begin{pmatrix}0\\0\\0\\1\end{pmatrix}, \begin{pmatrix}1\\0\\-1\\-1\end{pmatrix}, \begin{pmatrix}-1\\0\\1\\0\end{pmatrix}$	$\pm(0,1,0,0)$, $\pm(0,0,0,1)$	$(-1,0,-1,0)$, $(0,0,-1,0)$, $(0,0,-1,1)$	$\begin{pmatrix}\frac{98900}{650251}\\0\\\frac{219420}{650251}\\0\end{pmatrix}$
I-14	$\pm\begin{pmatrix}1\\0\\0\\0\end{pmatrix}, \begin{pmatrix}0\\1\\0\\0\end{pmatrix}, \begin{pmatrix}0\\0\\1\\0\end{pmatrix}, \begin{pmatrix}0\\0\\0\\1\end{pmatrix}, \begin{pmatrix}0\\-1\\0\\1\end{pmatrix}, \begin{pmatrix}1\\0\\-1\\-1\end{pmatrix}, \begin{pmatrix}-1\\0\\1\\0\end{pmatrix}$	$\pm(0,1,0,0)$	$(-1,0,-1,0)$, $(0,-1,0,-1)$, $(0,0,-1,0)$, $(0,0,0,-1)$	$\begin{pmatrix}\frac{21660926805}{218756440813}\\0\\\frac{90160555275}{218756440813}\\\frac{112125015975}{218756440813}\end{pmatrix}$
I-15	$\pm\begin{pmatrix}1\\0\\0\\0\end{pmatrix}, \begin{pmatrix}0\\1\\0\\0\end{pmatrix}, \begin{pmatrix}0\\0\\1\\0\end{pmatrix}, \begin{pmatrix}0\\0\\0\\1\end{pmatrix}, \begin{pmatrix}-1\\-1\\0\\0\end{pmatrix}, \begin{pmatrix}2\\0\\-1\\-1\end{pmatrix}, \begin{pmatrix}-1\\0\\1\\0\end{pmatrix}$	$\pm(0,1,0,0)$, $\pm(0,0,0,1)$	$(0,0,-1,0)$, $(0,0,-1,1)$	$\begin{pmatrix}\frac{329975235}{739116617}\\0\\\frac{184724865}{739116617}\\0\end{pmatrix}$
J-1	$\pm\begin{pmatrix}1\\0\\0\\0\end{pmatrix}, \begin{pmatrix}0\\1\\0\\0\end{pmatrix}, \begin{pmatrix}0\\0\\1\\0\end{pmatrix}, \begin{pmatrix}0\\0\\0\\1\end{pmatrix}, \begin{pmatrix}-1\\-1\\1\\1\end{pmatrix}, \begin{pmatrix}1\\0\\-1\\-1\end{pmatrix}, \begin{pmatrix}0\\0\\-1\\-1\end{pmatrix}$	$\pm(0,1,0,0)$, $\pm(0,0,1,-1)$		$\begin{pmatrix}-\frac{78240}{450983}\\0\\\frac{156480}{450983}\\\frac{156480}{450983}\end{pmatrix}$
J-2	$\pm\begin{pmatrix}1\\0\\0\\0\end{pmatrix}, \begin{pmatrix}0\\1\\0\\0\end{pmatrix}, \begin{pmatrix}0\\0\\1\\0\end{pmatrix}, \begin{pmatrix}0\\0\\0\\1\end{pmatrix}, \begin{pmatrix}-1\\-1\\1\\1\end{pmatrix}, \begin{pmatrix}0\\0\\-1\\-1\end{pmatrix}, \begin{pmatrix}-1\\0\\-1\\-1\end{pmatrix}$	$\pm(0,1,0,0)$, $\pm(0,0,1,-1)$	$(1,-1,0,0)$, $(1,0,0,0)$	$\begin{pmatrix}-\frac{88407285}{67936574}\\0\\\frac{23804655}{67936574}\\\frac{23804655}{67936574}\end{pmatrix}$
K-1	$\pm\begin{pmatrix}1\\0\\0\\0\end{pmatrix}, \begin{pmatrix}0\\1\\0\\0\end{pmatrix}, \begin{pmatrix}0\\0\\1\\0\end{pmatrix}, \begin{pmatrix}0\\0\\0\\1\end{pmatrix}, \begin{pmatrix}1\\-1\\0\\0\end{pmatrix}, \begin{pmatrix}1\\0\\-1\\0\end{pmatrix}, \begin{pmatrix}1\\0\\0\\-1\end{pmatrix}$	$\pm(0,1,0,0)$, $\pm(0,0,1,0)$, $\pm(0,0,0,1)$	$(-1,-1,-1,-1)$, $(-1,-1,-1,0)$, $(-1,-1,0,-1)$, $(-1,-1,0,0)$, $(-1,0,-1,-1)$, $(-1,0,-1,0)$, $(-1,0,0,-1)$, $(-1,0,0,0)$	$\begin{pmatrix}\frac{3150}{1577}\\0\\0\\0\end{pmatrix}$
K-2	$\pm\begin{pmatrix}1\\0\\0\\0\end{pmatrix}, \begin{pmatrix}0\\1\\0\\0\end{pmatrix}, \begin{pmatrix}0\\0\\1\\0\end{pmatrix}, \begin{pmatrix}0\\0\\0\\1\end{pmatrix}, \begin{pmatrix}1\\-1\\0\\0\end{pmatrix}, \begin{pmatrix}1\\-1\\-1\\0\end{pmatrix}, \begin{pmatrix}1\\-1\\0\\-1\end{pmatrix}$	$\pm(0,0,1,0)$, $\pm(0,0,0,1)$	$(-1,-1,0,0)$, $(-1,0,-1,-1)$, $(-1,0,-1,0)$, $(-1,0,0,-1)$, $(-1,0,0,0)$, $(0,1,-1,-1)$, $(0,1,-1,0)$, $(0,1,0,-1)$, $(0,1,0,0)$	$\begin{pmatrix}\frac{3878228}{2637399}\\-\frac{2370692}{2637399}\\0\\0\end{pmatrix}$
K-3	$\pm\begin{pmatrix}1\\0\\0\\0\end{pmatrix}, \begin{pmatrix}0\\1\\0\\0\end{pmatrix}, \begin{pmatrix}0\\0\\1\\0\end{pmatrix}, \begin{pmatrix}0\\0\\0\\1\end{pmatrix}, \begin{pmatrix}1\\-1\\0\\0\end{pmatrix}, \begin{pmatrix}1\\-1\\-1\\0\end{pmatrix}, \begin{pmatrix}0\\0\\1\\-1\end{pmatrix}$	$\pm(0,0,0,1)$	$(-1,-1,0,0)$, $(-1,0,-1,-1)$, $(-1,0,-1,0)$, $(-1,0,0,0)$, $(0,0,-1,-1)$, $(0,0,-1,0)$, $(0,1,-1,-1)$, $(0,1,-1,0)$, $(0,1,0,0)$	$\begin{pmatrix}\frac{6334289640}{6291985913}\\-\frac{2670575400}{6291985913}\\\frac{2239779600}{6291985913}\\0\end{pmatrix}$

P	$\mathcal{V}(P)$	$R_s(P)$	$R_u(P)$	$e(P)$
K-4	$\pm\begin{pmatrix}1\\0\\0\\0\end{pmatrix}, \pm\begin{pmatrix}0\\1\\0\\0\end{pmatrix}, \begin{pmatrix}0\\0\\1\\0\end{pmatrix}, \begin{pmatrix}0\\0\\0\\1\end{pmatrix}, \begin{pmatrix}0\\-1\\-1\\0\end{pmatrix}, \begin{pmatrix}0\\-1\\0\\-1\end{pmatrix}$	$\pm(1,0,0,0)$, $\pm(0,0,1,0)$, $\pm(0,0,0,1)$	$(0,1,-1,-1)$, $(0,1,-1,0)$, $(0,1,0,-1)$, $(0,1,0,0)$	$\begin{pmatrix}0\\-\frac{260}{219}\\0\\0\end{pmatrix}$
K-5	$\pm\begin{pmatrix}1\\0\\0\\0\end{pmatrix}, \begin{pmatrix}0\\1\\0\\0\end{pmatrix}, \begin{pmatrix}0\\0\\1\\0\end{pmatrix}, \begin{pmatrix}0\\0\\0\\1\end{pmatrix}, \begin{pmatrix}1\\-1\\0\\0\end{pmatrix}, \begin{pmatrix}1\\0\\-1\\0\end{pmatrix}, \begin{pmatrix}0\\0\\1\\-1\end{pmatrix}$	$\pm(0,1,0,0)$, $\pm(0,0,0,1)$	$(-1,-1,-1,-1)$, $(-1,-1,-1,0)$, $(-1,-1,0,0)$, $(-1,0,-1,-1)$, $(-1,0,-1,0)$, $(-1,0,0,0)$, $(0,0,-1,-1)$, $(0,0,-1,0)$	$\begin{pmatrix}\frac{26763240}{20356871}\\0\\\frac{7896000}{20356871}\\0\end{pmatrix}$
K-6	$\pm\begin{pmatrix}1\\0\\0\\0\end{pmatrix}, \pm\begin{pmatrix}0\\1\\0\\0\end{pmatrix}, \begin{pmatrix}0\\0\\1\\0\end{pmatrix}, \begin{pmatrix}0\\0\\0\\1\end{pmatrix}, \begin{pmatrix}0\\-1\\-1\\0\end{pmatrix}, \begin{pmatrix}0\\0\\1\\-1\end{pmatrix}$	$\pm(1,0,0,0)$, $\pm(0,0,0,1)$	$(0,0,-1,-1)$, $(0,0,-1,0)$, $(0,1,-1,-1)$, $(0,1,-1,0)$, $(0,1,0,0)$	$\begin{pmatrix}0\\-\frac{1450}{2541}\\\frac{7600}{17787}\\0\end{pmatrix}$
K-7	$\pm\begin{pmatrix}1\\0\\0\\0\end{pmatrix}, \begin{pmatrix}0\\1\\0\\0\end{pmatrix}, \pm\begin{pmatrix}0\\0\\1\\0\end{pmatrix}, \begin{pmatrix}0\\0\\0\\1\end{pmatrix}, \begin{pmatrix}1\\-1\\0\\0\end{pmatrix}, \begin{pmatrix}0\\0\\1\\-1\end{pmatrix}$	$\pm(0,1,0,0)$, $\pm(0,0,0,1)$	$(-1,-1,0,0)$, $(-1,0,0,0)$, $(0,0,-1,-1)$, $(0,0,-1,0)$	$\begin{pmatrix}\frac{6}{11}\\0\\\frac{6}{11}\\0\end{pmatrix}$
K-8	$\pm\begin{pmatrix}1\\0\\0\\0\end{pmatrix}, \pm\begin{pmatrix}0\\1\\0\\0\end{pmatrix}, \pm\begin{pmatrix}0\\0\\1\\0\end{pmatrix}, \pm\begin{pmatrix}0\\0\\0\\1\end{pmatrix}$	$\pm(1,0,0,0)$, $\pm(0,1,0,0)$, $\pm(0,0,1,0)$, $\pm(0,0,0,1)$		$\begin{pmatrix}0\\0\\0\\0\end{pmatrix}$
K-9	$\pm\begin{pmatrix}1\\0\\0\\0\end{pmatrix}, \pm\begin{pmatrix}0\\1\\0\\0\end{pmatrix}, \pm\begin{pmatrix}0\\0\\1\\0\end{pmatrix}, \begin{pmatrix}0\\0\\0\\1\end{pmatrix}, \begin{pmatrix}0\\0\\1\\-1\end{pmatrix}$	$\pm(1,0,0,0)$, $\pm(0,1,0,0)$, $\pm(0,0,0,1)$	$(0,0,-1,-1)$, $(0,0,-1,0)$	$\begin{pmatrix}0\\0\\\frac{6}{11}\\0\end{pmatrix}$
K-10	$\pm\begin{pmatrix}1\\0\\0\\0\end{pmatrix}, \begin{pmatrix}0\\1\\0\\0\end{pmatrix}, \begin{pmatrix}0\\0\\1\\0\end{pmatrix}, \begin{pmatrix}0\\0\\0\\1\end{pmatrix}, \begin{pmatrix}1\\-1\\0\\0\end{pmatrix}, \begin{pmatrix}1\\-1\\-1\\0\end{pmatrix}, \begin{pmatrix}0\\1\\0\\-1\end{pmatrix}$	$\pm(0,0,1,0)$, $\pm(0,0,0,1)$	$(-1,-1,0,-1)$, $(-1,-1,0,0)$, $(-1,0,-1,0)$, $(-1,0,0,0)$	$\begin{pmatrix}\frac{17040}{15439}\\0\\0\\0\end{pmatrix}$
K-11	$\pm\begin{pmatrix}1\\0\\0\\0\end{pmatrix}, \pm\begin{pmatrix}0\\1\\0\\0\end{pmatrix}, \begin{pmatrix}0\\0\\1\\0\end{pmatrix}, \begin{pmatrix}0\\0\\0\\1\end{pmatrix}, \begin{pmatrix}0\\-1\\-1\\0\end{pmatrix}, \begin{pmatrix}0\\1\\0\\-1\end{pmatrix}$	$\pm(1,0,0,0)$, $\pm(0,0,1,0)$, $\pm(0,0,0,1)$		$\begin{pmatrix}0\\0\\0\\0\end{pmatrix}$
K-12	$\pm\begin{pmatrix}1\\0\\0\\0\end{pmatrix}, \begin{pmatrix}0\\1\\0\\0\end{pmatrix}, \begin{pmatrix}0\\0\\1\\0\end{pmatrix}, \begin{pmatrix}0\\0\\0\\1\end{pmatrix}, \begin{pmatrix}1\\-1\\0\\0\end{pmatrix}, \begin{pmatrix}-1\\0\\-1\\0\end{pmatrix}, \begin{pmatrix}0\\0\\1\\-1\end{pmatrix}$	$\pm(0,1,0,0)$, $\pm(0,0,0,1)$	$(0,0,-1,-1)$, $(0,0,-1,0)$	$\begin{pmatrix}\frac{5940}{118907}\\0\\\frac{58080}{118907}\\0\end{pmatrix}$

P	$\mathcal{V}(P)$	$R_s(P)$	$R_u(P)$	$e(P)$
K-13	$\pm\begin{pmatrix}1\\0\\0\\0\end{pmatrix}, \begin{pmatrix}0\\1\\0\\0\end{pmatrix}, \begin{pmatrix}0\\0\\1\\0\end{pmatrix}, \begin{pmatrix}0\\0\\0\\1\end{pmatrix}, \begin{pmatrix}1\\-1\\0\\0\end{pmatrix}, \begin{pmatrix}1\\0\\-1\\0\end{pmatrix}, \begin{pmatrix}-1\\0\\0\\-1\end{pmatrix}$	$\pm(0,1,0,0)$, $\pm(0,0,1,0)$, $\pm(0,0,0,1)$		$\begin{pmatrix}\frac{110}{203}\\0\\0\\0\end{pmatrix}$
L-1	$\pm\begin{pmatrix}1\\0\\0\\0\end{pmatrix}, \begin{pmatrix}0\\1\\0\\0\end{pmatrix}, \pm\begin{pmatrix}0\\0\\1\\0\end{pmatrix}, \pm\begin{pmatrix}0\\0\\0\\1\end{pmatrix}, \begin{pmatrix}-1\\-1\\1\\1\end{pmatrix}$	$\pm(0,1,0,0)$	$(0,-1,-1,0)$, $(0,-1,0,-1)$, $(0,0,-1,0)$, $(0,0,0,-1)$, $(1,-1,0,0)$, $(1,0,0,0)$	$\begin{pmatrix}-\frac{1824900}{2853121}\\0\\\frac{1824900}{2853121}\\\frac{1824900}{2853121}\end{pmatrix}$
L-2	$\pm\begin{pmatrix}1\\0\\0\\0\end{pmatrix}, \begin{pmatrix}0\\1\\0\\0\end{pmatrix}, \begin{pmatrix}0\\0\\1\\0\end{pmatrix}, \begin{pmatrix}0\\0\\0\\1\end{pmatrix}, \begin{pmatrix}-1\\-1\\1\\1\end{pmatrix}, \begin{pmatrix}1\\0\\-1\\0\end{pmatrix}, \begin{pmatrix}1\\0\\0\\-1\end{pmatrix}$	$\pm(0,1,0,0)$	$(-1,-1,-1,-1)$, $(-1,0,-1,-1)$, $(-1,0,-1,0)$, $(-1,0,0,-1)$, $(0,-1,-1,0)$, $(0,-1,0,-1)$, $(0,0,-1,0)$, $(0,0,0,-1)$	$\begin{pmatrix}\frac{2555701260}{4944865931}\\0\\\frac{2483476860}{4944865931}\\\frac{2483476860}{4944865931}\end{pmatrix}$
L-3	$\pm\begin{pmatrix}1\\0\\0\\0\end{pmatrix}, \begin{pmatrix}0\\1\\0\\0\end{pmatrix}, \begin{pmatrix}0\\0\\1\\0\end{pmatrix}, \begin{pmatrix}0\\0\\0\\1\end{pmatrix}, \begin{pmatrix}-1\\-1\\1\\1\end{pmatrix}, \begin{pmatrix}1\\0\\-1\\0\end{pmatrix}, \begin{pmatrix}1\\0\\-1\\-1\end{pmatrix}$	$\pm(0,1,0,0)$	$(-1,0,-1,0)$, $(-1,0,0,-1)$, $(0,-1,0,-1)$, $(0,0,0,-1)$, $(0,0,1,-1)$	$\begin{pmatrix}\frac{4151273350320}{16440690503203}\\0\\\frac{4524155739540}{16440690503203}\\\frac{10150485305940}{16440690503203}\end{pmatrix}$
L-4	$\pm\begin{pmatrix}1\\0\\0\\0\end{pmatrix}, \begin{pmatrix}0\\1\\0\\0\end{pmatrix}, \begin{pmatrix}0\\0\\1\\0\end{pmatrix}, \pm\begin{pmatrix}0\\0\\0\\1\end{pmatrix}, \begin{pmatrix}-1\\-1\\1\\1\end{pmatrix}, \begin{pmatrix}1\\0\\-1\\0\end{pmatrix}$	$\pm(0,1,0,0)$	$(-1,0,-1,0)$, $(0,-1,-1,0)$, $(0,-1,0,-1)$, $(0,0,-1,0)$, $(0,0,0,-1)$	$\begin{pmatrix}-\frac{2793968096940}{13060549547443}\\0\\\frac{8560548364140}{13060549547443}\\\frac{8232843774060}{13060549547443}\end{pmatrix}$
M-1	$\begin{pmatrix}1\\0\\0\\0\end{pmatrix}, \begin{pmatrix}0\\1\\0\\0\end{pmatrix}, \begin{pmatrix}0\\0\\1\\0\end{pmatrix}, \begin{pmatrix}0\\0\\0\\1\end{pmatrix}, \begin{pmatrix}0\\-1\\0\\-1\end{pmatrix}, \begin{pmatrix}0\\0\\-1\\1\end{pmatrix}, \begin{pmatrix}1\\0\\0\\-1\end{pmatrix}, \begin{pmatrix}-1\\0\\1\\0\end{pmatrix}$	$\pm(0,1,0,0)$	$(-1,0,-1,-1)$, $(-1,0,-1,0)$, $(-1,0,0,0)$, $(-1,1,-1,-1)$, $(0,0,-1,0)$	$\begin{pmatrix}\frac{2003379647360}{5748810067571}\\0\\\frac{2700009652340}{5748810067571}\\-\frac{1155042818020}{5748810067571}\end{pmatrix}$
N-1	$\pm\begin{pmatrix}1\\0\\0\\0\end{pmatrix}, \pm\begin{pmatrix}0\\1\\0\\0\end{pmatrix}, \begin{pmatrix}0\\0\\1\\0\end{pmatrix}, \begin{pmatrix}0\\0\\0\\1\end{pmatrix}, \pm\begin{pmatrix}1\\-1\\0\\0\end{pmatrix}, \begin{pmatrix}2\\0\\-1\\-1\end{pmatrix}$	$\pm(0,0,1,0)$, $\pm(0,0,1,-1)$, $\pm(0,0,0,1)$		$\begin{pmatrix}\frac{27300}{16571}\\0\\0\\0\end{pmatrix}$

P	$\mathcal{V}(P)$	$R_s(P)$	$R_u(P)$	$e(P)$
N-2	$\pm\begin{pmatrix}1\\0\\0\\0\end{pmatrix}, \pm\begin{pmatrix}0\\1\\0\\0\end{pmatrix}, \begin{pmatrix}0\\0\\1\\0\end{pmatrix}, \begin{pmatrix}0\\0\\0\\1\end{pmatrix}, \pm\begin{pmatrix}1\\-1\\0\\0\end{pmatrix}, \begin{pmatrix}1\\1\\-1\\-1\end{pmatrix}$	$\pm(0,0,1,0)$, $\pm(0,0,1,-1)$, $\pm(0,0,0,1)$		$\begin{pmatrix}\frac{2950}{3751}\\\frac{2950}{3751}\\0\\0\end{pmatrix}$
N-3	$\pm\begin{pmatrix}1\\0\\0\\0\end{pmatrix}, \pm\begin{pmatrix}0\\1\\0\\0\end{pmatrix}, \begin{pmatrix}0\\0\\1\\0\end{pmatrix}, \begin{pmatrix}0\\0\\0\\1\end{pmatrix}, \pm\begin{pmatrix}1\\-1\\0\\0\end{pmatrix}, \begin{pmatrix}1\\0\\-1\\-1\end{pmatrix}$	$\pm(0,0,1,0)$, $\pm(0,0,1,-1)$, $\pm(0,0,0,1)$		$\begin{pmatrix}\frac{25050}{35413}\\0\\0\\0\end{pmatrix}$
N-4	$\pm\begin{pmatrix}1\\0\\0\\0\end{pmatrix}, \pm\begin{pmatrix}0\\1\\0\\0\end{pmatrix}, \begin{pmatrix}0\\0\\1\\0\end{pmatrix}, \begin{pmatrix}0\\0\\0\\1\end{pmatrix}, \pm\begin{pmatrix}1\\-1\\0\\0\end{pmatrix}, \begin{pmatrix}0\\0\\-1\\-1\end{pmatrix}$	$\pm(0,0,1,0)$, $\pm(0,0,1,-1)$, $\pm(0,0,0,1)$		$\begin{pmatrix}0\\0\\0\\0\end{pmatrix}$
O-1	$\pm\begin{pmatrix}1\\0\\0\\0\end{pmatrix}, \begin{pmatrix}0\\1\\0\\0\end{pmatrix}, \begin{pmatrix}0\\0\\1\\0\end{pmatrix}, \begin{pmatrix}0\\0\\0\\1\end{pmatrix}, \pm\begin{pmatrix}1\\-1\\0\\0\end{pmatrix}, \begin{pmatrix}0\\1\\-1\\0\end{pmatrix}, \begin{pmatrix}0\\1\\0\\-1\end{pmatrix}$	$\pm(0,0,1,0)$, $\pm(0,0,0,1)$	$(-1,-1,-1,-1)$, $(-1,-1,-1,0)$, $(-1,-1,0,-1)$, $(-1,-1,0,0)$, $(0,-1,-1,-1)$, $(0,-1,-1,0)$, $(0,-1,0,-1)$, $(0,-1,0,0)$	$\begin{pmatrix}0\\\frac{363545}{266403}\\0\\0\end{pmatrix}$
O-2	$\pm\begin{pmatrix}1\\0\\0\\0\end{pmatrix}, \begin{pmatrix}0\\1\\0\\0\end{pmatrix}, \begin{pmatrix}0\\0\\1\\0\end{pmatrix}, \begin{pmatrix}0\\0\\0\\1\end{pmatrix}, \pm\begin{pmatrix}1\\-1\\0\\0\end{pmatrix}, \begin{pmatrix}0\\1\\-1\\0\end{pmatrix}, \begin{pmatrix}0\\0\\1\\-1\end{pmatrix}$	$\pm(0,0,0,1)$	$(-1,-1,-1,-1)$, $(-1,-1,-1,0)$, $(-1,-1,0,0)$, $(0,-1,-1,-1)$, $(0,-1,-1,0)$, $(0,-1,0,0)$, $(0,0,-1,-1)$, $(0,0,-1,0)$	$\begin{pmatrix}0\\\frac{211857525}{237174992}\\\frac{80463375}{237174992}\\0\end{pmatrix}$
O-3	$\pm\begin{pmatrix}1\\0\\0\\0\end{pmatrix}, \begin{pmatrix}0\\1\\0\\0\end{pmatrix}, \begin{pmatrix}0\\0\\1\\0\end{pmatrix}, \begin{pmatrix}0\\0\\0\\1\end{pmatrix}, \pm\begin{pmatrix}1\\-1\\0\\0\end{pmatrix}, \begin{pmatrix}1\\0\\-1\\0\end{pmatrix}, \begin{pmatrix}1\\0\\0\\-1\end{pmatrix}$	$\pm(0,0,1,0)$, $\pm(0,0,0,1)$	$(-1,-1,-1,-1)$, $(-1,-1,-1,0)$, $(-1,-1,0,-1)$, $(-1,-1,0,0)$, $(0,-1,0,0)$	$\begin{pmatrix}\frac{269739295}{201232113}\\\frac{20572710}{67077371}\\0\\0\end{pmatrix}$
O-4	$\pm\begin{pmatrix}1\\0\\0\\0\end{pmatrix}, \begin{pmatrix}0\\1\\0\\0\end{pmatrix}, \begin{pmatrix}0\\0\\1\\0\end{pmatrix}, \begin{pmatrix}0\\0\\0\\1\end{pmatrix}, \pm\begin{pmatrix}1\\-1\\0\\0\end{pmatrix}, \begin{pmatrix}0\\1\\-1\\0\end{pmatrix}, \begin{pmatrix}1\\0\\0\\-1\end{pmatrix}$	$\pm(0,0,1,0)$, $\pm(0,0,0,1)$	$(-1,-1,-1,-1)$, $(-1,-1,-1,0)$, $(-1,-1,0,-1)$, $(-1,-1,0,0)$, $(0,-1,-1,0)$, $(0,-1,0,0)$	$\begin{pmatrix}\frac{51832575}{73313456}\\\frac{58443525}{73313456}\\0\\0\end{pmatrix}$
O-5	$\pm\begin{pmatrix}1\\0\\0\\0\end{pmatrix}, \begin{pmatrix}0\\1\\0\\0\end{pmatrix}, \begin{pmatrix}0\\0\\1\\0\end{pmatrix}, \begin{pmatrix}0\\0\\0\\1\end{pmatrix}, \pm\begin{pmatrix}1\\-1\\0\\0\end{pmatrix}, \begin{pmatrix}1\\0\\-1\\0\end{pmatrix}, \begin{pmatrix}0\\0\\1\\-1\end{pmatrix}$	$\pm(0,0,0,1)$	$(-1,-1,-1,-1)$, $(-1,-1,1,0)$, $(-1,-1,0,0)$, $(0,-1,0,0)$, $(0,0,-1,-1)$, $(0,0,-1,0)$	$\begin{pmatrix}\frac{394596985345}{573603226348}\\\frac{48137018910}{143400806587}\\\frac{234003144575}{573603226348}\\0\end{pmatrix}$
O-6	$\pm\begin{pmatrix}1\\0\\0\\0\end{pmatrix}, \begin{pmatrix}0\\1\\0\\0\end{pmatrix}, \pm\begin{pmatrix}0\\0\\1\\0\end{pmatrix}, \begin{pmatrix}0\\0\\0\\1\end{pmatrix}, \pm\begin{pmatrix}1\\-1\\0\\0\end{pmatrix}, \begin{pmatrix}0\\1\\0\\-1\end{pmatrix}$	$\pm(0,0,1,0)$, $\pm(0,0,0,1)$	$(-1,-1,0,-1)$, $(-1,-1,0,0)$, $(0,-1,0,-1)$, $(0,-1,0,0)$	$\begin{pmatrix}0\\\frac{17020}{19651}\\0\\0\end{pmatrix}$

P	$\mathcal{V}(P)$	$R_s(P)$	$R_u(P)$	$e(P)$
O-7	$\pm\begin{pmatrix}1\\0\\0\\0\end{pmatrix},\begin{pmatrix}0\\1\\0\\0\end{pmatrix},\begin{pmatrix}0\\0\\1\\0\end{pmatrix},\begin{pmatrix}0\\0\\0\\1\end{pmatrix},\ \pm\begin{pmatrix}1\\-1\\0\\0\end{pmatrix},\begin{pmatrix}1\\0\\-1\\0\end{pmatrix},\begin{pmatrix}-1\\1\\0\\-1\end{pmatrix}$	$\pm(0,0,1,0)$, $\pm(0,0,0,1)$	$(-1,-1,-1,0)$, $(-1,-1,0,0)$, $(0,-1,0,-1)$, $(0,-1,0,0)$	$\begin{pmatrix}0\\ \frac{176055}{198188}\\0\\0\end{pmatrix}$
O-8	$\pm\begin{pmatrix}1\\0\\0\\0\end{pmatrix},\begin{pmatrix}0\\1\\0\\0\end{pmatrix},\pm\begin{pmatrix}0\\0\\1\\0\end{pmatrix},\begin{pmatrix}0\\0\\0\\1\end{pmatrix},\ \pm\begin{pmatrix}1\\-1\\0\\0\end{pmatrix},\begin{pmatrix}1\\0\\0\\-1\end{pmatrix}$	$\pm(0,0,1,0)$, $\pm(0,0,0,1)$	$(-1,-1,0,-1)$, $(-1,-1,0,0)$, $(0,-1,0,0)$	$\begin{pmatrix}\frac{1663200}{2735927}\\ \frac{982960}{2735927}\\0\\0\end{pmatrix}$
O-9	$\pm\begin{pmatrix}1\\0\\0\\0\end{pmatrix},\begin{pmatrix}0\\1\\0\\0\end{pmatrix},\begin{pmatrix}0\\0\\1\\0\end{pmatrix},\begin{pmatrix}0\\0\\0\\1\end{pmatrix},\ \pm\begin{pmatrix}1\\-1\\0\\0\end{pmatrix},\begin{pmatrix}1\\0\\-1\\0\end{pmatrix},\begin{pmatrix}1\\-1\\0\\-1\end{pmatrix}$	$\pm(0,0,1,0)$, $\pm(0,0,0,1)$	$(-1,-1,-1,0)$, $(-1,-1,0,0)$	$\begin{pmatrix}\frac{230614890}{197727059}\\ -\frac{53988825}{395454118}\\0\\0\end{pmatrix}$
O-10	$\pm\begin{pmatrix}1\\0\\0\\0\end{pmatrix},\begin{pmatrix}0\\1\\0\\0\end{pmatrix},\pm\begin{pmatrix}0\\0\\1\\0\end{pmatrix},\ \begin{pmatrix}0\\0\\0\\1\end{pmatrix},\pm\begin{pmatrix}1\\-1\\0\\0\end{pmatrix},\begin{pmatrix}0\\0\\1\\-1\end{pmatrix}$	$\pm(0,0,0,1)$	$(-1,-1,0,0)$, $(0,-1,0,0)$, $(0,0,-1,-1)$, $(0,0,-1,0)$	$\begin{pmatrix}0\\ \frac{168}{409}\\ \frac{6}{11}\\0\end{pmatrix}$
O-11	$\pm\begin{pmatrix}1\\0\\0\\0\end{pmatrix},\begin{pmatrix}0\\1\\0\\0\end{pmatrix},\pm\begin{pmatrix}0\\0\\1\\0\end{pmatrix},\ \pm\begin{pmatrix}0\\0\\0\\1\end{pmatrix},\pm\begin{pmatrix}1\\-1\\0\\0\end{pmatrix}$	$\pm(0,0,1,0)$, $\pm(0,0,0,1)$	$(-1,-1,0,0)$, $(0,-1,0,0)$	$\begin{pmatrix}0\\ \frac{168}{409}\\0\\0\end{pmatrix}$
O-12	$\pm\begin{pmatrix}1\\0\\0\\0\end{pmatrix},\begin{pmatrix}0\\1\\0\\0\end{pmatrix},\begin{pmatrix}0\\0\\1\\0\end{pmatrix},\begin{pmatrix}0\\0\\0\\1\end{pmatrix},\ \pm\begin{pmatrix}1\\-1\\0\\0\end{pmatrix},\begin{pmatrix}0\\1\\-1\\0\end{pmatrix},\begin{pmatrix}1\\-1\\0\\-1\end{pmatrix}$	$\pm(0,0,1,0)$, $\pm(0,0,0,1)$	$(-1,-1,-1,0)$, $(-1,-1,0,0)$	$\begin{pmatrix}\frac{115654710}{196251577}\\ \frac{66679950}{196251577}\\0\\0\end{pmatrix}$
O-13	$\pm\begin{pmatrix}1\\0\\0\\0\end{pmatrix},\begin{pmatrix}0\\1\\0\\0\end{pmatrix},\begin{pmatrix}0\\0\\1\\0\end{pmatrix},\begin{pmatrix}0\\0\\0\\1\end{pmatrix},\ \pm\begin{pmatrix}1\\-1\\0\\0\end{pmatrix},\begin{pmatrix}1\\-1\\-1\\0\end{pmatrix},\begin{pmatrix}1\\-1\\0\\-1\end{pmatrix}$	$\pm(0,0,1,0)$, $\pm(0,0,0,1)$	$(-1,-1,0,0)$	$\begin{pmatrix}\frac{38978775}{35875483}\\ -\frac{23199825}{35875483}\\0\\0\end{pmatrix}$
O-14	$\pm\begin{pmatrix}1\\0\\0\\0\end{pmatrix},\begin{pmatrix}0\\1\\0\\0\end{pmatrix},\begin{pmatrix}0\\0\\1\\0\end{pmatrix},\ \begin{pmatrix}0\\0\\0\\1\end{pmatrix},\pm\begin{pmatrix}1\\-1\\0\\0\end{pmatrix},\ \begin{pmatrix}1\\-1\\-1\\0\end{pmatrix},\begin{pmatrix}0\\0\\1\\-1\end{pmatrix}$	$\pm(0,0,0,1)$	$(-1,-1,0,0)$, $(0,0,-1,-1)$, $(0,0,-1,0)$	$\begin{pmatrix}\frac{1233812146125}{2450627729348}\\ -\frac{151909753125}{2450627729348}\\ \frac{1142554261875}{2450627729348}\\0\end{pmatrix}$
O-15	$\pm\begin{pmatrix}1\\0\\0\\0\end{pmatrix},\begin{pmatrix}0\\1\\0\\0\end{pmatrix},\pm\begin{pmatrix}0\\0\\1\\0\end{pmatrix},\ \begin{pmatrix}0\\0\\0\\1\end{pmatrix},\pm\begin{pmatrix}1\\-1\\0\\0\end{pmatrix},\begin{pmatrix}1\\-1\\0\\-1\end{pmatrix}$	$\pm(0,0,1,0)$, $\pm(0,0,0,1)$	$(-1,-1,0,0)$	$\begin{pmatrix}\frac{42144}{78995}\\ -\frac{7936}{78995}\\0\\0\end{pmatrix}$

P	$\mathcal{V}(P)$	$R_s(P)$	$R_u(P)$	$e(P)$
O-16	$\pm\begin{pmatrix}1\\0\\0\\0\end{pmatrix}, \begin{pmatrix}0\\1\\0\\0\end{pmatrix}, \begin{pmatrix}0\\0\\1\\0\end{pmatrix}, \begin{pmatrix}0\\0\\0\\1\end{pmatrix}, \pm\begin{pmatrix}1\\-1\\0\\0\end{pmatrix}, \begin{pmatrix}1\\0\\-1\\0\end{pmatrix}, \begin{pmatrix}-1\\0\\0\\-1\end{pmatrix}$	$\pm(0,0,1,0)$, $\pm(0,0,0,1)$	$(0,-1,0,0)$	$\begin{pmatrix}\frac{1544575}{111477439}\\ \frac{43444950}{111477439}\\ 0\\ 0\end{pmatrix}$
O-17	$\pm\begin{pmatrix}1\\0\\0\\0\end{pmatrix}, \begin{pmatrix}0\\1\\0\\0\end{pmatrix}, \begin{pmatrix}0\\0\\1\\0\end{pmatrix}, \begin{pmatrix}0\\0\\0\\1\end{pmatrix}, \pm\begin{pmatrix}1\\-1\\0\\0\end{pmatrix}, \begin{pmatrix}1\\-1\\-1\\0\end{pmatrix}, \begin{pmatrix}-1\\0\\0\\-1\end{pmatrix}$	$\pm(0,0,1,0)$, $\pm(0,0,0,1)$		$\begin{pmatrix}0\\ -\frac{17940}{162187}\\ 0\\ 0\end{pmatrix}$
P-1	$\begin{pmatrix}1\\0\\0\\0\end{pmatrix}, \begin{pmatrix}0\\1\\0\\0\end{pmatrix}, \begin{pmatrix}0\\0\\1\\0\end{pmatrix}, \pm\begin{pmatrix}0\\0\\0\\1\end{pmatrix}, \pm\begin{pmatrix}1\\1\\-1\\-1\end{pmatrix}, \begin{pmatrix}0\\0\\-1\\-1\end{pmatrix}, \begin{pmatrix}-1\\-1\\1\\0\end{pmatrix}$	$\pm(1,-1,0,0)$	$(-1,0,-1,0)$, $(0,-1,-1,0)$, $(0,0,-1,1)$	$\begin{pmatrix}\frac{451036894560}{2028544442569}\\ \frac{451036894560}{2028544442569}\\ \frac{701882402340}{2028544442569}\\ -\frac{984670344060}{2028544442569}\end{pmatrix}$
P-2	$\begin{pmatrix}1\\0\\0\\0\end{pmatrix}, \begin{pmatrix}0\\1\\0\\0\end{pmatrix}, \begin{pmatrix}0\\0\\1\\0\end{pmatrix}, \pm\begin{pmatrix}0\\0\\0\\1\end{pmatrix}, \pm\begin{pmatrix}1\\1\\-1\\-1\end{pmatrix}, \begin{pmatrix}-1\\-1\\0\\0\end{pmatrix}, \begin{pmatrix}-1\\-1\\1\\0\end{pmatrix}$	$\pm(1,-1,0,0)$	$(-1,0,-1,0)$, $(0,-1,-1,0)$, $(0,0,-1,0)$, $(0,0,-1,1)$	$\begin{pmatrix}-\frac{54894540}{324135871}\\ -\frac{54894540}{324135871}\\ \frac{238278060}{324135871}\\ 0\end{pmatrix}$
P-3	$\begin{pmatrix}1\\0\\0\\0\end{pmatrix}, \begin{pmatrix}0\\1\\0\\0\end{pmatrix}, \pm\begin{pmatrix}0\\0\\1\\0\end{pmatrix}, \pm\begin{pmatrix}0\\0\\0\\1\end{pmatrix}, \pm\begin{pmatrix}1\\1\\-1\\-1\end{pmatrix}, \begin{pmatrix}-1\\-1\\1\\0\end{pmatrix}$	$\pm(1,-1,0,0)$	$(-1,0,-1,0)$, $(0,-1,-1,0)$	$\begin{pmatrix}\frac{91335300}{369479321}\\ \frac{91335300}{369479321}\\ \frac{142965700}{369479321}\\ 0\end{pmatrix}$
Q-1	$\pm\begin{pmatrix}1\\0\\0\\0\end{pmatrix}, \begin{pmatrix}0\\1\\0\\0\end{pmatrix}, \begin{pmatrix}0\\0\\1\\0\end{pmatrix}, \pm\begin{pmatrix}0\\0\\0\\1\end{pmatrix}, \begin{pmatrix}1\\-1\\-1\\0\end{pmatrix}, \begin{pmatrix}1\\-1\\-1\\-1\end{pmatrix}, \begin{pmatrix}-1\\1\\0\\0\end{pmatrix}$		$(-1,-1,0,0)$, $(0,-1,1,0)$, $(0,0,1,0)$, $(0,0,0,1)$	$\begin{pmatrix}\frac{74703180}{153124261}\\ 0\\ -\frac{57721380}{153124261}\\ -\frac{74703180}{153124261}\end{pmatrix}$
R-1	$\pm\begin{pmatrix}1\\0\\0\\0\end{pmatrix}, \pm\begin{pmatrix}0\\1\\0\\0\end{pmatrix}, \pm\begin{pmatrix}0\\0\\1\\0\end{pmatrix}, \pm\begin{pmatrix}0\\0\\0\\1\end{pmatrix}, \begin{pmatrix}1\\1\\1\\1\end{pmatrix}$		$(-1,0,0,0)$, $(0,-1,0,0)$, $(0,0,-1,0)$, $(0,0,0,-1)$	$\begin{pmatrix}\frac{1436760}{2766841}\\ \frac{1436760}{2766841}\\ \frac{1436760}{2766841}\\ \frac{1436760}{2766841}\end{pmatrix}$
S-1	$\pm\begin{pmatrix}1\\0\\0\\0\end{pmatrix}, \pm\begin{pmatrix}0\\1\\0\\0\end{pmatrix}, \begin{pmatrix}0\\0\\1\\0\end{pmatrix}, \begin{pmatrix}0\\0\\0\\1\end{pmatrix}, \pm\begin{pmatrix}1\\-1\\0\\0\end{pmatrix}, \begin{pmatrix}1\\0\\-1\\0\end{pmatrix}, \begin{pmatrix}1\\0\\0\\-1\end{pmatrix}$	$\pm(0,0,1,0)$, $\pm(0,0,0,1)$		$\begin{pmatrix}\frac{3850}{3389}\\ 0\\ 0\\ 0\end{pmatrix}$

P	$\mathcal{V}(P)$	$R_s(P)$	$R_u(P)$	$e(P)$
S-2	$\pm\begin{pmatrix}1\\0\\0\\0\end{pmatrix}, \pm\begin{pmatrix}0\\1\\0\\0\end{pmatrix}, \begin{pmatrix}0\\0\\1\\0\end{pmatrix}, \begin{pmatrix}0\\0\\0\\1\end{pmatrix}, \pm\begin{pmatrix}1\\-1\\0\\0\end{pmatrix}, \begin{pmatrix}1\\0\\-1\\0\end{pmatrix}, \begin{pmatrix}0\\1\\0\\-1\end{pmatrix}$	$\pm(0,0,1,0)$, $\pm(0,0,0,1)$		$\begin{pmatrix}\frac{11175}{19936}\\\frac{11175}{19936}\\0\\0\end{pmatrix}$
S-3	$\pm\begin{pmatrix}1\\0\\0\\0\end{pmatrix}, \pm\begin{pmatrix}0\\1\\0\\0\end{pmatrix}, \begin{pmatrix}0\\0\\1\\0\end{pmatrix}, \begin{pmatrix}0\\0\\0\\1\end{pmatrix}, \pm\begin{pmatrix}1\\-1\\0\\0\end{pmatrix}, \begin{pmatrix}1\\0\\-1\\0\end{pmatrix}, \begin{pmatrix}-1\\1\\0\\-1\end{pmatrix}$	$\pm(0,0,1,0)$, $\pm(0,0,0,1)$		$\begin{pmatrix}0\\\frac{139}{258}\\0\\0\end{pmatrix}$
S-4	$\pm\begin{pmatrix}1\\0\\0\\0\end{pmatrix}, \pm\begin{pmatrix}0\\1\\0\\0\end{pmatrix}, \pm\begin{pmatrix}0\\0\\1\\0\end{pmatrix}, \pm\begin{pmatrix}0\\0\\0\\1\end{pmatrix}, \pm\begin{pmatrix}1\\-1\\0\\0\end{pmatrix}$	$\pm(0,0,1,0)$, $\pm(0,0,0,1)$		$\begin{pmatrix}0\\0\\0\\0\end{pmatrix}$
S-5	$\pm\begin{pmatrix}1\\0\\0\\0\end{pmatrix}, \pm\begin{pmatrix}0\\1\\0\\0\end{pmatrix}, \begin{pmatrix}0\\0\\1\\0\end{pmatrix}, \pm\begin{pmatrix}0\\0\\0\\1\end{pmatrix}, \pm\begin{pmatrix}1\\-1\\0\\0\end{pmatrix}, \begin{pmatrix}1\\0\\-1\\0\end{pmatrix}$	$\pm(0,0,1,0)$, $\pm(0,0,0,1)$		$\begin{pmatrix}\frac{36}{67}\\0\\0\\0\end{pmatrix}$
S-6	$\pm\begin{pmatrix}1\\0\\0\\0\end{pmatrix}, \pm\begin{pmatrix}0\\1\\0\\0\end{pmatrix}, \pm\begin{pmatrix}0\\0\\1\\0\end{pmatrix}, \begin{pmatrix}0\\0\\0\\1\end{pmatrix}, \pm\begin{pmatrix}1\\-1\\0\\0\end{pmatrix}, \begin{pmatrix}0\\0\\-1\\-1\end{pmatrix}$	$\pm(0,0,0,1)$	$(0,0,1,-1)$, $(0,0,1,0)$	$\begin{pmatrix}0\\0\\-\frac{6}{11}\\0\end{pmatrix}$
S-7	$\pm\begin{pmatrix}1\\0\\0\\0\end{pmatrix}, \pm\begin{pmatrix}0\\1\\0\\0\end{pmatrix}, \begin{pmatrix}0\\0\\1\\0\end{pmatrix}, \begin{pmatrix}0\\0\\0\\1\end{pmatrix}, \pm\begin{pmatrix}1\\-1\\0\\0\end{pmatrix}, \begin{pmatrix}1\\0\\-1\\0\end{pmatrix}, \begin{pmatrix}1\\0\\-1\\-1\end{pmatrix}$	$\pm(0,0,0,1)$	$(0,0,1,-1)$, $(0,0,1,0)$	$\begin{pmatrix}\frac{453705}{462191}\\0\\-\frac{409005}{924382}\\0\end{pmatrix}$
S-8	$\pm\begin{pmatrix}1\\0\\0\\0\end{pmatrix}, \pm\begin{pmatrix}0\\1\\0\\0\end{pmatrix}, \begin{pmatrix}0\\0\\1\\0\end{pmatrix}, \begin{pmatrix}0\\0\\0\\1\end{pmatrix}, \pm\begin{pmatrix}1\\-1\\0\\0\end{pmatrix}, \begin{pmatrix}1\\0\\-1\\0\end{pmatrix}, \begin{pmatrix}-1\\0\\0\\-1\end{pmatrix}$	$\pm(0,0,1,0)$, $\pm(0,0,0,1)$		$\begin{pmatrix}0\\0\\0\\0\end{pmatrix}$
T-1	$\pm\begin{pmatrix}1\\0\\0\\0\end{pmatrix}, \pm\begin{pmatrix}0\\1\\0\\0\end{pmatrix}, \pm\begin{pmatrix}0\\0\\1\\0\end{pmatrix}, \pm\begin{pmatrix}0\\0\\0\\1\end{pmatrix}, \pm\begin{pmatrix}1\\1\\1\\1\end{pmatrix}$			$\begin{pmatrix}0\\0\\0\\0\end{pmatrix}$
U-1	$\pm\begin{pmatrix}1\\0\\0\\0\end{pmatrix}, \pm\begin{pmatrix}0\\1\\0\\0\end{pmatrix}, \pm\begin{pmatrix}0\\0\\1\\0\end{pmatrix}, \pm\begin{pmatrix}0\\0\\0\\1\end{pmatrix}, \begin{pmatrix}-1\\-1\\0\\0\end{pmatrix}, \begin{pmatrix}0\\0\\-1\\-1\end{pmatrix}$		$(1,0,0,0)$, $(0,1,0,0)$, $(0,0,1,0)$, $(0,0,0,1)$	$\begin{pmatrix}-\frac{168}{409}\\-\frac{168}{409}\\-\frac{168}{409}\\-\frac{168}{409}\end{pmatrix}$

P	$\mathcal{V}(P)$	$R_s(P)$	$R_u(P)$	$e(P)$
V-1	$\pm\begin{pmatrix}1\\0\\0\\0\end{pmatrix}, \pm\begin{pmatrix}0\\1\\0\\0\end{pmatrix}, \pm\begin{pmatrix}0\\0\\1\\0\end{pmatrix}, \pm\begin{pmatrix}0\\0\\0\\1\end{pmatrix}, \pm\begin{pmatrix}0\\0\\1\\1\end{pmatrix}, \begin{pmatrix}-1\\-1\\0\\0\end{pmatrix}$		$(1, 0, 0, 0)$, $(0, 1, 0, 0)$	$\begin{pmatrix}-\frac{168}{409}\\-\frac{168}{409}\\0\\0\end{pmatrix}$
W-1	$\pm\begin{pmatrix}1\\0\\0\\0\end{pmatrix}, \pm\begin{pmatrix}0\\1\\0\\0\end{pmatrix}, \pm\begin{pmatrix}0\\0\\1\\0\end{pmatrix}, \pm\begin{pmatrix}0\\0\\0\\1\end{pmatrix}, \pm\begin{pmatrix}1\\1\\0\\0\end{pmatrix}, \pm\begin{pmatrix}0\\0\\1\\1\end{pmatrix}$			$\begin{pmatrix}0\\0\\0\\0\end{pmatrix}$
X-1	$\begin{pmatrix}1\\0\\0\\0\end{pmatrix}, \begin{pmatrix}0\\1\\0\\0\end{pmatrix}, \begin{pmatrix}0\\0\\1\\0\end{pmatrix}, \begin{pmatrix}0\\0\\0\\1\end{pmatrix}, \begin{pmatrix}-1\\-1\\-1\\-1\end{pmatrix}, \begin{pmatrix}-1\\-1\\0\\0\end{pmatrix}, \begin{pmatrix}0\\0\\-1\\-1\end{pmatrix}, \begin{pmatrix}1\\0\\1\\0\end{pmatrix}, \begin{pmatrix}0\\1\\0\\1\end{pmatrix}$			$\begin{pmatrix}0\\0\\0\\0\end{pmatrix}$

Acknowledgments

This work was supported in part by the Grant-in-Aid for Scientific Research (No. 08211205), The Ministry of Education, Science and Culture, Japan, and also by the Inamori Foundation.

The author wishes to thank Professor Toshiki Mabuchi for introducing him to this topic and also for constant encouragements.

References

1. M. Audin, "The topology of torus actions on symplectic manifolds", Progress in Math., 93, Birkhäuser Verlag, Basel, Boston, Berlin, 1991.
2. V. V. Batyrev, *Toroidal Fano 3-folds*, Math. USSR-Izv. **19** (1982), pp. 13–25; Izv. Akad. Nauk SSSR **45** (1981), pp. 704–717.
3. V. V. Batyrev, *Higher-dimensional toric varieties with ample anticanonical class*, Thesis, Moscow State University, 1985 (in Russian).
4. V. V. Batyrev, *On the classification of toric Fano 4-folds*, preprint.
5. R. Bott, *Vector fields and characteristic numbers*, Michigan Math. J. **14** (1967), pp. 231–244.
6. E. Calabi, *Extremal Kähler metrics*, in "Seminar on differential geometry", Annals of Math. Studies **102**, Princeton Univ. Press, Princeton, 1982, pp. 259–290.
7. E. Calabi, *Extremal Kähler metrics* II, in "Differential geometry and complex analysis", Springer-Verlag, Berlin, Heidelberg, New York, 1985, pp. 95–114.
8. D. A. Cox, *The homogeneous coordinate ring of a toric variety*, J. Alg. Geom. **4** (1995), pp. 17–50.

9. T. Delzant, *Hamiltoniens périodiques et image convexe de l'application moment*, Bull. Soc. Math. France **116** (1988), pp. 315–339.
10. M. Demazure, *Sous-groupes algébriques de rang maximum du groupe de Cremona*, Ann. Sci. École Norm. Sup. **3** (1970), pp. 507–588.
11. W.-Y. Ding and G. Tian, *Kähler-Einstein metrics and the generalized Futaki invariant*, Invent. Math. **110** (1992), pp. 315–335.
12. A. Futaki, *An obstruction to the existence of Kähler Einstein metrics*, Invent. Math. **73** (1983), pp. 437–443.
13. A. Futaki and T. Mabuchi, *Bilinear forms and extremal Kähler vector fields associated with Kähler classes*, Math. Ann. **301** (1995), pp. 199–210.
14. A. Futaki, T. Mabuchi and Y. Sakane, *Einstein-Kähler metrics with positive Ricci curvature*, in "Kähler metric and moduli spaces", Adv. Stud. in Pure Math. **18-II**, Kinokuniya and Academic Press, Tokyo and Boston, 1990, pp. 339–358.
15. A. Futaki and S. Morita, *Invariant polynomials on compact complex manifolds*, Proc. Japan Acad. **60** (1984), pp. 369–372.
16. V. Guillemin, "Moment maps and combinatorial invariants of Hamiltonian T^n-spaces", Progress in Math., 122, Birkhäuser Verlag, Basel, Boston, Berlin, 1994.
17. G. Kempf, F. Knudsen, D. Mumfold and B. Saint-Donat, "Toroidal embeddings I", Lecture Notes in Mathematics, 339, Springer-Verlag, Berlin, Heidelberg, New York, 1973.
18. S. Kobayashi, "Transformation groups in differential geometry", Springer-Verlag, Berlin, Heidelberg, New York, 1972.
19. R. Koelman, *The number of moduli of families of curves on toric surfaces*, Thesis, Univ. Nijmegen, 1991.
20. T. Mabuchi, *Einstein-Kähler forms, Futaki invariants and convex geometry on toric Fano varieties*, Osaka J. Math. **24** (1987), pp. 705–737.
21. T. Mabuchi, *An algebraic character associated with the Poisson brackets*, in "Recent Topics in Differential and Analytic Geometry", Adv. Stud. in Pure Math. **18-I**, Kinokuniya and Academic Press, Tokyo and Boston, 1990, pp. 339–358.
22. Y. Matsushima, *Sur la structure du groupe d'homomorphismes analitique d'une certaine variété kählerienne*, Nagoya Math. J. **46** (1957), pp. 145–150.
23. A. M. Nadel, *Multiplier ideal sheaves and Kähler-Einstein metrics of positive scalar curvature*, Ann. of Math. **132** (1990), pp. 549–596.
24. Y. Nakagawa, *Einstein-Kähler toric Fano fourfolds*, Tôhoku Math. J. **45** (1993), pp. 297–310.

25. Y. Nakagawa, *Classification of Einstein-Kähler toric Fano fourfolds*, Tôhoku Math. J. **46** (1994), pp. 125–133
26. T. Oda, "Convex bodies and algebraic geometry : An introduction to the theory of toric varieties", Ergebnisse der Math. (3) 15, Springer-Verlag, Berlin, Heidelberg, New York, 1988.
27. Y. Sakane, *Example of compact Einstein Kähler manifolds with positive Ricci tensor*, Osaka J. Math. **23** (1986), pp. 585–616.
28. H. Sakuma, *Structure of automorphisms for toric Fano threefolds*, (Revision of Master thesis, Kyoto University, 1994), preprint.
29. I. Satake, *The Gauss-Bonnet theorem for V-manifolds*, J. Math. Soc. Japan **9** (1957), pp. 464–492.
30. H. Sato, *Toward the classification of higher-dimensional toric Fano varieties*, preprint.
31. Y.-T. Siu, *The existence of Kähler-Einstein metrics on manifolds with positive anticanonical line bundle and a suitable finite symmetry group*, Ann. of Math. **127** (1988), pp. 585–627.
32. G. Tian, *Kähler-Einstein metrics with positive scalar curvature*, Invent. Math. **137** (1997), pp. 1–37.
33. G. Tian and S.-T. Yau, *Kähler-Einstein metrics on complex surfaces with* $C_1 > 0$, Commun. Math. Phys. **112** (1987), pp. 175–203.
34. K. Watanabe and M. Watanabe, *The classification of Fano 3-folds with torus embeddings*, Tokyo J. Math. **5** (1982), pp. 37–48.
35. S.-T. Yau, *On the Ricci curvature of a compact Kähler manifold and the complex Monge-Ampère equation* I, Pure and Appl. Math. **31** (1978), pp. 339–411.

NATURALITY OF FLOER HOMOLOGY OF OPEN SUBSETS IN LAGRANGIAN INTERSECTION THEORY

YONG-GEUN OH
Department of Mathematics, University of Wisconsin, Maidson, WI 53706, USA
E-mail: oh@math.wisc.edu

In this paper, naturality of Floer homology of open subsets, which was previously constructed by Kasturirangan and the author, is proved and some of its consequences are explored.

1 Introduction

In a recent paper R. Kasturirangan and the author [5] associated a singular Lagrangian submanifold in T^*M, denoted by $\nu^*\overline{U}$, to each open subset U of M with smooth boundary ∂U, which they call the *co-normal* to $\overline{U}$:

$$\nu^*\overline{U} := o_M|_U \coprod \nu^*_-(\partial U)$$

where

$$\nu^*_-(\partial U) = \{\alpha \in \nu^*(\partial U) \mid \alpha(\vec{n}) \leq 0, \ \vec{n} \quad \text{outward normal to} \quad \partial U\}.$$

From now on, we will call this paper [5] by [KO].

When a Hamiltonian $H : T^*M \times [0,1] \to \mathbf{R}$ and so a Lagrangian submanifold $L = \phi^1_H(o_M)$ is given, they constructed a (filtered) Floer homology, denoted by $HF^\lambda_*(H, U : M)$, for each given real number λ. They call it *Floer homology* of the open subset U, compute $HF^\infty_*(H, U : M) := HF_*(H, U : M)$ and prove that there exists a canonical isomorphism

$$F_{(H,U)} : H_*(U, \mathbf{Z}) \to HF_*(H, U : M) \tag{1. 1}$$

As a consequence, they proved a micro-local version of Lagrangian intersection theorem,

$$\#(\phi^1_H(o_M) \cap \nu^*\overline{U}) \geq \ \text{rank}\ H_*(U, \mathbf{Z}). \tag{1. 2}$$

This refines the global version

$$\#(\phi^1_H(o_M) \cap o_M) \geq \ \text{rank}\ H_*(M, \mathbf{Z})$$

that was conjectured by Arnol'd [1] and proven by Hofer [4].

The purpose of the present paper is two-fold. First, we prove *naturality* of the (filtered) Floer homology $HF_*^\lambda(H, U : M)$ by proving that for any given pair $U \subset \overline{U} \subset V$ (with smooth boundaries) there exists a canonical homomorphism

$$i_{UV}^H : HF_*^\lambda(H, U : M) \to HF_*^\lambda(H, V : M) \tag{1. 3}$$

for which the following diagram commutes

$$\begin{array}{ccc} HF_*^\lambda(H, U : M) & \xrightarrow{i_{UV}^H} & HF_*^\lambda(H, V : M) \\ \uparrow F_{(H,U)} & & \uparrow F_{(H,V)} \\ H_*(U, \mathbf{Z}) & \xrightarrow{i_*} & H_*(V, \mathbf{Z}) \end{array} \tag{1. 4}$$

Note that the co-normals $\nu^*\overline{V}$ and $\nu^*\overline{U}$ have a priori no inclusion relation and so it is not at all a trivial matter to construct a canonical inclusion-induced homomorphism, especially when taking filtrations into account. This is a contrast to the case of Floer-Hofer's symplectic homology[3] (of open subsets of *symplectic manifold*) for which defining an inclusion-induced homomorphism is a straightforward matter.

The second purpose is then to define the Floer homology for *arbitrary* open subsets of M and to set the stage of quantization of Eilenberg-Steenrod Axioms. Once we construct the above inclusion-induced homomorphism, it is straightforward to define the Floer homology for arbitrary open subsets by taking a limiting process. By taking the dual construction, we also define a "Cêch"-version of the Floer cohomology for arbitrary locally closed subset $C \subset M$. For any given locally closed subset $C \subset M$, we take the direct system of neighborhoods of C and give a partial order by the opposite inclusion. We define

$$\check{H}F_\lambda^*(H, C : M) := \varinjlim_{U \supset C} HF_\lambda^*(H, U : M)$$

and prove that there exists a canonical isomorphism

$$\check{H}^*(C, \mathbf{Z}) \underset{\check{F}_{(H,C)}}{\to} \check{H}F_\infty^*(H, C : M). \tag{1. 5}$$

Recall from Oh[8] that when S is a closed submanifold of M, by considering the co-normal bundle $\nu^*S \subset T^*M$, we defined another Floer cohomology $HF_\lambda^*(H, S : M)$ which can be considered as the "singular" version. By a construction similar to (1.3), we can define a canonical homomorphism respecting the filtration.

$$\check{H}F_\lambda^*(H, S : M) \to HF_\lambda^*(H, S : M) \tag{1. 6}$$

for which the diagram

$$\begin{array}{ccc} \check{H}F^*_\lambda(H,S:M) & \longrightarrow & HF^*_\lambda(H,S:M) \\ \uparrow & & \uparrow \\ \check{H}^*(S,\mathbf{Z}) & \longrightarrow & H^*(S,\mathbf{Z}) \end{array} \qquad (1.\ 7)$$

commutes. By the isomorphisms (1.5) and the bottom row and the commutativity of (1.7), it follows that

$$\check{H}F^*_\infty(H,S:M) \to HF^*_\infty(H,S:M)$$

is an isomorphism. However, we will further prove elsewhere that (1.6) is an isomorphism for all $\lambda \in \mathbf{R}$.

Note that our definition of $\nu^*\overline{U}$ in [KO] [5] had been chosen so that certain positivity property required in the proofs of all the main theorems [KO] is satisfied but the intrinsic meaning of the correspondence had not been clear to us. After finishing writing [KO] , we learned that the singular Lagrangian variety $\nu^*\overline{U}$ had been naturally introduced on the one hand in micro-local analysis (see Kashiwara & Schapira [6]) and on the other hand in the stratified Morse theory [7] . Kashiwara and Schapira [6] had defined the concept of the *co-normal cone* (denoted by N^*S) for arbitrary locally closed subset $S \subset M$. Our definition $\nu^*\overline{U}$ turns out to be essentially the *co-normal cone* to the closure $\overline{U}$ in their definition. We also learned from the lectures by MacPherson [7] that this co-normal $\nu^*\overline{U}$ is also the *characteristic Lagrangian cycle* associated to the characteristic function $\chi_{\overline{U}} : M \to \{0,1\}$ of the open subset $U \subset M$. "Micro-localization" or "symplectification" is by now well-known to be an essential process both in the micro-local analysis and in the stratified Morse theory. Our work in [KO] [5] and the present paper should be considered as a part of "quantization" program of the classical homology theory and the Morse theory. At this stage, there arise the following natural questions.

Questions

1. Quantize the "homology" of the constructible functions of the Whitney stratification on M and study its geometric meaning in the context of symplectic geometry of T^*M.

2. Prove the cup-length estimate

$$\#(\phi^1_H(o_M) \cap \nu^*C) \geq \quad \text{cup-length} \quad \check{H}^*(C,\mathbf{Z}) \qquad (1.\ 8)$$

for any locally closed subset $C \subset M$ and for any Hamiltonian H.

The organization of the paper is now in order: In Section 2, for any given Hamiltonian H, we construct another Hamiltonian $\overset{\approx}{H}$ such that the Hofer's norm $\|H - \overset{\approx}{H}\|$ can be made arbitrarily close to 0 and construct a natural map

$$i_{UV}^{\lambda} : CF_*^{\lambda}(H, U : M) \to CF_*^{\lambda+\delta}(\overset{\approx}{H}, V : M) \tag{1. 9}$$

where δ can be chosen as small as we want. In Section 3, we prove the chain property of (1.9) and establish the homomorphism (1.3). Interestingly enough, to establish the chain property of (1.9), it is crucial to look at the kind of Hamiltonians $\overset{\approx}{H}$ such that

$$\overset{\approx}{H} \equiv 0 \quad \text{near} \quad t = 1.$$

Such Hamiltonians have been used in the literature of symplectic geometry, which also naturally enter in our Floer theory. In Section 4, we use our constructions to define some symplectic invariants of $L = \phi_H^1(o_M)$ relative to the open subset $U \subset M$ and prove their basic monotonicity properties under the inclusion.

The present paper should be considered as a continuation of our previous work[5] with Kasturirangan. We assume that readers are familiar to [KO][5] and refer thereto for unexplained notations and details. We would like to thank A. Weinstein for pointing out the reference Kashiwara & Schapira[6] and R. MacPherson for his interest in the present work. This research is partially supported by NSF grant # DMS 950-4455.

2 Construction of the map

Let (H, J) be a given pair of Hamiltonian H and a one-parameter family of almost complex structure $J = \{J_t\}_{0 \le t \le 1}$ such that

$$H \equiv \text{ const.} \quad \text{and} \quad J_t \equiv J_g \tag{2. 1}$$

outside the compact subset

$$D_R := \{(q, p) \in T^*M | \ |p|_g \le R\}.$$

And let $U \subset \overline{U} \subset V$ be a pair of open subsets of M that have smooth boundaries ∂V and ∂U respectively. We will denote by Υ^U (resp. Υ^V) the approximating smooth Lagrangian submanifold of $\nu^*\overline{U}$ (resp. $\nu^*\overline{V}$) such that

$$\Upsilon^U \equiv \nu_-^*(\partial U) \tag{2. 2}$$

outside a small neighborhood of $o_M|_{\partial U}$ and in particular outside D_R. See [KO][5] for the details of this approximation. The following uniform C^0-estimate of the Cauchy-Riemann equation

$$\begin{cases} \frac{\partial u}{\partial \tau} + J(\frac{\partial u}{\partial t} - X_H(u)) = 0 \\ u(\tau, 0) \in o_M, u(\tau, 1) \in \Upsilon^U \end{cases} \tag{2. 3}$$

over $u \in \mathcal{M}(o_M, \Upsilon^U : H, J)$ independent of U, H and J, is important for later purpose. We refer to Oh [8] for its proof where a uniform C^0-estimate has been proven in a similar context.

Lemma 2.1 *Let (H, J) and Υ^U be as above satisfying (2.1) and (2.2) respectively. Then for any solution u of (2.3), we have*

$$\text{Image } u \subset D_R \tag{2. 4}$$

for all such (H, J) and U.

We now consider $\nu^*\overline{U} \cap D_R$ and construct a Hamiltonian flow that leaves o_M and $\nu^*_-(\partial U)$ invariant and maps $\nu^*_-(\partial U) \cap D_R$ into a small neighborhood of $o_M|_{\partial U}$,

$$B_\delta = \{(q, p) \in T^*M \mid q \in Tb(\partial U) \subset V, \quad |p| < \delta\}.$$

Such a flow will be constructed explicitly using the identification

$$Tb(\partial U) \simeq \partial U \times (-1, 1)$$
$$T^*M|_{Tb(\partial U)} \simeq T^*(\partial U) \times ((-1, 1) \times \mathbf{R})$$

and

$$\nu^*_-(\partial U) \cap \pi^{-1}(Tb(\partial U)) \simeq T^*(\partial U) \times \{(0, s) | s \leq 0\} \subset T^*(\partial U) \times ((-1, 1) \times \mathbf{R}).$$

As a first step, we will find an isotopy in $\mathbf{R}^2$ that leaves $\mathbf{R} \times \{0\}$ invariant and that maps $\{(0, s)| - R \leq s \leq 0\}$ into $\{(0, s)| - \delta \leq s \leq 0\}$. This will be constructed by considering appropriate scalar multiples of the Hamiltonian vector field

$$X(r, s) = (r, -s) \tag{2. 5}$$

whose generating Hamiltonian is given by the function

$$rs.$$

Note that the flow of (2.5) contracts the s-axis and expands the r-axis. When $R, \delta > 0$ are given, we can choose appropriate positive number $\lambda > 0$ so that the time one-map of λX satisfies the requirement. We now cut-off the Hamiltonian λrs

$$k_{R,\delta,\epsilon}(r,s) = \chi_R(|s|)\chi_\epsilon(|r|)(\lambda rs) \tag{2. 6}$$

where $\chi : \mathbf{R}_+ \to [0,1]$ is a function such that

$$\chi(t) = \begin{cases} 0 & \text{for } t \geq 2 \\ 1 & \text{for } 0 \leq t \leq 1 \,, \end{cases}$$

and $\chi_K(t) = \chi(\frac{t}{K})$. The constant $\epsilon > 0$ will be determined later. Note that by choosing $\epsilon > 0$ small, we can make the C^0-norm of $k_{R,\delta,\epsilon}$ can be made arbitrarily small, provided R and δ (and so λ) are fixed. It follows that the Hamiltonian flow of $k_{R,\delta,\epsilon}$ will satisfy all the requirements if we choose λ sufficiently big, and in turn the C^0-norm of $k_{R,\delta,\epsilon}$ can be made arbitrarily small by choosing ϵ sufficiently small.

For higher dimensions, we define our Hamiltonian to be

$$K(q,p) = K_{R,\delta,\epsilon}(q,p) := \chi_R(|p'|_g)k_{R,\delta,\epsilon}(r,s) \tag{2. 7}$$

in coordinates $(q,p) = (q',r,p',s)$ on $T^*(Tb(\partial U))$ and trivially extending to the whole T^*M. It is easy to check that the Hamiltonian flow of K will satisfy the following properties:

1. It leaves o_M and $\nu^*_-(\partial U)$ invariant,
2. maps $\nu^*_-(\partial U) \cap D_R$ into $\nu^*_-(\partial U) \cap D_\delta$ and
3. leaves $(T^*M \backslash D_{2R}) \cup (T^*M \backslash T^*(Tb(\partial U))$ fixed.

   Furthermore, by choosing $\epsilon > 0$ sufficiently small, we can make
4. C^0-norm of $K_{R,\delta,\epsilon}$ arbitrarily small.

   Finally for later purposes, we will adjust K into a time-dependent Hamiltonian, again denoted by K, so that
5. $K \equiv 0$ near $t = 1$ (See Lemma 3.1 below for relevant remarks).

We go back to the equations

$$\begin{cases} \dot{z} = X_H(z) \\ z(0) \in o_M, \quad z(1) \in \Upsilon^U \end{cases} \tag{2. 8}$$

and (2.3) i.e.

$$\begin{cases} \frac{\partial u}{\partial \tau} + J\Big(\frac{\partial u}{\partial t} - X_H(u)\Big) = 0 \\ u(\tau,0) \in o_M, \quad u(\tau,1) \in \Upsilon^U \end{cases}$$

and study how these equations are transformed under the map

$$z \mapsto \tilde{z}\ ;\ \tilde{z}(t) := \phi_K^t(z(t)) \tag{2. 9}$$

on the path space

$$\Omega(o_M, \Upsilon^U) = \{z : [0,1] \to T^*M \mid z(0) \in o_M, z(1) \in \Upsilon^U\}.$$

Note that

$$\tilde{z}(0) \in o_M, \quad \tilde{z}(1) \in \phi_K^1(\Upsilon^U)$$

where the manifolds

$$\widetilde{\Upsilon}^U := \phi_K^1(\Upsilon^U)$$

are again approximations of $\nu_-^*(\overline{U})$ in the sense of [Section 2, KO [5]] such that

$$\widetilde{\Upsilon}^U \equiv \Upsilon^U$$

outside $(T^*M\backslash D_{2R}) \cup (T^*M\backslash T^*(Tb(\partial U))$. Furthermore, it is easy to check that

$$\widetilde{\Upsilon}_1^U \le \widetilde{\Upsilon}_2^U \quad \text{if } \Upsilon_1^U \le \Upsilon_2^U.$$

And the map (2.9) maps

$$CF(o_M, \Upsilon^U : H) \to CF(o_M, \widetilde{\Upsilon}^U : \widetilde{H})$$

$$\mathcal{M}(o_M, \Upsilon^U : H, J) \to \mathcal{M}(o_M, \widetilde{\Upsilon}^U : \widetilde{H}, \widetilde{J})$$

where

$$\widetilde{H} = K\#H\ ;\quad (K\#H)_t = K_t + H_t \circ (\phi_K^t)^{-1}$$

$$\widetilde{J} = \{\widetilde{J}_t\}\ ;\ \widetilde{J}_t = (\phi_K^t)_* J_t.$$

This can be proven by straightforward computations (see Oh [8] for details). Now, we estimate the changes of the action under (2.9). We recall from [KO] [5] the definition of the *effective action functional* of Υ^U. Denote by $h_{\Upsilon^U} : \Upsilon^U \to \mathbf{R}$ the potential of $i_{\Upsilon^U}^*\theta = dh_{\Upsilon^U}$ defined as in [KO]. Then the effective action functional is defined by

$$\mathcal{A}_H^{\Upsilon^U}(\gamma) = \mathcal{A}_H(\gamma) - h_{\Upsilon^U}(\gamma(1)) \tag{2. 10}$$

in [KO]. Recall that solutions of (2.8) are critical points of $\mathcal{A}_H^{\Upsilon^U}$ and those of (2.3) are gradient trajectories of $\mathcal{A}_H^{\Upsilon^U}$ with respect to the metric

$$\langle\langle , \rangle\rangle_J = \int_0^1 g_{J_t}(\cdot,\cdot)dt.$$

Lemma 2.2 *For any $z \in CF(o_M, \Upsilon^U : H)$, we have the inequality*

$$\mathcal{A}_{\widetilde{H}}^{\widetilde{\Upsilon}^U}(\tilde{z}) - \mathcal{A}_H^{\Upsilon^U}(z) \leq -\int_0^1\int_0^1 K(\phi_{sK}^t(z(t)),t)dtds. \qquad (2.\ 11)$$

Proof As in [O2], we consider the families

$$s \mapsto (sK)\#H$$
$$s \mapsto \phi_{sK}^t(z(t)) := \gamma^s(t).$$

Note that

$$\gamma^s(0) = z(0), \quad \gamma^s(1) = \phi_K^s(z(1)).$$

We have

$$\mathcal{A}_{K\#H}(\tilde{z}) - \mathcal{A}_H(z) = \int_0^1 \frac{d}{ds}\Big(\mathcal{A}_{(sK)\#H}(\gamma^s)\Big)ds$$

and

$$\frac{d}{ds}\Big(\mathcal{A}_{(sK)\#H}(\gamma^s)\Big) = d\mathcal{A}_{(sK)\#H}(\gamma^s)\frac{\partial\gamma^s}{\partial s} - \int_0^1 K(\gamma^s(t),t)dt$$

Note that for each fixed t, the curve $s \mapsto \gamma^s(t)$ is a (re-parametrized) Hamiltonian orbit of K with initial point at $z(t)$. On the other hand, from the general variation formula

$$d\mathcal{A}_H(\gamma)(\xi) = \int_0^1 (\omega(\dot{\gamma},\xi) - dH(\gamma)\xi)dt + \langle\xi(1),\theta(\gamma(1))\rangle - \langle\xi(0),\theta(\gamma(0))\rangle \qquad (2.\ 12)$$

and from the boundary condition we have

$$\begin{aligned} d\mathcal{A}_{(sK)\#H}(\gamma^s)\frac{\partial\gamma^s}{\partial s} &= dh_{\phi_K^s(\Upsilon^U)}(\gamma^s(1))\frac{\partial\gamma^s}{\partial s}(1) \\ &= \frac{d}{ds}\Big(h_{\phi_K^s(\Upsilon^U)}(\gamma^s(1))\Big) - \frac{\partial}{\partial s}\Big(h_{\phi_K^s(\Upsilon^U)}\Big) \\ &= \frac{d}{ds}\Big(h_{\phi_K^s(\Upsilon^U)}(\gamma^s(1))\Big) - \frac{\partial}{\partial s}\Big(h_{\phi_K^s(\Upsilon^U)}\Big)(\gamma^s(1)) \end{aligned}$$

Note from the construction that

$$\frac{\partial}{\partial s}\Big(h_{\phi^s_K(\Upsilon^U)}\Big) \geq 0 \quad \text{for all} \quad 0 \leq s \leq 1 \tag{2. 13}$$

if we choose Υ^U sufficiently close to $\nu^*\overline{U}$. Combining these, we have

$$\begin{aligned}
\mathcal{A}_{K\#H}(\tilde{z}) - \mathcal{A}_H(z) &= \int_0^1 \tfrac{d}{ds}\Big(h_{\phi^s_K(\Upsilon^U)}(\gamma^s(1)\Big)ds - \int_0^1 \tfrac{\partial}{\partial s}\Big(h_{\phi^s_K(\Upsilon^U)}\Big)(\gamma^s(1))ds \\
&\qquad - \int_0^1\int_0^1 K(\gamma^s(t),t)dt\, ds \\
&\leq h_{\phi^1_K(\Upsilon^U)}(\gamma^1(1)) - h_{\phi^0_K(\Upsilon^U)}(\gamma^0(1)) - \int_0^1\int_0^1 K(\gamma^s(t),t)dt\, ds \\
&= h_{\widetilde{\Upsilon}^U}(\tilde{z}(1)) - h_{\Upsilon^U}(z(1)) - \int_0^1\int_0^1 K(\gamma^s(t),t)dt\, ds,
\end{aligned}$$

and hence

$$\Big(\mathcal{A}_{K\#H}(\tilde{z}) - h_{\widetilde{\Upsilon}^U}(\tilde{z}(1))\Big) - \Big(\mathcal{A}_H(z) - h_{\Upsilon^U}(z(1)\Big) \leq -\int_0^1\int_0^1 K(\gamma^s(t),t)dt\, ds$$

which is exactly (2.11) from the definition of $\mathcal{A}^{\Upsilon}_H$. □

(2.11) obviously implies that the assignment (2.9) restricts to a map

$$HF^{\lambda}(o_M, \Upsilon^U : H) \to HF^{\lambda+(-\min K)}(o_M, \widetilde{\Upsilon}^U : \widetilde{H}).$$

By choosing $\epsilon > 0$ sufficiently small and taking the approximation Υ^U close enough to $\nu^*\overline{U}$, we can make the change of filtration as small as we want. We now note that for any $u \in \mathcal{M}(o_M, \Upsilon^U : H, J)$, $\tilde{u}$ satisfies

$$\tilde{u}(\tau, 1) \in \widetilde{\Upsilon}^U \cap B_\delta \subset \{(q', p', r, s) \mid -\delta \leq s \leq 0, \qquad |p'|_{g'} < \delta\}.$$

Next, we will find the second Hamiltonian isotopy that maps $\widetilde{\Upsilon}^U \cap B_\delta$ into $o_M \cap Tb(\partial U) \subset o_V$. We again start with the one-dimensional case. We isotope $\widetilde{\Upsilon}^U \cap B_{2\delta}$ so that $\widetilde{\Upsilon}^U \cap B_\delta$ is mapped into $o_{Tb(\partial U)} \subset o_V$ and $\widetilde{\Upsilon}^U \backslash B_{2\delta}$ remains fixed. We need to define this isotopy in the following way: We rotate $(\widetilde{\Upsilon}^U \cap B_\delta) \cap i\mathbf{R}$ around the origin *counterclockwise* with constant angular velocity $\pi/2$ so that at time 1 the y-axis is mapped onto the x-axis. Then we extend this isotopy to the whole $\widetilde{\Upsilon}^U$ so that $\widetilde{\Upsilon}^U \backslash B_{2\delta}$ is fixed and $\widetilde{\Upsilon}^U \cap B_\delta$ is mapped into the x-axis. Such an isotopy is not difficult to construct. Then we extend this (Lagrangian isotopy) to an ambient Hamiltonian isotopy in $\mathbf{R}^2$. Note that because we made a counterclockwise rotation, the image of $\widetilde{\Upsilon}^U \cap B_\delta$ will lie on the *outside* of U. We denote by $R_\delta = \{R^\delta_t\}_{0\leq t\leq 1}$ the corresponding Hamiltonian. By construction, we can choose

$$R^\delta_t \equiv r^2 + s^2 \tag{2. 14}$$

on a neighborhood of the image of $(\widetilde{\Upsilon}^U \cap B_\delta) \cap i\mathbf{R}$ under the isotopy. We pull-back R_δ to $T^*(Tb(\partial U)) \simeq T^*(\partial U) \times (\mathbf{R} \times (-1,1)$ in an obvious way. Then we extend R_δ from $T^*(Tb(\partial U))$ to the whole T^*M so that $R_\delta \equiv 0$ outside $B_{2\delta}$. Now, we consider the assignment

$$\begin{aligned} \tilde{z} \mapsto \approx{z} \quad &; \quad \approx{z}(t) = \phi^t_{R_\delta} \circ \tilde{z}(t) = \phi^t_{R_\delta} \circ \phi^t_K(z(t)) \\ \tilde{u} \mapsto \approx{u} \quad &; \quad \approx{u}(\tau, t) = \phi^t_{R_\delta} \circ \phi^t_K(u(\tau, t)) \end{aligned} \tag{2. 15}$$

By construction, these map

$$\begin{aligned} CF(o_M, \widetilde{\Upsilon}^U : \widetilde{H}) &\to CF(o_M, \Upsilon^V : \approx{H}) \\ \mathcal{M}(o_M, \widetilde{\Upsilon}^U : \widetilde{H}, \widetilde{J}) &\to \mathcal{M}(o_M, \Upsilon^V : \approx{H}, \approx{J}) \end{aligned} \tag{2. 16}$$

respectively, where

$$\approx{H} = R_\delta \# K \# H$$
$$\approx{J} = \{\approx{J}_t\}; \quad \approx{J}_t = (\phi^t_{R_\delta} \circ \phi^t_K)_* J_t$$

Note also that if we choose $\epsilon > 0$ sufficiently small and Υ^U sufficiently close to $\nu^* \overline{U}$ (and so δ sufficiently small), then we can make the C^0-norm of R_δ as small as we want. By the similar calculations as we did above for K, we can make the change of filtration levels for the map (2.16) as small as we want.

Finally composing (2.9) and (2.15), we have constructed the map

$$\psi_{UV} : CF(o_M, \Upsilon^U : H) \to CF(o_M, \Upsilon^V : \approx{H})$$

and

$$\mathcal{M}(o_M, \Upsilon^U : H, J) \to \mathcal{M}(o_M, \Upsilon^V : \approx{H}, \approx{J})$$

whose filtration change can be made as small as we want.

At this stage, it remains to prove the chain property of ψ_{UV} , i.e., the identity

$$\partial^V_{(\approx{H}, \approx{J})} \circ \psi_{UV} = \psi_{UV} \circ \partial^U_{(H,J)} \tag{2. 17}$$

which will be proven in the next section.

3 Chain property of the map ψ_{UV}

It turns out that the chain property (2.17) cannot be proven (and probably not true), unless we perturb (H, J) to (H', J') in a way that

$$H' \equiv 0 \quad \text{near} \quad t = 1 \quad \text{and}$$

$$\widetilde{\widetilde{J'}} \equiv J_g \qquad \text{near } t = 1 \text{ and near the zero section } o_M \subset T^*M.$$

As we mentioned in the introduction, such a perturbation can be achieved so that both

$$\Big| \int_0^1 \max_x (H' - H)\, dt \Big| \quad \text{and} \quad \Big| \int_0^1 \min_x (H' - H)\, dt \Big|$$

are as small as we want. We would like to emphasize that this perturbation cannot be made in a way that the slightly different norms $\Big| \max_{x,t}(H' - H) \Big|$ and $\Big| \min_{x,t}(H' - H) \Big|$ become small, which have been also used in the literature in relation to Hofer's geometry. Because of this, in our (filtered) Floer theory, it is crucial to use Hofer's norm defined as

$$\|H\| = \int_0^1 (\max_x H - \min_x H) dt \tag{3. 1}$$

not as

$$\max_{x,t} H - \min_{x,t} H. \tag{3. 2}$$

This indicates that the correct Hofer's norm to be used should be (3.1) but not (3.2) in general. Here is the precise perturbation lemma that we need.

Lemma 3.1 *Let H be a given Hamiltonian $H : T^*M \times [0, 1] \to \mathbf{R}$ and $\phi = \phi_H^1$ be its time-one map. Then we can perturb H so that the perturbed Hamiltonian H' has the property*

1. $\phi_{H'}^1 = \phi_H^1$
2. $H' \equiv 0$ *near* $t = 0$ *and* 1
3. *Both* $|\int_0^1 \max_x (H' - H)\, dt|$ *and* $|\int_0^1 \min_x (H' - H)\, dt|$ *can be made as small as we want.*

Proof We first reparametrize ϕ_H^t in the following way: We choose a smooth function $\zeta : [0, 1] \to [0, 1]$ such that

$$\zeta(t) = \begin{cases} 0 & \text{for } 0 \le t \le \epsilon/2 \\ 1 & \text{for } 1 - \frac{\epsilon}{2} \le t \le 1 \end{cases}$$

and

$$\zeta'(t) \geq 0 \quad \text{for all} \quad t \in [0,1],$$

and consider the isotopy

$$\psi^t := \phi_H^{\zeta(t)}.$$

It is easy to check that the Hamiltonian generating the isotopy $\{\psi^t\}_{0\leq t\leq 1}$ is $H' = \{H'_t\}_{0\leq t\leq 1}$ with $H'_t = \zeta'(t)H_{\zeta(t)}$. By definition, it follows that H' satisfies (1) and (2). For (3), we compute

$$\begin{aligned}\int_0^1 \max_x \left(H' - H\right)dt &= \int_0^1 \max_x(\zeta'(t)H_{\zeta(t)} - H_t)dt \\ &\leq \int_0^1 \max_x \Big(\zeta'(t)(H_{\zeta(t)} - H_t)\Big)dt + \int_0^1 \max_x \Big((\zeta'(t) - 1)H_t\Big)dt.\end{aligned}$$

For the first term,

$$\begin{aligned}\int_0^1 \max_x \Big(\zeta'(t)(H_{\zeta(t)} - H_t)\Big)dt &= \int_0^1 \zeta'(t) \max_x(H_{\zeta(t)} - H_t)dt \\ &\leq \int_0^1 \zeta'(t) \max_{x,t} |H_{\zeta(t)} - H_t|dt \\ &= \max_{x,t} |H_{\zeta(t)}(x) - H_t(x)|\end{aligned}$$

which can be made arbitrarily small by choosing ζ so that $\|\zeta - t\|_{C^0}$ become sufficiently small. For the second term,

$$\begin{aligned}\int_0^1 \max_x \Big((\zeta'(t) - 1)H_t\Big)dt &\leq \textstyle\int_0^1 |\zeta'(t) - 1|dt \cdot \max_{x,t} H(x,t) \\ &\leq \|H\|_{C^0} \textstyle\int_0^1 |\zeta'(t) - 1|dt.\end{aligned}$$

Again by appropriately choosing ζ, we can make

$$\int_0^1 |\zeta'(t) - 1|dt$$

as small as we want. Combining these two, we have verified $|\int_0^1 \max_x(H' - H)\,dt|$ can be made as small as we want. Similar consideration applies to $|\int_0^1 \min_x(H' - H)\,dt|$ and hence we have finished the proof of (3). □

By the same calculation as we did in [O1], we can prove that the canonical chain homomorphism

$$h_{HH'} : HF_*(H, U : M) \to HF_*(H', U : M)$$

restricts to

$$h_{HH'} : HF_*^{\lambda}(H,U : M) \to HF_*^{\lambda+\epsilon_{HH'}}(H',U : M)$$

where

$$\epsilon_{HH'} = -\int_0^1 \min(H - H')dt$$

and so Lemma 3.1 implies that we can make the filtration change as small as we want. Because of this, we will concern only those Hamiltonians H that have the property (2), i.e.

$$H_t \equiv 0 \qquad \text{near} \quad t = 1 \tag{3. 3}$$

and J such that

$$\widetilde{\widetilde{J}} \equiv J_g \qquad \text{near } t = 1 \text{ and near the zero section } o_M \subset T^*M. \tag{3. 4}$$

The up-shot is that for such H's and J's, we can prove the chain property of the map ψ_{UV}. What we will have to show is that the image of $CF(o_M, \Upsilon^U : H)$ of the map ψ_{UV} becomes a sub-complex of $CF(o_M, \Upsilon^V : \widetilde{\widetilde{H}})$ under the boundary map $\partial^V_{(\widetilde{\widetilde{H}},\widetilde{\widetilde{J}})}$. More precisely, we will prove

Proposition 3.2 *Let H and J be as above satisfying (3.3) and (3.4) respectively. Then there exists some $\delta > 0$ and λ sufficiently large such that for any $u \in \mathcal{M}(o_M, \Upsilon^V : \widetilde{\widetilde{H}}, \widetilde{\widetilde{J}})$ with $u(-\infty, 1) \in U_\delta$*

$$u(\tau, 1) \in U_\delta \quad \textit{for all} \quad \tau \in \mathbf{R}$$

where U_ϵ is a ϵ-neighborhood of U in M, i.e.,

$$U_\epsilon = U \cup \{(q', r) \in Tb(\partial U) \mid q' \in \partial U, 0 \leq r < \epsilon\}.$$

In other words, the local Floer complex

$$\mathcal{M}(o_M, \Upsilon^V : \widetilde{\widetilde{H}}, \widetilde{\widetilde{J}}, U_\delta) := \{u \in \mathcal{M}(o_M, \Upsilon^V : \widetilde{\widetilde{H}}, \widetilde{\widetilde{J}}) \mid \text{Image } u \subset U_\delta\}$$

is a sub-complex of $\mathcal{M}(o_M, \Upsilon^V : \widetilde{\widetilde{H}}, \widetilde{\widetilde{J}})$ under the boundary $\partial^V_{(\widetilde{\widetilde{H}},\widetilde{\widetilde{J}})}$. Furthermore the map (2.17) is a bijective map from $\mathcal{M}(o_M, \Upsilon^U : H, J)$ to the local Floer complex.

Assuming this proposition for the moment, we proceed with the proof of chain property of ψ_{UV}. We choose $\lambda > 0$ so that

$$\phi_K^1(u(\tau,1)) \subset U_\delta$$

for all $u \in \mathcal{M}(o_M, \Upsilon^U : H, J)$. To prove the chain property, it will be enough to prove

$$n_{(H,J)}(z^\alpha, z^\beta : U) = n_{(\approx{H},\approx{J})}(\approx{z^\alpha}, \approx{z^\beta} : V) \tag{3. 5}$$

where

$$n_{(H,J)}(z^\alpha, z^\beta : U) = \#\mathcal{M}(o_M, \Upsilon^U : H, J)$$

and similar for $n_{(\approx{H},\approx{J})}(\approx{z^\alpha}, \approx{z^\beta} : V)$. The proof of (3.5) immediately follows from Proposition 3.2.

Now, it remains to prove Proposition 3.2.

Proof of Proposition 3.2 The last statement is immediate from our construction: Just recall the fact that any $u' \in \mathcal{M}(o_M, \Upsilon^V : \approx{H}, \approx{J})$ with $u'(\tau,1) \in U_\delta$ comes from $\mathcal{M}(o_M, \Upsilon^U : H, J)$ since

$$u(\tau,t) := (\phi_{R_\delta}^t \circ \phi_K^t)^{-1} u'(\tau,t)$$

satisfies $u(\tau,1) \in \Upsilon^U$ and (2.3). Following Floer [2] and [KO], we rewrite the equation

$$\begin{cases} \frac{\partial u}{\partial \tau} + \approx{J}\Big(\frac{\partial u}{\partial t} - X_{\approx{H}}(u)\Big) = 0 \\ u(\tau,0) \in o_M, u(\tau,1) \in o_V \subset \Upsilon^V \end{cases} \tag{3. 6}$$

in the canonical coordinate $u = (q,p)$ near $t = 1$. Here we recall that near $t = 1$, (H,J) satisfies (3.3) and (3.4) and that K also satisfies

$$K \equiv 0 \quad \text{near} \quad t = 1.$$

Therefore we have

$$\approx{H} = R_\delta \# K \# H \equiv R_\delta \quad \text{near } t = 1.$$

Now thanks to (3.3) and (3.4), one can check that (3.6) becomes

$$\begin{cases} \frac{\partial q}{\partial \tau} - \nabla_t p - \pi_q(J_g X_{R_\delta}) = 0 \\ \frac{\partial q}{\partial t} + \nabla_\tau p - \pi_q^*(J_g X_{R_\delta}) = 0 \end{cases} \tag{3. 7}$$

near $t = 1$, where π_q (resp. π_q^*) is the horizontal (resp. vertical) projection on $T_{(q,p)}(T^*M)$ with respect to the splitting induced from the Levi-Civita connection of the given metric g. We now prove Proposition 3.2 by contradiction. Suppose the contrary that $u(\tau, 1)$ escapes from U_δ and denote by τ_0 the first exit time at which $u(\tau_0, 1) \in \partial \overline{U}_\delta$. Using the coordinates $(q,p) = (q', r, p', s)$ near o_M, we compute

$$\frac{\partial}{\partial s} R_\delta = 2s, \quad \frac{\partial}{\partial r} R_\delta = 2r \quad \text{and } d_{(q',p')} R_\delta \equiv 0.$$

Therefore if $u(\tau_0, 1) \in \partial U_\delta$, then we have

$$\frac{\partial}{\partial s} R_\delta(u(\tau_0, 1)) = 0.$$

Hence the second equation of (3.7) becomes

$$\frac{\partial q}{\partial t}(\tau_0, 1) = 0 \tag{3. 8}$$

since $p \equiv 0$ and so $\nabla_\tau p(\tau_0, 1) = 0$ for $-\infty < \tau \leq \tau_0$. Substituting (3.8) and $\frac{\partial}{\partial r} R_\delta(U(\tau_0, 1)) = 2r$ into the first equation of (3.7), we have

$$\begin{cases} \frac{\partial q'}{\partial \tau} = 0 \\ \frac{\partial r}{\partial \tau} + 2r = 0 \quad \text{at } (\tau, t) = (\tau_0, 1) \ . \end{cases}$$

If $r(\tau_0, 1) = \delta$, we have

$$\frac{\partial r}{\partial \tau}(\tau_0, 1) = -2\delta < 0.$$

On the other hand by the assumption that τ_0 is the first exit time from U_δ, we must have

$$\frac{\partial r}{\partial \tau}(\tau_0, 1) \geq 0$$

which gives rise to a contradiction. This finishes the proof.

□

Now, we have defined a canonical homomorphism

$$\psi_{UV} : HF_*^\lambda(o_M, \Upsilon^U : H, J) \to HF_*^{\lambda+\delta_1}(o_M, \Upsilon^V : \approx{H}, \approx{J})$$

for δ_1 arbitrarily small. By combining this with the canonical continuation homomorphism

$$h : HF_*^{\lambda+\delta_1}(o_M, \Upsilon^V : \approx{H}, \approx{J}) \to HF_*^{\lambda+\delta_1+\epsilon_{\approx{H}H}}(o_M, \Upsilon^V : H, J),$$

we have defined a homomorphism

$$i_{UV}^{(H,J)} : HF_*^{\lambda}(o_M, \Upsilon^U : H, J) \to HF_*^{\lambda+\delta_2}(o_M, \Upsilon^V : H, J)$$

where δ_2 can be made arbitrarily small. Now by taking the direct limits of both sides using the algebraic lemma [Lemma 4.6, KO], we can define

$$i_{UV}^H : HF_*^{\lambda}(H, U : M) \to HF_*^{\lambda}(H, V : M) \tag{3. 9}$$

for any $\lambda \in \mathbf{R}\backslash\mathrm{Spec}\ (H, U : M)$. It is also easy to check that (3.9) satisfies the following commutativity: For $\lambda_0 < \lambda_1$, the diagram

$$\begin{array}{ccc} HF_*^{\lambda_0}(H, V : M) & \stackrel{j_*^{\lambda_0\lambda_1}}{\longrightarrow} & HF_*^{\lambda_1}(H, V : M) \\ \uparrow i_{UV}^H & & \uparrow i_{UV}^H \\ HF_*^{\lambda_0}(H, U : M) & \stackrel{j_*^{\lambda_0\lambda_1}}{\longrightarrow} & HF_*^{\lambda_1}(H, U : M) \end{array}$$

commutes where the horizontal rows are the canonical homomorphism (4.1) in [KO]. This finally finishes the construction of the inclusion-induced homomorphism preserving the filtration. We summarize the basic properties of the homomorphism (3.9) in the following theorem.

Theorem 3.3 *Let H be an arbitrary Hamiltonian such that $\phi_H^1(o_M)$ is transverse to o_M, and $U \subset \overline{U} \subset V$ be a pair of open subsets of M with smooth boundary. Suppose that $\phi_H^1(o_M)$ is transverse to $\nu^*\overline{U}$ and $\nu^*\overline{V}$. Then for each $\lambda \in \mathbf{R}\backslash\mathrm{Spec}\ (H, U : M)$ there exists a canonical homomorphism*

$$i_{UV}^H : HF_*^{\lambda}(H, U : M) \to HF_*^{\lambda}(H, V : M)$$

that satisfies the following properties.

1. *The diagram*

$$\begin{array}{ccc} HF_*^{\lambda}(H, U : M) & \stackrel{i_{UV}^H}{\longrightarrow} & HF_*^{\lambda}(H, V : M) \\ \uparrow F_{(H,U)} & & \uparrow F_{(H,V)} \\ H_*(U, \mathbf{Z}) & \underset{i_*}{\longrightarrow} & H_*(V, \mathbf{Z}) \end{array} \tag{3. 10}$$

   commutes.

2. *For the triple $U \subset V \subset W$ with smooth boundary, we have*

$$h_{UW}^H = h_{VW}^H \circ h_{UV}^H \tag{3. 11}$$

Now, we extend the definition to arbitrary open subsets whether they have smooth boundary or not. Let U be a given open subset of M, and denote

$$\mathcal{A}_U^H = \{U_\alpha \subset U \mid U_\alpha \quad \text{open subset with smooth boundary } \partial U_\alpha \text{ and } \phi_H^1(o_M) \text{ transverse to } \nu^*\overline{U}_\alpha\}$$

We give the partial ordering by inclusion. Using (3.11), we now take the direct limit of $HF_*(H, U_\alpha : M)$ to define

Definition 3.4. Let $U \subset M$ be a given open subset and $\mathcal{A}_U^H$ be the above direct system. We define

$$HF_*(H, U : M) = \varinjlim_{\alpha} HF_*(H, U_\alpha : M)$$

It follows that when $U \subset \overline{U} \subset V$ this Floer homology has the natural inclusion homomorphism, again denoted by i_{UV}^H, such that the following diagram commutes

$$\begin{array}{ccc} HF_*^\lambda(H, U : M) & \xrightarrow{i_{UV}^H} & HF_*^\lambda(H, V : M) \\ \uparrow F_{(H,U)} & & \uparrow F_{(H,V)} \\ H_*(U, \mathbf{Z}) & \xrightarrow[i_*]{} & H_*(V : \mathbf{Z}) \end{array} \tag{3. 12}$$

and the homomorphism

$$F_{(H,U)} : H_*(U, \mathbf{Z}) \to HF_*^\infty(H, U : M)$$

is an isomorphism.

By taking the dual construction, we can define Floer cohomology and the canonical inclusion-induced homomorphism

$$i_H^{UV} : HF_\lambda^*(H, V : M) \to HF_\lambda^*(H, U : M) \tag{3. 13}$$

whenever $U \subset \overline{U} \subset V$. Now for any locally closed subset $C \subset M$, we choose the direct system

$$\mathcal{B}_C^H = \{U \supset C \mid U \quad \text{open subset with smooth } \partial U \text{ and } \phi_H^1(o_M) \text{ transverse to } \nu^*\overline{U}\}$$

whose partial order is given by the opposite inclusion. We take the direct limit to define the Cêch-version of the Floer cohomology

$$\check{H}F_\lambda^*(H, C : M) = \varinjlim_{U \supset C} HF_\lambda^*(H, U : M).$$

It then follows that

1. there exists a canonical isomorphism

$$\check{F}^{(H,C)} : \check{H}^*(C, \mathbf{Z}) \to \check{H}F^*_\infty(H, C; M)$$

2. the following diagram commutes

$$\begin{array}{ccc} \check{H}F^*_\lambda(H, D : M) & \xrightarrow{i_H^{CD}} & \check{H}F^*_\lambda(H, C : M) \\ \uparrow F^{(H,D)} & & \uparrow F^{(H,C)} \\ \check{H}^*(D, \mathbf{Z}) & \xrightarrow{i^*_{CD}} & \check{H}^*(C, \mathbf{Z}) \end{array}$$

whenever $C \subset D$ are two locally closed subsets of M.

4 Construction of symplectic invariants

In this section, we imitate the construction from Oh [8,9] to define various symplectic invariants of given $L = \phi^1_H(o_M)$ relative to open subsets and study the basic properties of them. We start with the most primitive form of the invariants.

Definition 4.1 Let $U \subset M$ be an open subset with smooth boundary ∂U, and H such that $\phi^1_H(o_M)$ is transverse both to $\nu^*\overline{U}$ and o_M. We define

1. $\rho_+(H, U) = \inf_\lambda \{\lambda \in \mathbf{R} \mid j^\lambda_* : HF^\lambda_*(H, U : M) \to HF_*(H, U : M) \text{ is surjective } \}$
2. $\rho_-(H, U) = \sup_\lambda \{\lambda \in \mathbf{R} \mid j^\lambda_* : HF^\lambda_*(H, U : M) \to HF_*(H, U : M) \text{ is trivial}\}$
3. $\gamma(H, U) = \rho_+(H, U) - \rho_-(H, U)$.

Theorem 4.2 *Let H, U be as in Definition 4.1.*

1. *The values of $\rho_\pm(H, U)$ lie in* $\mathrm{Spec}(H, U : M)$ *i.e., values of some Hamiltonian orbits z with*

$$z(0) \in o_M, \quad z(1) \in \nu^*\overline{U}.$$

2. *$\rho_+(H, U) \geq \rho_-(H, U)$ and so $\gamma(H, U) \geq 0$*
3. *For each fixed U, we have*

$$-\int_0^1 \max_x (H_\alpha - H_\beta) dt \leq \rho_\pm(H_\alpha, U) - \rho_\pm(H_\beta, U) \leq -\int_0^1 \min_x (H_\alpha - H_\beta) dt$$

In particular, the function $H_\alpha \mapsto \rho(H_\alpha, U)$ can be extended as a continuous function in C^0-topology to arbitrary H's. We still denote this extension by $\rho_\pm(H, U)$

4. *For $H \equiv 0$, $\rho_\pm(H,U) = 0$ for any open subset U with smooth boundary. Therefore, we have from (3)*

$$- \int_0^1 \max_x H dt \leq \rho_\pm(H,U) \leq - \int_0^1 \min_x H dt$$

5. $\gamma(H,U) \leq \|H\|$(= *Hofer's norm of* H)

6. $\gamma(H,U)$ *depends only on* $L = \phi_H^1(o_M)$ *after the normalization given in [O1].*

Now, we study the dependence of $\rho(H,U)$ on the open subsets U.

Theorem 4.3 *Let $U \overset{i}{\subset} V$ be open subsets of M with smooth boundary. Then we have the following:*

1. *If $i_* : H_*(U, \mathbf{Z}) \to H_*(V, \mathbf{Z})$ is surjective, then*

$$\rho_+(H,V) \leq \rho_+(H,U)$$

2. *If $i_* : H_*(U, \mathbf{Z}) \to H_*(V, \mathbf{Z})$ is injective, then*

$$\rho_-(H,U) \leq \rho_-(H,V)$$

Therefore if $i_* : H_*(U, \mathbf{Z}) \to H_*(V, \mathbf{Z})$ is an isomorphism, then we have

$$\gamma(H,V) \leq \gamma(H,U).$$

Now we want to extend our definition of these invariants to arbitrary open subsets. Let $U \subset M$ be an open subset of M and $\{U_j\}$ be an increasing sequence such that

$$i_* : H_*(U_j, \mathbf{Z}) \to H_*(U_{j+1}, \mathbf{Z})$$

becomes an isomorphism and such that $U = \cup_{j=1}^\infty U_j$.

Then $j \mapsto \rho_+(H, U_j)$ (resp. $j \mapsto \rho_-(H, U_j)$) is a bounded decreasing (resp. bounded increasing) sequence and so has the limit. One can easily show that the limit is independent of the sequence chosen.

Definition 4.4. Let U and $\{U_j\}$ as above. We define

$$\rho_\pm(H,U) := \lim_{j \to \infty} \rho_\pm(H, U_j)$$

and

$$\gamma(H,U) = \rho_+(H,U) - \rho_-(H,U).$$

One natural question occurring at this stage is to study the geometric meaning of these invariants, which will be investigated elsewhere.

References

1. Arnol'd, V. I., C. R. Acad. Sci. Paris **261**, 3719 (1965).
2. Floer, A., *J. Differ. Geom.* **30**, 207 (1989).
3. Floer, A. and Hofer, H., Math. Z. **215**, 37 (1984).
4. Hofer, H., Ann. Inst. H. Poincaré, Anal. Non Linéaire **2**, 407 (1985).
5. Kasturirangan, R. and Oh, Y. -G., submitted.
6. Kashiwara, M. and Schapira, P., *Sheaves on Manifolds*, A Series of Comp. Studies in Math. 292, Springer-Verlag, New York, 1990.
7. MacPherson, R., Lectures in IAS-Park City Summer School for Symplectic Geometry and Topology, July 1997.
8. Oh, Y. -G., *J. Differ. Geom.* **46**, 499 (1997).
9. Oh, Y. -G., *Comm. Anal. Geom.*, (to appear).

AN EXTENSION OF CHERN-LASHOF THEOREM TO OTHER SPACE FORMS

TAKASHI OKAYASU
3190 Gofuku, Toyama, 930
Faculty of Education, Toyama University
JAPAN
E-mail: okayasu@edu.toyama-u.ac.jp

The famous Chern-Lashof inequality shows that the total absolute curvature of a compact submanifold in the Euclidean space is estimated from below by the total Betti number. In this paper we extend this result to other space forms.

1 Introduction

The total absolute curvature of a closed curve in the Euclidean space is the integral of the curvature of the curve. Fenchel([8])proved that the total absolute curvature of a closed curve in $\mathbf{R}^3$ is at least 2π. Borsuk([1]) extended this to closed curves in $\mathbf{R}^n$.

In 1957 Chern and Lashof ([6]) made a beautiful generalization of this result. Let $f : M^n \to \mathbf{R}^m$ be an isometric immersion of a compact n-dimensional Riemannian manifold M. Let $\nu(f)^1$ be the unit normal bundle of M in $\mathbf{R}^m$. Note that $\nu(f)^1$ has a canonical Riemannian metric (see section 2). For $e \in \nu(f)^1$, set A_e to be the shape operator with respect to e. The total absolute curvature of f is defined by

$$\tau(f) = \frac{1}{\omega_{m-1}} \int_{e\in\nu(f)^1} |\det A_e| d\sigma, \tag{1}$$

where ω_{m-1} is the volume of the unit $(m-1)$-sphere, and $d\sigma$ the volume element of $\nu(f)^1$. Let $b_i(M)$ be the i-th Betti number of M. The famous Chern-Lashof theorem is

$$\tau(f) \geq \sum_{i=0}^{n} b_i(M). \tag{2}$$

This inequality shows that the total absolute curvature is closely linked with the topology of the manifold. This is the start point of the study of the so called "tight immersion" ([10]).

Since the discovery of Chern-Lashof, it becomes a natural question whether one can extend their result to other spaces. Several people defined various "total absolute curvatures" and get results similar to (2) ([4],[14],[18]). But all their notion of total absolute curvature are not directly calculable from the second fundamental form.

In 1964 Willmore and Saleemi([19])defined a total absolute curvature of a closed manifold in a simply-connected Riemannian manifold of nonpositive curvature by (1), and claimed that the inequality (2) still holds. But unfortunately their proof contained a mistake in the last two paragraphs of page 159 (their Morse function was not correct). To remedy this situation Chen ([4])defined a different total absolute curvature and proved an inequality similar to (2).

In 1984 Teufel([15]) defined a total absolute curvature of a closed manifold in the hyperbolic space $H^m(-1)$ by (1), and showed the following result. Let M^n be an n-dimensional compact Riemannian manifold. Suppose $f : M^n \to H^m(-1)$ is an isometric immersion with $f(M) \subset B_R$, where B_R is a geodesic ball of $H^m(-1)$. Then, $\tau(f) > \frac{1}{\cosh R}\sum_{i=0}^{n} b_i(M)$. This gives a fairly good positive answer to Willmore-Saleemi conjecture in the hyperbolic space case.

In this paper we also define a total absolute curvature of a closed manifold in the spheres or the hyperbolic spaces by (1), and show that inequalities similar to (2) hold (theorem 1, 4). Our method of proof is totally different from Teufel's. In the sphere case our result is new. In the hyperbolic space case our estimate is weaker than Teufel's, but from the proof we can relate the total absolute curvature of the manifold to that of some manifold in the Euclidean space. By using this relation we can give many interesting applications. For example we can give an extension of Fenchel-Borsuk inequality to the spheres and hyperbolic spaces. We can extend a theorem of Fary-Milnor ([7], [12]) about the total curvature of a knot in $\mathbf{R}^3$ to spheres and hyperbolic spaces. Moreover we can extend an inequality of Chen about total mean curvature ([5]) to other space forms. Other applications are explained in section 5.

2 Preliminaries

First we review a well-known fact (for more details, see Ferus [9]). Let $\pi : E \to M$ be a vector bundle on a compact Riemannian manifold M. Suppose E has a fiber metric $<,>$ and let ∇^E be a metric connection on E. Then we can construct a canonical Riemannian metric on E as follows. Let TE be the tangent bundle of E. Let $e \in E$. We set $V_e = \ker(\pi_{*e} : T_eE \to T_{\pi(e)}M)$ (the vertical space). We denote by $\Gamma(E)$ the set of all cross sections of E. We choose $s \in \Gamma(E)$ such that $s(\pi(e)) = e$, $\nabla^E_X s = 0$ for all $X \in T_{\pi(e)}M$. Now we set $H_e = s_*(T_{\pi(e)}M)$ (the horizontal space). Then $T_eE = V_e \bigoplus H_e$. By the parallel translation we can identify V_e with $E_{\pi(e)}$. Also we can identify H_e with $T_{\pi(e)}M$ by π_*. We give a Riemannian metric to E so that $T_eE =$

$V_e \bigoplus H_e \cong E_{\pi(e)} \bigoplus T_{\pi(e)}M$ becomes an isometry.

Let $f : M^n \to \overline{M}^m$ be an isometric immersion. Let $\nabla, \overline{\nabla}$ be the Riemannian connection of $M, \overline{M}$ respectively. Then the second fundamental form h can be defined by

$$h(X,Y) = \overline{\nabla}_X Y - \nabla_X Y \quad \text{for } \forall X, Y \in TM.$$

For any normal vector e, we define the shape operator with respect to e as

$$< A_e(X), Y > = < h(X,Y), e > \qquad \text{for } \forall X, Y \in TM.$$

where $<,>$ denotes the Riemannian metric of $\overline{M}$. Let $\nu(f)$ be the normal bundle of M in $\overline{M}$ and $\pi : \nu(f) \to M$ be its natural projection. Since $\nu(f)$ has a normal connection $\nabla^{\perp}$, we can give a Riemannian metric to $\nu(f)$ as above. The unit normal bundle $\nu(f)^1$ is a submanifold of $\nu(f)$, so we get an induced metric on $\nu(f)^1$.

Now we define the central projection which is a main tool of our proof. Let $S^m(1) = \{x = (x_1, \ldots, x_{m+1}) \in \mathbf{R}^{m+1} | |x| = 1\}$ and $S^m_+ = \{s \in S^m(1) | x_{m+1} > 0\}$ the upper hemisphere. Set $N = (0, \ldots, 1) \in S^m(1)$ and identify $\mathbf{R}^m$ with $T_N S^m(1) = \{x \in \mathbf{R}^{m+1} | x_{m+1} = 1\}$. We define the **central projection** (or Bertrami map [3]) $\beta : S^m_+ \to \mathbf{R}^m$ as

$$\beta(x) = \frac{1}{x_{m+1}}(x_1, \ldots, x_m).$$

We have the following elementary but important lemma.

Lemma 1 *Let $f : M^n \to S^m_+$ be an isometric immersion. We set $\tilde{f} = \beta \circ f : M^n \to \mathbf{R}^m$ and give M^n the induced metric. We denote M with this Riemannian structure as $\tilde{M}$.*

(i)

$$f(x) = \left(\frac{\tilde{f}(x)}{\sqrt{1 + |\tilde{f}(x)|^2}}, \frac{1}{\sqrt{1 + |\tilde{f}(x)|^2}} \right),$$

(ii) *We take a local coordinate $(u_1, \ldots, u_n)$ on M and set $f_i = \frac{\partial f}{\partial u_i}, \tilde{f}_i = \frac{\partial \tilde{f}}{\partial u_i}$ $(i = 1, \ldots, n)$, then*

$$f_i = \left(\frac{(1 + |\tilde{f}|^2)\tilde{f}_i - < \tilde{f}, \tilde{f}_i > \tilde{f}}{(1 + |\tilde{f}|^2)^{3/2}}, -\frac{< \tilde{f}, \tilde{f}_i >}{(1 + |\tilde{f}|^2)^{3/2}} \right).$$

(iii) *If $\tilde{e}$ is a unit normal vector to $\tilde{M}$, then*

$$e = \left(\frac{\tilde{e}}{\sqrt{1+ < \tilde{f}, \tilde{e} >^2}}, -\frac{< \tilde{f}, \tilde{e} >}{\sqrt{1+ < \tilde{f}, \tilde{e} >^2}} \right)$$

is a unit normal vector of M. Conversely, any unit normal vector e of M can be written in this way.

(iv) *We define a map*

$$\varphi : \nu(f)^1 \to S^{m-1}(1)$$

as follows. Any element $u \in \nu(f)^1$ can be written uniquely as $u = (\overline{u}, \hat{u}) \in \mathbf{R}^m \oplus \mathbf{R}$. Since $f(M) \subset S^m_+$, $\overline{u} \neq 0$. We set

$$\varphi(u) = \frac{1}{|\overline{u}|}\overline{u}.$$

Then

$$\varphi(\nu_x(f)^1) = \nu_x(\tilde{f})^1 \quad \textit{for } \forall x \in M$$

Proof. (i), (ii) and (iii) are elementary. (iv) can be derived easily from (iii). □

3 Submanifolds in the spheres

In this section we prove the following theorem.

Theorem 1 *Let M^n be an n-dimensional compact Riemannian manifold. Suppose $f : M^n \to S^m(1)$ is an isometric immersion with $f(M) \subset B_R(N)$ $(R < \pi/2)$, where N is the north pole of $S^m(1)$ and $B_R(N)$ is the geodesic ball of radius R centered at N. Then*

$$\frac{1}{\omega_{m-1}} \int_{e \in \nu(f)^1} |\det A_e| dv_{\nu(f)^1} \geq \cos^{m-1} R \cdot \sum_{i=0}^{n} b_i(M).$$

The equality holds if and only if M is the small hypersphere of radius $\sin R$.

Let $dv_{S^{m-1}}$ be the volume form of $S^{m-1}(1)$ and $dv_{\nu(f)^1}$ the volume form of $\nu(f)^1$. Using the spherical Gauss map $\varphi : \nu(f)^1 \to S^{m-1}(1)$ (see lemma 1), we pull back the form $dv_{S^{m-1}(1)}$. We get

$$\varphi^* dv_{S^{m-1}} = G dv_{\nu(f)^1}.$$

for some function G on $\nu(f)^1$. We define the **spherical total absolute curvature** as

$$\tau_S(f) = \frac{1}{\omega_{m-1}} \int_{e \in \nu(f)^1} |G(e)| dv_{\nu(f)^1}. \tag{3}$$

Theorem 2 *Let M^n be an n-dimensional compact Riemannian manifold. Suppose $f : M^n \to S^M(1)$ is an isometric immersion with $f(M) \subset B_R(N)$ $(R < \pi/2)$. Then*

$$\tau_S(f) \geq \sum_{i=0}^{n} b_i(M).$$

Proof. For each $e^* \in S^{m-1}(1)$, we define a hight function $h_{e^*} : M \to \mathbf{R}$ as

$$h_{e^*} = < \beta(f(x)), e^* > .$$

Suppose $p \in M$ is a critical point of h_{e^*}. Then

$$dh_{e^*}(p) = < d(\beta \circ f)(p), e^* > = 0.$$

Thus e^* is a normal vector of $\beta \circ f$, hence from lemma 1, there is a unique unit normal vector $e \in \nu_p(f)^1$ such that $\varphi(e) = e^*$. Conversely, let $e \in \nu_p(f)^1$ and $\varphi(e) = e^*$. Thus from (iv) of lemma 1

$$dh_{\varphi(e)}(p) = < d(\beta \circ f)(p), \varphi(e) > = 0.$$

Thus p is a critical point of $h_{\varphi(e)} = h_{e^*}$. Consequently, we see that the number of critical points of h_e* (which is denoted by $n(h_{e^*})$) is equal to the number of $e \in \nu(f)^1$ such that $\varphi(e) = e^*$. Hence we obtain

$$\int_{\nu(f)^1} |\varphi^* dv_{s^{m-1}}| = \int_{e^* \in s^{m-1}} n(h_{e^*}) dv_{s^{m-1}} \tag{4}$$

(for more detailed argument, see the proof of theorem 3.14 of [9]).
By Sard's theorem, for almost all $e^* \in s^{m-1}$, h_{e^*} is a non-degenerate function. Therefore $n(h_{e^*})$ is well-defined and is finite for a.e. $e^* \in s^{m-1}(1)$. For such a non-degenerate function h_{e^*}, we have the following Morse inequality ([9]).

$$n(h_{e^*}) \geq \sum_{i=0}^{n} b_i(M). \tag{5}$$

Combining (4) and (5) we get the conclusion. □

Lemma 2 *Under the hypothesis of theorem 2,*

$$|G(e)| \leq \frac{1}{\cos^{m-1} R} |\det A_e| \quad \textit{for } \forall e \in \nu(f)^1.$$

Proof. Set $\pi(e) = p$. We take an orthonormal basis $Y_1, \ldots, Y_{m-1}$ of $T_e(\nu(f)^1)$ such that $Y_1, \ldots, Y_n$ are horizontal and $Y_{n+1}, \ldots, Y_{m-1}$ are vertical. We choose a cross section $s \in \Gamma(\nu(f)^1)$ such that $s(p) = e$ and $(\nabla^\perp s)_p = 0$. Set $X_i = \pi_*(Y_i) \in T_pM$, then $\{X_1, \ldots, X_n\}$ is an orthonormal basis of T_pM and $Y_i = s_*X_i$. We fix $i \in \{1, \ldots, n\}$. We choose a curve $c_i(t)$ on M such that $c_i'(o) = X_i$. Then

$$\varphi_* Y_i = \frac{d}{dt}\varphi(s(c_i(t)))|_{t=0} = \frac{d}{dt}\Big|_{t=0} \frac{\overline{s}(t)}{|\overline{s}(t)|}, \tag{6}$$

where $s(x) = (\overline{s}(x), \hat{s}(x)) \in \mathbf{R}^m \oplus \mathbf{R}$ and we abbreviate $\overline{s}(c_i(t)) = \overline{s}(t)$. Since $(\nabla^\perp s)_p = 0$ and $|s| = 1$,

$$\begin{aligned} \frac{d}{dt}\Big|_{t=0} s(t) &= \overline{\nabla}_{X_i} s \ \ (\textit{the covariant derivative of } \mathbf{R}^{m+1}) \\ &= -A_s X_i \\ &= \sum_j < h(X_i, X_j), s(p) > X_j. \end{aligned}$$

We may assume that we have chosen $Y_1, \ldots Y_n$ such that $< h(X_i, X_j), e >$ $= \delta_{i,j} < h(X_i, X_i), e >$. Then

$$\begin{aligned} \frac{d}{dt}\Big|_{t=0} \frac{\overline{s}(t)}{|\overline{s}(t)|} &= - < h(X_i, X_i), e > \left\{ \frac{\overline{X}_i}{|\overline{s}(0)|} - \frac{< \overline{X}_i, \overline{s}(0) >}{|\overline{s}(0)|^3} \overline{s}(0) \right\} \\ &= - < h(X_i, X_i), e > \left\{ \frac{\overline{X}_i}{|\overline{e}|} - \frac{< \overline{X}_i, \overline{s}(0) >}{|\overline{e}|^3} \overline{e} \right\} \end{aligned} \tag{7}$$

From (6) and (7)

$$< \varphi_* Y_i, \varphi_* Y_j >= \frac{< h(X_i, X_i), e >< h(X_j, X_j), e >}{|\overline{e}|^2} \left\{ \delta_{i,j} - \frac{\hat{X}_i \hat{X}_j}{|\overline{e}|^2} \right\}.$$

Since $f(M) \subset B_R(N)$, $|\overline{e}| \geq \cos R$. Thus

$$\begin{aligned} &(\textit{ the volume of the parallelotope generated by } \varphi_* Y_1, \ldots, \varphi_* Y_n)^2 \\ &= \det < \varphi_* Y_i, \varphi_* Y_j > \\ &= < h(X_1, X_1), e >^2 \cdots < h(X_n, X_n), e >^2 |\overline{e}|^{-2n} \det \left(\delta_{i,j} - \frac{\hat{X}_i \hat{X}_j}{|\overline{e}|^2} \right) \end{aligned}$$

$$= < h(X_1, X_1), e >^2 \cdots < h(X_n, X_n), e >^2 |\bar{e}|^{-2n} \left(1 - \frac{\hat{X}_1^2 + \cdots + \hat{X}_n^2}{|\bar{e}|^2}\right)$$

$$\leq \frac{|\det A_e|^2}{|\bar{e}|^{2n}} \leq \frac{|\det A_e|^2}{\cos^{2n} R}. \tag{8}$$

For each Y_k $(k = n+1, \ldots m-1)$, we take a generating curve $\gamma_k(t)$ in $\nu_(f)^1$. Then

$$\varphi_* Y_k = \frac{d}{dt}\Big|_{t=0} \varphi(\gamma_k(t)) = \frac{d}{dt}\Big|_{t=0} \frac{\overline{\gamma_k}(t)}{|\overline{\gamma_k}(t)|}$$

$$= \frac{|\overline{\gamma_k}(0)|\overline{\gamma_k'}(0) - < \overline{\gamma_k'}(0), \overline{\gamma_k}(0) > |\overline{\gamma_k}(0)|^{-1}\overline{\gamma_k}(0)}{|\overline{\gamma_k}(0)|^2}$$

Therefore

$$|\varphi_* Y_k|^2 \leq \frac{|\overline{\gamma_k}(0)|^2 |\overline{\gamma_k'}(0)|^2 - < \overline{\gamma_k'}(0), \overline{\gamma_k}(0) >^2}{|\overline{\gamma_k}(0)|^4}$$

$$\leq \frac{1}{|\overline{\gamma_k}(0)|^2} \leq \frac{1}{\cos^2 R}. \tag{9}$$

Combining (8) and (9), we obtain

$$|G(e)| \leq \frac{1}{\cos^{m-1} R} |\det A_e|.$$

□

Theorem 1 is a direct consequence of theorem 2 and lemma 2.

4 Submanifolds in the hyperbolic spaces

Let $\mathbf{R}_L^{m+1}$ be the Lorentz space, that is, $\mathbf{R}^{m+1}$ equipped with the Lorentz metric

$$< x, x >_L = (x_1)^2 + \cdots + (x_m)^2 - (x_{m+1})^2.$$

The Lorentz model of hyperbolic space $H^m(-1)$ is

$$H^m(-1) = \{x \in \mathbf{R}_L^{m+1} | |x|_L^2 = -1, x_{m+1} > 0\}.$$

Set $N = (0, \ldots, 0, 1) \in H^m(-1)$. We can identify $\mathbf{R}^m$ with $T_N H^m(-1) = \{x \in \mathbf{R}_L^{m+1} | x_{m+1} = 1\}$. We define the **central projection** $\beta : H^m(-1) \to \mathbf{R}^m$ as

$$\beta(x) = \frac{1}{x_{m+1}}(x_1, \ldots, x_m).$$

We have the following lemma similar to lemma 1 (whose proof is the same as lemma 1, so we omit it).

Lemma 3 *Let $f : M^n \to H^m(-1)$ be an isometric immersion. We set $\tilde{f} = \beta \circ f : M^n \to \mathbf{R}^m$ and give M^n the induced metric. We denote M with this Riemannian structure as $\tilde{M}$.*

(i)

$$f(x) = \left(\frac{\tilde{f}(x)}{\sqrt{1 - |\tilde{f}(x)|_L^2}}, \frac{1}{\sqrt{1 - |\tilde{f}(x)|_L^2}} \right).$$

(ii) *We take a local coordinate $(u_1, \ldots, u_n)$ on M and set $f_i = \frac{\partial f}{\partial u_i}, \tilde{f}_i = \frac{\partial \tilde{f}}{\partial u_i}$ $(i = 1, \ldots, n)$, then*

$$f_i = \left(\frac{(1 - |\tilde{f}|_L^2)\tilde{f}_i + < \tilde{f}, \tilde{f}_i >_L \tilde{f}}{(1 - |\tilde{f}|_L^2)^{3/2}}, \frac{< \tilde{f}, \tilde{f}_i >_L}{(1 - |\tilde{f}|_L^2)^{3/2}} \right).$$

(iii) *If $\tilde{e}$ is a unit normal vector of $\tilde{M}$ at x, then*

$$e = \left(\frac{\tilde{e}}{\sqrt{1 - < \tilde{f}, \tilde{e} >_L^2}}, \frac{< \tilde{f}, \tilde{e} >_L}{\sqrt{1 - < \tilde{f}, \tilde{e} >_L^2}} \right)$$

is a unit normal vector of M at x.

(iv) *We define a map $\varphi : \nu(f)^1 \to S^{m-1}(1)$ as follows. Any $u \in \nu(f)^1$ can be written uniquely as $u = (\overline{u}, \hat{u}) \in \mathbf{R}^m \oplus \mathbf{R}$ $(\overline{u} \neq 0)$. We set*

$$\varphi(u) = \frac{1}{|\overline{u}|}\overline{u}.$$

Then

$$\varphi(\nu_x(f)^1) = \nu_x(\tilde{f})^1 \quad \textit{for } \forall\, x \in M.$$

Let M^n be an n-dimensional compact Riemannian manifold. Suppose $f : M^n \to H^m(-1)$ is an isometric immersion. Let $dv_{S^{m-1}}$ be the volume form of S^{m-1} and $dv_{\nu(f)^1}$ the volume form of $\nu(f)^1$. We define a function $G : M \to \mathbf{R}$ as

$$\varphi^* dv_{S^{m-1}} = G dv_{\nu(f)^1}.$$

The **hyperbolic total absolute curvature** is defined as

$$\tau_H(f) = \frac{1}{\omega_{m-1}} \int_{e\in\nu(f)^1} |G(e)| dv_{\nu(f)^1}.$$

By using lemma 3, we can prove the following theorem (whose proof is the same as theorem 2).

Theorem 3 *Let $f : M^n \to H^m(-1)$ be an isometric immersion with $f(M) \subset B_R(N)$. Then*

$$\tau_H(f) \geq \sum_{i=0}^{n} b_i(M).$$

Lemma 4 *Under the hypothesis of theorem 3,*

$$|G(e)| \leq \sqrt{1 + n\sinh^2 R} \cdot (\cosh R)^{m-n+1} |\det A_e|.$$

Proof. We do the same calculation as lemma 2, but we must be careful that the inner product is indefinite.

Set $\pi(e) = p$. We take an orthonormal basis $Y - 1, \ldots, Y_{m-1}$ of $T_e(\nu(f)^1)$ such that $Y_1, \ldots, Y_n$ are horizontal and $Y_{n+1}, \ldots, Y_{m-1}$ are vertical. We choose a cross section $s \in \Gamma(\nu(f)^1$ such that $s(p) = e$ and $(\nabla^\perp)_p = 0$. Set $X_i = \pi_* Y_i$, then $\{X_1, \ldots, X_n\}$ is an orthonormal basis of T_pM. We may assume that $X_1, \ldots, X_n$ are chosen so that $< h(X_i, X_j), e >$ is diagonalized. As in lemma 2 we have

$$\varphi_* Y_i = - < h(X_i, X_j), e > \left\{ \frac{\overline{X}_i}{|\overline{e}|_L} - \frac{< \overline{X}_i, \overline{e} >_L}{|\overline{e}|_L^3} \overline{e} \right\}.$$

Therefore

$$< \varphi_* Y_i, \varphi_* Y_j > = \frac{< h(X_i, X_i), e >< h(X_j, X_j), e >}{|\overline{e}|_L^2} \left\{ \delta_{i,j} + \frac{\hat{X}_i \hat{X}_j}{|\overline{e}|_L^2} \right\}.$$

$$\begin{aligned} &(\text{ the volume of the parallelotope generated by } \varphi_* Y_1, \ldots, \varphi_* Y_n)^2 \\ &= \det < \varphi_* Y_i, \varphi_* Y_j > \\ &= \frac{< h(X_1, X_1), e >^2 \cdots < h(X_n, X_n), e >^2}{|\overline{e}|_L^{2n}} \left(1 + \frac{\hat{X}_1^2 + \cdots + \hat{X}_n^2}{|\overline{e}|_L^2} \right). \quad (10) \end{aligned}$$

Since $1 = |e|_L^2 = |\bar{e}|^2 - \hat{e}^2$, $|\bar{e}| \geq 1$. From $f(M) \subset B_R(M)$, we obtain $\hat{X}^2 \leq \sinh^2 R$ for all $X \in T_pM$. So from (10),

$$\begin{aligned}&(\textit{the volume of the parallelotope generated by } \varphi_* Y_1, \ldots, \varphi_* Y_n)^2 \\ &\qquad \leq |\det A_e|^2 (1 + n \sinh^2 R).\end{aligned}$$

For each Y_k $(k = n+1, \ldots .m-1)$, we take a generating curve $\gamma_k(t)$ in $\nu(f)^1$. Then as in lemma 2

$$|\varphi_* Y_k|^2 \leq \frac{|\overline{\gamma'}_k(0)|^2}{|\overline{\gamma}_k(0)|^2} \leq \cosh^2 R. \tag{11}$$

Combining (10) and (11) we get the conclusion. □

From theorem 3 and lemma 4 we get the following theorem.

Theorem 4 *Let M^n be an n-dimensional compact Riemannian manifold. Suppose $f : M^n \to H^m(-1)$ is an isometric immersion with $f(M) \subset B_R(N)$. Then*

$$\begin{aligned}&\frac{1}{\omega_{m-1}} \int_{\nu(f)^1} |\det A_e| dv_{\nu(f)^1} \\ &> \quad (1 + n \sinh^2 R)^{-1/2} (\cosh R)^{-m+n+1} (\sum_{i=0}^{n} b_i(M)).\end{aligned}$$

Remark Since the estimate in lemma 4 is not sharp, the equality in theorem 4 never holds. This contrasts with the sphere case.

5 Applications

Let $f : M^n \to S^m(1)$ be an isometric immersion with $f(M) \subset B_R(N)$ $(R < \pi/2)$. From (4) and theorem 3.14 of [9], we see that

$$\tau_S(f) = \frac{1}{\omega_{m-1}} \int_{S^{m-1}} n(h_{e^*}) dv_{S^{m-1}} = \tau(\tilde{f}). \tag{12}$$

Note that $\tilde{f} : \tilde{M}^n \to \mathbf{R}^m$. (The same equation holds for the hyperbolic case). Many known results about total curvature of submanifolds in the Euclidean spaces are derived from the interpretation of total curvature as the last equality of (12). So using (12), we can extend many known results in the Euclidean spaces to the other space forms. In this section we give some of these results. All our results are obtained by applying theorem 1 or 4 to the left hand side of

(12) and estimating the right hand side of (12) by the corresponding theorems. So we omit the proofs.

First we give an extension of Fary-Milnor theorem ([7],[12]) about a knot (see [2],[13],[15],[16],[17] for different estimates).

Theorem 5 *Let $f : S^1 \to S^3(1)$ be a knot with $f(S^1) \subset B_R(N)$ $(R < \pi/2)$. Then*

$$\int_{S^1} \kappa \geq 4\cos^2 R \cdot \pi.$$

where κ is the geodesic curvature of $f(S^1)$ in $S^3(1)$.

Theorem 6 *Let $f : S^1 \to H^3(-1)$ be a knot with $f(S^1) \subset B_R(N)$. Then*

$$\int_{S^1} \kappa > 4\pi(\cosh R)^{-2}.$$

where κ is the geodesic curvature of $f(S^1)$ in $H^3(-1)$.

Langevin and Rosenberg ([11]) extended Fary-Milnor theorem to a knotted torus in $\mathbf{R}^3$.

Theorem 7 ([11]) *Let $f : T^2 \to \mathbf{R}^3$ be a knotted torus, then*

$$\frac{1}{2\pi}\int_M |K_M| dv \geq 8,$$

where K_M is the Gaussian curvature of M.

From the proof of this theorem, we immediately get the following results.

Theorem 8 *Let $f : T^2 \to S^3(1)$ be a knotted torus with $f(T^2) \subset B_R(N)$ $(R < \pi/2)$, then*

$$\frac{1}{2\pi}\int_M |K_M - 1| dv \geq 8\cos^2 R.$$

Theorem 9 *Let $f : T^2 \to H^3(-1)$ be a knotted torus with $f(T^2) \subset B_R(N)$, then*

$$\frac{1}{2\pi}\int_M |K_M + 1| dv \geq 8(1 + 2\sinh^2 R)^{-1/2}.$$

Chen ([5]) derived an inequality of total mean curvature from (2).

Theorem 10 ([5]) Let $f : M^n \to \mathbf{R}^m$ be an isometric immersion of a compact manifold. Let H be the mean curvature vector. Then

$$\int_M |H|^n dv \geq \omega_n.$$

The same proof as theorem 10 gives thc following theorems.

Theorem 11 *Let $f : M^n \to S^m(1)$ be an isometric immersion of a compact manifold with $f(M) \subset B_R(N)$ $(R < \pi/2)$. Let H be the mean curvature vector. Then*

$$\int_M |H|^n \geq \cos^{m-1} R \cdot \omega_n.$$

Theorem 12 *Let $f : M^n \to H^m(-1)$ be an isometric immersion of a compact manifold with $f(M) \subset B_R(N)$. Let H be the mean curvature vector. Then*

$$\int_M |H|^n dv \geq (1 + n \sinh^2 R)^{-1/2} (\cosh R)^{-m+n+1} \omega_n.$$

There are many more applications. These will be discussed in the future.

6 Acknowledgments

This paper is dedicated to Professor S. Tanno on his 60th birthday.

The author is partially supported by Grant-in-Aid for Scientific Research (No.09640099), Ministry of Education, Science and Culture, Japan.

References

1. K. Borsuk, *Sur la courbure totale des courbes fermées, Ann. Soc. Polon.* **20**, 251–265 (1948).
2. F. Brickell and C. C. Hsiung, *The total absolute curvature of closed curves in Riemannian manifolds, J. Diff. Geom.* **9**, 177–193 (1974)
3. M. P. do Carmo and F. W. Warner, *Rigidity and convexity of hypersurfaces in spheres, J. Diff. Geom.* **4**, 133–144 (1970)
4. B.Y. Chen, *On the total curvature of manifolds immersed in Riemannian manifolds, Kōdai Math. Sem. Rep.* **19**, 299–311 (1967)
5. B. Y. Chen, *On total curvature of immersed manifolds, Amer. J. Math.* **93**, 148–162 (1971)

6. S. S. Chern and R. K. Lashof, *On the total curvature of immersed manifolds, Amer. J. Math.* **79**, 306–313 (1957)
7. I. Fary, *Sur la courbure totale d'une courbe gauche faisant un noeud, Bull. Soc. Math. France* **78**, 128–138 (1949)
8. W. Fenchel, *Über Krümmung und Windung geschlossener Raumkurven, Math. Ann.* **101**, 238–252 (1929)
9. D. Ferus, *Total Absolutkrümmung in Differentialgeometrie und topologie, Lect. Notes in Math.* **66** (Springer, Berlin-Heidelberg-New York, 1968)
10. N. H. Kuiper, *Minimal total absolute curvature for immersions, Inv. Math.* **10**, 209–238 (1970)
11. R. Langevin and H. Rosenberg, *On curvature integrals and knots , Topology* **15**, 405–416 (1976)
12. J. Milnor, *On the total curvature of knots, Ann. of Math.* **52**, 248–260 (1950)
13. J. Szenthe, *On the total curvature of closed curves in Riemannian manifolds, Publ. Math. Debrecen* **15**, 99–105 (1968)
14. E. Teufel, *Eine Differentialtopologische berechnung der totalen Krümmung und totalen Absolutkrümmung in der sphärichen Differentialgeometrie, Manuscripta Math.* **31**, 119–147 (1980)
15. E. Teufel, *On the total absolute curvature of immersions into hyperbolic spaces, Coll. Math. Soc. J.Bolyai* **46**, 1201–1209 (1984)
16. E. Teufel, *The isoperimetric inequality and the total absolute curvature of closed curves in spheres, Manuscripta Math.* **75**, 43–48 (1992)
17. Y. Tsukamoto, *On the total absolute curvature of closed curves in manifolds of negative curvature, Math. Ann.* **21**, 313–319 (1974)
18. J. L. Weiner, *Total curvature and total absolute curvature of immersed submanifolds of spheres, J. Diff. Geom.* **9**, 391–400 (1974)
19. T. J. Willmore and B. A. Saleemi, *The total absolute curvature of immersed manifolds, J. London Math. Soc.* **41**, 153–160 (1966)

Topological entropy for geodesic flows under Ricci curvature and conjugate radius bounded below

Seong-Hun Paeng
Korea Institute for Advanced Study,
207-43, Cheongryangri-dong, Dongdaemun-gu
Seoul 130-012, Korea
e-mail:shpaeng@kias.kaist.ac.kr

It is known that the topological entropy for the geodesic flow on a Riemannian manifold M is bounded if the absolute value of sectional curvature $|K_M|$ is bounded. We will prove this boundedness under Ricci curvature and conjugate radius bounded below. Furthermore we can find the increasing rate of conjugate radius such that topological entropy converges to 0 when $\mathrm{Ric}_{M_i} \geq -\epsilon_i \to 0$.

1 Introduction

The topological entropy is the most important invariant related to the orbit growth of a dynamical system. It represents the exponential growth rate for the number of orbit segments. In a sense, topological entropy describes the total exponential complexity of the orbit structure.

Manning shows that topological entropy for geodesic flows on manifolds with $K_M \leq 0$ is the same as the volme growth rate, $\lim_{r\to\infty} r^{-1} \log V(x,r)$, where $V(x,r)$ is the volume of the r-ball $B(x,r)$ contained in the universal covering space [Ma1]. This result was extended to the case of manifolds without conjugate points [FM]. So the topological entropy for geodesic flows is closely related to the curvatures of manifolds since the volume growth rates depend on the curvatures of manifolds. Let ϕ_t be a geodesic flow on a Riemannian manifold M and $h(\phi_t)$ be the topological entropy for ϕ_t.

It was known that if the absolute value of sectional curvature satisfies $|K_M| \leq k$, then the topological entropy for geodesic flows satisfies $h(\phi_t) \leq (n-1)\sqrt{k}$ by [Ma2]. In the case of conjugate radius $\mathrm{conj}_M = \infty$, the topological entropy for geodesic flows is the same as the volume growth rate [FM]. So Bishop's comparison theorem [GHL] implies that if Ricci curvature satisfies $\mathrm{Ric}_M \geq -(n-1)k^2$ and the conjugate radius of the universal covering space of M is infinite, then $h(\phi_t) \leq (n-1)k$. Thus if $|K_{M_i}| \leq \epsilon_i \to 0$ or $\mathrm{Ric}_{M_i} \geq -\epsilon_i \to 0$ and $\mathrm{conj}_{M_i} = \infty$, then topological entropy for geodesic flows of M_i, $h_i(\phi_t) \to 0$.

Then the followings are questions;
(1)Let $\mathrm{Ric}_M \geq -k$. Even if $\mathrm{conj}_M \neq \infty$, is it possible to find an upper bound of $h(\phi_t)$?

(2)If $\mathrm{Ric}_{M_i} \geq -(n-1)\epsilon_i^2 \to 0$ and $\mathrm{conj}_{M_i} \geq c_i \to \infty$, then $h_i(\phi_t) \to 0$?

The significant difference between the volume growth rate and the topological entropy for geodesic flows occurs when geodesics pass through conjugate points, i.e. for the volume growth rate, we only compute the volume form along geodesics before they meet conjugate points. So we know intuitively that without any assumption on conjugate points, the question (1) is not true.

If the topological entropy for geodesic flows is continuous with respect to metrics of the base manifold, then we easily know that the question (2) is true. But the topological entropy for geodesic flows is not continuous under C^∞-topology for metrics of base manifold. So (2) is not a trivial problem. We prove the following theorem;

Theorem 1.1 [a] *Let k, c_0 be positive real numbers. Then there exists a constant $C(c_0, n, k)$ such that for every n-dimensional compact Riemannian manifold M with $\mathrm{Ric}_M \geq -k$, $\mathrm{conj}_M \geq c_0$, the topological entropy for geodesic flow of M is bounded by $C(c_0, n, k)$.*

For answering to (2), we need some additional assumption on conjugate radius. In fact, the condition of $|K_M| \leq \epsilon^2$ implies the condition of $\mathrm{conj}_M \geq \dfrac{1}{\epsilon}$. So it is also a natural assumption that $\mathrm{conj}_M \geq \dfrac{C}{\epsilon}$ for some constant C in our case.

Theorem 1.2 *Let $\{M_i\}$ be a sequence of compact Riemannian manifolds with $\mathrm{Ric}_{M_i} \geq -(n-1)\epsilon_i^2 \to 0$ and $\mathrm{conj}_{M_i} \geq \dfrac{C}{\epsilon_i} \to \infty$ as $i \to \infty$ for some constant C. Then $h_i(\phi_t) \to 0$ as $i \to \infty$.*

For the proof, we will use Brocks' estimate for the Laplacian of the distance function, which will play an important role on estimating the norm of Jacobi fields.

2 Preliminaries

Let X be a compact metric space with the distance function d, and let $\phi = \{\phi_t : X \to X\}$ be a flow, i.e. 1-parameter subgroup of homeomorphisms.

Define

$$d_T^\phi(x,y) := \max\{d(\phi_t(x), \phi_t(y)) | 0 \leq t \leq T\}.$$

Let $S_d(\phi, \epsilon, T)$ be the minimal number of balls of radius ϵ in the metric d_T^ϕ

[a] This theorem appears in Proc. Amer. Math. Soc. under the injectivity radius condition instead of conjugate radius condition but the proof is the same.

which cover X. Define

$$h_d(\phi, \epsilon) := \lim\sup_{T\to\infty} \frac{1}{T} \log S_d(\phi, \epsilon, T).$$

Then the *topological entropy* for ϕ is defined as follows;

$$h(\phi) := \lim_{\epsilon\to 0} h_d(\phi, \epsilon)$$

The following theorem is Brocks' estimate on the Laplacian of the distance function.

Theorem 2.1 (Brocks[1]) *Let M^n be a complete Riemannian manifold and c be a geodesic on M^n with $\mathrm{conj}_M \geq c_0$ and $\mathrm{Ric}_M \geq -(n-1)k^2$. Let $A(t)$ be a Hessian of distance function along $c(t)$ from $c(0)$. Let $\alpha(t) = k^2 t + 2k \coth k(c_0 - t)/2$, $\gamma(t) = 2\sqrt{(n-1)\alpha(t)}\sqrt{t}$ for $0 \leq t \leq c_0$. Then*

$$\int_0^t ||A(\tau) - \frac{I}{\tau}|| d\tau \leq 2\gamma(t)$$

for all t such that $\gamma(t) \leq 1/7$.

The following lemma is important to estimate the upper bound of the topological entropy. It is proved in the discrete time case.(See Katok and Hasselblatt[7].) In this case, almost the same proof can be applied. For the precise proof, see Paeng[11].

Lemma 2.2 *Let X be a compact metric space and $\phi = \{\phi_t\}$ be a flow on M with $d(\phi_t(x), \phi_t(y)) \leq e^{tC} d(x,y)$ for some constant C. Then $h(\phi) \leq |C| D(X)$ where $D(X)$ is the ball dimension defined by $D(X) := \lim_{\epsilon\to 0} \dfrac{\log b(\epsilon)}{|\log(\epsilon)|}$, and $b(\epsilon)$ is the minimum number of a covering of X by ϵ-balls.*

It is known that $D(X) = \dim X$, when X is a topological manifold.

3 Proof of Theorem 1.1

In the calculation of the entropy, it is an important step to define a metric on the unit tangent bundle, T_1M. Let $\gamma_v(t)$ be the geodesic on M with $\gamma_v'(0) = v_p$.

In Manning's paper [8], a metric on T_1M is defined as following;

$$d_1(v_p, w_q) := \sup_{0\leq t\leq 1} d(\gamma_v(t), \gamma_w(t)).$$

In the book[7], they used the following metric;

$$d_2(v_p, w_q) := d(p,q) + ||P_p^q(v_p) - w_q||,$$

where $P_p^q(v_p)$ is the parallel translation of v_p along the geodesic from p to q.

Now we construct a metric on T_1M as following. Since $\mathrm{conj}_M \geq c_0$, it is possible to identify $v_p \in T_1M$ with the geodesic $\gamma_v(t)$, $t \in [0, c_0/4]$ with $\gamma_v'(0) = v_p$. Then we may consider a Jacobi field J along $\gamma_v(t)$ as a tangent vector at $v_p \in T_1M$, where J is generated by a C^∞-rectangle $Q(t,s)$ such that $\frac{\partial Q}{\partial s}(t,0) = J(t), ||\frac{\partial Q}{\partial t}(t,s)|| = 1$ for any s and $0 \leq t \leq c_0/4$ and $Q(t, s_0)$ is a geodesic for any fixed s_0. In this identification, we consider a C^∞-rectangle, $Q(t,s)$, where $0 \leq s \leq 1$ and $0 \leq t \leq c_0/4$ as a C^∞-curve on T_1M from $\frac{\partial Q}{\partial t}(0,0)$ to $\frac{\partial Q}{\partial t}(0,1)$.

Then we define an inner product of tangent vectors and a distance on T_1M as following;

Definition 3.1 Let J_1, J_2 be tangent vectors on T_1M, i.e. Jacobi fields with above property. Then

$$\begin{aligned} (J_1, J_2) &= \textstyle\int_0^{c_0/4} \langle J_1(t), J_2(t)\rangle dt \\ d(v_p, w_q) &= \inf \textstyle\int_0^1 \left(\frac{\partial Q}{\partial s}, \frac{\partial Q}{\partial s}\right)^{\frac{1}{2}} ds = \inf \int_0^1 \int_0^{c_0/4} ||\frac{\partial Q}{\partial s}|| dt ds, \end{aligned} \tag{3.1}$$

where the inf is taken over piecewise C^∞-curves Q on T_1M from v_p to w_q, i.e. $\frac{\partial Q}{\partial t}(0,0) = v_p$ and $\frac{\partial Q}{\partial t}(0,1) = w_q$.

Since T_1M is a compact Hausdorff space, above metrics induce the same topology. The remaining proof of Theorem 1.1 is almost the same as Paeng[11] but for the proof of Theorem 1.2, we will survey the proof.

The following lemma is a key lemma. (See Brocks[1] or Paeng[11].)

Lemma 3.2 *Let M be a complete Riemannian manifold with $\mathrm{Ric}_M \geq -k$, $\mathrm{conj}_M \geq c_0$ and $\gamma(t)$ be a minimal geodesic starting from q and $J(t)$ is a Jacobi field along γ such that $J(0) = 0$ and $\langle J'(0), \gamma'(0)\rangle = 0$. Then*

$$||J||(t) \leq ||J'(0)|| t \exp\left(\int_0^t ||A(\tau) - \frac{I}{\tau}|| d\tau\right). \tag{3.2}$$

Proof See Paeng[11].□

Let

$$D(c_0, n, k) := \int_0^{c_0/4} ||A(\tau) - \frac{I}{\tau}|| d\tau.$$

From Lemma 5.2 in Dai, Shen and Wei[3], we know that $||J(t)|| \leq \frac{t}{t_0} e^D ||J(t_0)||$, $0 \leq t \leq t_0 < c_0$. Then $||J'(0)|| = \lim_{t\to 0} \frac{||J(t)||}{t} \leq \frac{e^D ||J(t_0)||}{t_0}$.

From Lemma 3.2 and above inequality, we obtain the following inequality

$$e^{-D}||J'(0)||t \leq ||J(t)|| \leq e^{D}||J'(0)||t. \tag{3.3}$$

Also we know that

$$K_1||J(c_0/4)|| \leq ||J'(0)|| \leq K_2||J(c_0/4)||, \tag{3.4}$$

for some constants $K_1,\ K_2$ depending only on c_0, n, k. Consequently,

$$K_3||J(c_0/4)||t \leq ||J(t)|| \leq K_4||J(c_0/4)||t. \tag{3.5}$$

If not $\langle J'(0), \gamma'(0)\rangle = 0$, we can decompose J into tangential and perpendicular components and obtain the above boundedness (3), since tangential component is linear in t (See Dai, Shen and Wei[3].)

Remark 3.3 Theorem 1.1, 1.2 said that geodesic flows under Ricci curvature and conjugate radius bounded below is not too much complicated. By the above consequence, Ricci curvature and conjugate radius conditions play a similar role as $|K_M| \leq k$. Using the above lemma with explicit computation, we can obtain the following another stable properties on sphere theorem and splitting theorem by Paeng[12].

Theorem 3.4 *Let M be an n-dimensional compact Riemannian manifold with* $\mathrm{Ric}_M \geq -(n-1)k$ *and* $\mathrm{inj}_M \geq i_0$. *For given real numbers $k, i_0 > 0$ and a positive integer n, there exists an $\epsilon(i_0, n, k) > 0$ depending only on i_0, n, k such that if* $\mathrm{Ric}_M \geq n-1$ *on $M - A$,* $\mathrm{vol}(A) \leq \epsilon$ *and either*
(1) $\mathrm{vol}(M) \geq \omega_n - \epsilon$,
or
(2) $\mathrm{diam}(M) \geq \pi - \epsilon$,
then M is diffeomorphic to S^n.

Theorem 3.5 *Let M an n-dimensional compact Riemannian manifold with* $\mathrm{Ric}_M \geq -(n-1)k$ *and* $\mathrm{inj}_M \geq c_0$ *with* $\mathrm{diam}(M) \leq d$. *Then there exists an $\epsilon(i_0, n, k, d) > 0$ depending only on i_0, n, k, d such that if* $\mathrm{Ric}_M \geq 0$ *on $M - A$,* $\mathrm{vol}(A) \leq \epsilon$, *then M is diffeomorphic to $T^k \times N$ up to finite cover, where N is a simply connected space.*

Theorem 3.5 is a consequence in splitting theorem for almost nonnegative Ricci curvature with injectivity radius bounded below. If we assume $\mathrm{conj}_M \geq c_0$ instead of $\mathrm{inj}_M \geq i_0$, we cannot prove this splitting theorem.

Now we compute Lipschitz continuity along geodesic flows.

Lemma 3.6 *If h is sufficiently small,*

$$d(\phi_h(v_p), \phi_h(w_q)) - d(v_p, w_q) \leq N(c_0, n, k)d(v_p, w_q)h,$$

for some constant $N(c_0, n, k)$.

Proof Let $Q(t,s)$ be a length minimizing curve from v_p to w_q, i.e. a piecewise C^∞-rectangle which realizes the distance from v_p to w_q. We may assume the existence of a length minimizing curve.

Then we know that

$$d(v_p, w_q) = \int_0^1 \int_0^{c_0/4} ||\frac{\partial Q}{\partial S}||dtds$$

and

$$d(\phi_h(v_p), \phi_h(w_q)) \leq \int_0^1 \int_h^{c_0/4+h} ||\frac{\partial Q}{\partial s}||dtds.$$

So we get

$$d(\phi_h(v_p), \phi_h(w_q)) - d(v_p, w_q) \leq \int_0^1 \int_0^h + \int_0^1 \int_{c_0/4}^{c_0/4+h} ||\frac{\partial Q}{\partial s}||dtds.$$

Let $\frac{\partial Q}{\partial s}(0,s) = V(s)$ and $\frac{\partial Q}{\partial s}(c_0/4, s) = W(s)$. Decompose J into J_1 and J_2 such that $J_1(0) = 0$, $J_2(c_0/4) = 0$. Then $||J_2(0)|| = ||V||$ and $||J_1(c_0/4)|| = ||W||$. If $||V|| = ||W|| = 0$, then $J(t) = J_1(t) = J_2(t) = 0$ for all $t \leq c_0/4$ and $v_p = w_q$. So we do not need consider this case. From now on we may assume $||V|| \neq 0$. Then by (3.5), we have

$$K_3||W||t \leq ||J_1(t)|| \leq K_4||W||t, \tag{3.6}$$

$$K_3||V||t \leq ||J_2(c_0/4 - t)|| \leq K_4||V||t. \tag{3.7}$$

Thus we have

$$\begin{aligned} d(\phi_h(v_p), \phi_h(w_q)) - d(v_p, w_q) &\leq \textstyle\int_0^h \int_0^1 + \int_{c_0/4}^{c_0/4+h} \int_0^1 ||J(t)||dsdt \\ &\leq (\textstyle\int_{c_0/4}^{c_0/4+h} + \int_0^h K_4 tdt)(\int_0^1 ||V|| + ||W||ds) \\ &\leq K(c_0, n, k)(\textstyle\int_0^1 ||V|| + ||W||ds)h, \end{aligned} \tag{3.8}$$

for some constant $K(c_0, n, k)$. Now compute the $d(v_p, w_q)$. Let

$$a = \frac{c_0 K_3||V||}{4(K_3||V|| + K_4||W||)}, \quad b = \frac{c_0 K_3||W||}{4(K_3||W|| + K_4||V||)}.$$

On $[0, a]$, we know that $||J_2(t)|| \geq ||J_1(t)||$ and on $[c_0/4 - b, c_0/4]$, $||J_1(t)|| \geq$

$||J_2(t)||$ by (3.6) and (3.7). So we obtain

$$\begin{aligned} d(v_p, w_q) &= \int_0^{c_0/4} \int_0^1 ||J||dsdt \\ &\geq \int_0^1 \int_0^a ||J_2|| - ||J_1||dtds \\ &+ \int_0^1 \int_{c_0/4-b}^{c_0/4} ||J_1|| - ||J_2||dtds \\ &\geq \int_0^1 \int_0^a K_3||V||(c_0/4 - t) - K_4||W||tdtds \\ &+ \int_0^1 \int_{c_0/4-b}^{c_0/4} K_3||W||t - K_4||V||(c_0/4 - t)dtds. \end{aligned}$$

The integral $\int_0^a K_3||V||(c_0/4 - t) - K_4||W||tdt$ is the area of $\{(x, y) \mid K_4||W||x \leq y \leq K_3||V||(c_0/4 - x), \quad x \leq a\}$. Hence we get

$$\int_0^a ||J_2|| - ||J_1|| \geq ac_0K_3||V||/8 = c_0^2K_3/32\frac{K_3||V||^2}{K_4||W|| + K_3||V||}. \tag{3.9}$$

Similarly we get

$$\int_{c_0/4-b}^{c_0/4} ||J_1|| - ||J_2|| \geq c_0^2K_3/32\frac{K_3||W||^2}{K_4||V|| + K_3||W||}. \tag{3.10}$$

It is an easy calculation that

$$\int_0^{c_0/4} ||J||dt \geq \frac{a_1(||V||^2||W|| + ||V||||W||^2) + a_2(||V||^3 + ||W||^3)}{b_1(||W||^2 + ||V||^2) + b_2||V||||W||},$$

for some positive constant a_i, b_i depending only on c_0, n, k.

Then we obtain

$$\begin{aligned} \frac{||V||+||W||}{\int_0^{c_0/4} ||J||dt} &\leq \frac{(||V|| + ||W||)(b_1(||V||^2 + ||W||^2) + b_2||V||||W||)}{a_1(||V||^2||W|| + ||V||||W||^2) + a_2(||V||^3 + ||W||^3)} \\ &\leq C_0(c_0, n, k), \end{aligned} \tag{3.11}$$

for some positive constants $C_0(c_0, n, k)$ since we assume $||V|| \neq 0$. (See Paeng[11].) Consequently, we have

$$\begin{aligned} d(\phi_h(v_p), \phi_h(w_q)) - d(v_p, w_q) &\leq K(\int_0^1 ||V|| + ||W||ds)h \\ &\leq KC_0(\int_0^1 \int_0^{c_0/4} ||J||dtds)h \\ &= KC_0d(v_p, w_q)h, \end{aligned} \tag{3.12}$$

which completes the proof. □

For fixed T, we have

$$\begin{aligned} d(\phi_T(v_p), \phi_T(w_q)) &\leq \lim_{h\to 0}(1 + KC_0h)^{T/h}d(v_p, w_q) \\ &\leq e^{TKC_0}d(v_p, w_q). \end{aligned}$$

Then by Lemma 2.2, the topological entropy for geodesic flows of compact Riemannian manifolds with $\mathrm{Ric}_M \geq -k$ and $\mathrm{conj}_M \geq c_0$ is bounded i.e. $h(\phi_t) \leq (2n-1)KC_0$. This completes the proof of Theorem 1.1.

4 Proof of Theorem 1.2

Theorem 1.2 is obtained by explicit computation of the Laplacian of distance function. We use the notation A_i, B_i for A, B of M_i as previous sections. For the proof of Theorem 1.2, we need the following more precise estimate of L^1-norm of $||B_i||$.

First estimate $||B_i||$ in Preliminaries in the case of $\mathrm{Ric}_M \geq -(n-1)\epsilon_i^2$, $\mathrm{conj}_M \geq c_i := C\epsilon_i^{-1}$. From the definition of $\gamma(t)$, we estimate $\alpha(t) = \epsilon_i^2 t + 2\epsilon_i \coth \epsilon_i (\frac{c_i - t}{2})$ in section 2. For $0 \leq t \leq c_i/2$, we have $\frac{\epsilon_i c_i}{4} \leq \frac{\epsilon_i(c_i - t)}{2} \leq \frac{\epsilon_i c_i}{2}$ and

$$\begin{aligned} 2\epsilon_i \coth \epsilon_i \frac{(c_i - t)}{2} &= \tfrac{2}{c_i - t} 2\epsilon_i \tfrac{c_i - t}{2} \coth \epsilon_i \tfrac{(c_i - t)}{2} \\ &\leq \tfrac{4}{c_i} 2\epsilon_i \tfrac{(c_i - t)}{2} \coth \epsilon_i \tfrac{(c_i - t)}{2} \\ &\leq \tfrac{H_1}{c_i} = \tfrac{H_1 \epsilon_i}{C}. \end{aligned}$$

for some constant $H_1 > 0$, since we know that $x \coth x$ is bounded for any x so is $\epsilon_i \frac{c_i - t}{2} \coth \epsilon_i \frac{(c_i - t)}{2}$.

Since $\gamma(t) = 2\sqrt{(n-1)\alpha(t)}\sqrt{t}$ for $0 \leq t \leq c_i$,

$$\begin{aligned} \gamma(t) &= 2\sqrt{(n-1)(\epsilon_i t)^2 + 2(\epsilon_i \coth \epsilon_i \tfrac{(c_i - t)}{2}) \cdot t} \\ &\leq 2\sqrt{(n-1)\{(\epsilon_i c_i)^2 (\tfrac{t}{c_i})^2 + \tfrac{H_1 \epsilon_i t}{C}\}}, \end{aligned}$$

so we can find $H_2(n, C) > 0$ such that if $0 < t < c_i/H_2$ then $\gamma(t) \leq 1/7$ and the assumptions of Theorem 2.1 are satified. Furthermore we may assume $\gamma(t) \leq \sqrt{H_3(n, C)\epsilon_i t}$ for some constant H_3 depending only n, C if $t \leq \min(c_i/2C, c_i/H_2)$ since $(\epsilon_i t)^2 \leq \epsilon_i t$.

Now we define an inner product and a distance on $T_1 M_i$ as that of Definition 3.1, i.e. we substitute c_i/H_2 for $c_0/4$. Then (3.3) is replaced by

$$e^{-\int_0^t ||B_i||} ||J'(0)|| t \leq ||J(t)|| \leq e^{\int_0^t ||B_i||} ||J'(0)|| t. \tag{4.1}$$

By Theorem 2.1,

$$\int_0^{c_i/H_2} ||B_i|| \leq 2\gamma(t) \leq \frac{2}{7}.$$

Then

$$e^{-\frac{2}{7}}||J'(0)||t \leq ||J(t)|| \leq e^{\frac{2}{7}}||J'(0)||t \tag{4.2}$$

on $[0, c_i/H_2]$. Also we know that (3.4) is replaced by

$$e^{-\frac{2}{7}}\frac{||J(c_i/H_2)||}{c_i/H_2} \leq ||J'(0)|| \leq e^{\frac{2}{7}}\frac{||J(c_i/H_2)||}{c_i/H_2}. \tag{4.3}$$

So

$$K_1'\frac{||J(c_i/H_2)||}{c_i} \leq ||J'(0)|| \leq K_2'\frac{||J(c_i/H_2)||}{c_i} \tag{4.4}$$

for some constant K_1', K_2' depending only on n, C.

Consequently,

$$\frac{K_3'}{c_i}||J(c_i/H_2)||t \leq ||J(t)|| \leq \frac{K_4'}{c_i}||J(c_i/H_2)||t \tag{4.5}$$

for some constant K_3', K_4' depending only on n, C. So K_3, K_4 in (3.5) are replaced by $K_3'/c_i, K_4'/c_i$ where K_3', K_4' depend only on n, C.

In fact, without the condition of $\text{conj}_{M_i} \geq C/\epsilon_i$, we cannot consider K_3', K_4' as constant independent of c_i and we cannot say $\int ||J_1|| - ||J_2|| \to \infty$ as $i \to \infty$. So we get the following integral as (3.9), (3.10),

$$\begin{aligned}\int_0^a ||J_2|| - ||J_1|| &\geq c_i^2(K_3'/c_i)\frac{(K_3'/c_i)||V||^2}{(K_4'/c_i)||W|| + (K_3'/c_i)||V||}\\ &= c_iK_3'\frac{K_3'||V||^2}{K_4'||W|| + K_3'||V||}.\end{aligned}$$

Similarly

$$\int_{c_i/H_2-b}^{c_i/H_2} ||J_1|| - ||J_2|| \geq c_iK_3'\frac{K_3'||W||^2}{K_4'||V|| + K_3'||W||},$$

where a, b are also obtained by substituting c_i/H_2 for $c_0/4$. So we get

$$\int_0^{c_i/H_2} ||J||dt \geq c_i\frac{a_1(||V||^2||W|| + ||V||||W||^2) + a_2(||V||^3 + ||W||^3)}{b_1(||W||^2 + ||V||^2) + b_2||V||||W||},$$

for some positive constant a_i, b_i depending only on n, C. Then (3.11) is replaced by

$$\frac{||V|| + ||W||}{\int_0^{c_i/H_2} ||J||dt} \leq \frac{1}{c_i}C_0'(n, C)$$

for some constant C_0' depending only on n, C. Now we estimate K in (3.8). From the inequality (3.8), we get immediately that

$$Kh = \int_{c_i/H_2}^{c_i/H_2+h} + \int_0^h \frac{K_4'}{c_i} t dt \leq \frac{K_4'}{2c_i}((\frac{c_i}{H_2}+h)^2 - \frac{c_i^2}{H_2^2} + h^2)$$
$$\leq \frac{K_4'}{2c_i}(2\frac{c_i}{H_2}h + 2h^2) \leq K_5(n, C)h,$$

for some constant K_5 depending only on n, C. So $K \leq K_5(n, C)$. Consequently, KC_0 in (3.12) has the value

$$(K_5(n, C))(\frac{1}{c_i}C_0'(n, C)) = \frac{K_5 C_0'}{c_i} \to 0$$

so we conclude that $h_i(\phi_t)$ converges to 0. This completes the proof of Theorem 1.2.

From Manning[9], we know that topological entropy for geodesic flows of almost flat manifolds are almost 0. Also we know that almost flat manifolds are infranilmanifolds by Gromov [6]. From Theorem 1.2, we obtain that manifolds with almost nonnegative Ricci curvature and large conjugate radius have almost 0 topological entropy for geodesic flows. So it is a question that these manifolds are infranilmanifolds. We obtain the following result about this question;

Theorem 4.1 (Paeng[13]) *Let M be an n-dimensional compact Riemannian manifold with $\mathrm{Ric}_M \geq -(n-1)$. Then there exists an $\epsilon > 0$ depending on n and c_0 such that if $\mathrm{conj}_M \geq c_0$ and $\mathrm{diam}(M) \leq \epsilon$, then a finite covering space of M is a nilmanifold.*

We obtain this result by Jacobi fields estimate and parallel translation estimate by Theorem 2.1.

From this result or Cheeger-Colding's result[2], we also know that manifolds with almost nonnegative Ricci curvature have almost nilpotent fundamental groups. So their volume growth rates are polynomial growth rates by Gromov[6]. So we have the following consequence immediately by Freire-Mañé [4];

Corollary 4.2 *Let M be an n-dimensional compact Riemannian manifold with no conjugate points and $\mathrm{diam}(M) \leq d$. Then there exists an $\epsilon > 0$ depending on n, d such that if $\mathrm{Ric}_M \geq -(n-1)\epsilon$, then $h(\phi_t) = 0$.*

The diameter condition in Corollary 4.2 is an essential assumption. In this case, we can obtain the same result under the rescaling of metrics.

Acknowledgments

The author is partially supported by the Basic Science Research Institute Program and in part supported by GARC-KOSEF. We would like to express our gratitude to Professor Hong-Jong Kim and Professor Dong-Pyo Chi for kind and helpful advices.

References

1. R. Brocks, *C. R. Acad. Sci. Paris* **t.319, série I**, 73 (1994).
2. J. Cheeger, T. B. Colding, *Ann. Math.* **144**, 189 (1996).
3. X. Dai, Z. Shen, G. Wei , *Duke Math. J.* **76**, 59 (1994).
4. A. Freire, R. Mañé, *Ann. Math.* **69**, 375 (1982).
5. S. Gallot, D. Hulin, J. Lafontaine, *Riemannian geometry*, (Springer-Verlag, 1987).
6. M. Gromov, *Publ. Math. IHES* **53**, 53 (1981).
7. A. B. Katok, B. Hasselblatt, *Introduction to the modern theory of dynamical systems*, (Cambridge university press, 1995).
8. A. Manning, *Ann. Math.* **110**, 567 (1979).
9. A. Manning, *Dynamical systems and turbulence, Warwick* , 243 (1980).
10. S.-H. Paeng, *Kyushu J. Math.* **50**, 459 (1996).
11. S.-H. Paeng, *Proc. Amer. Math. Soc.* **125**, 1873 (1997).
12. S.-H. Paeng, *Ph. D. Thesis.*
13. S.-H. Paeng, preprint.

Canonical flows of Einstein-Weyl manifolds

Hong Kyung Pak [a]
Department of Mathematics
Kyungsan University
Kyungsan, 712-240, Korea
E-mail : hkpak@kirin.ksucc.ac.kr

From the point of view generalizing results in [12], [14], [9], we study sufficient conditions that the canonical flow determined from the strict Gauduchon gauge on a compact Einstein-Weyl manifold admits a transversal Kähler structure. Besides, the existence problem of a strict Gauduchon gauge is investigated. We construct seversl examples of compact Einstein-Weyl manifolds which admit the strict Gauduchon gauge.

1 Introduction

Given a bundle-like (or equivalently metric) flow $\mathcal{F}$ on a compact Einstein manifold (M, g), the author ([12]) gave a sufficient condition that $\mathcal{F}$ admits a transversal Kähler structure. The following result obtained in [12] extended that of Bérard Bergery ([1]) for Einstein structures on principal S^1-bundles over compact Einstein bases.

Theorem 1.1 *Let (M, g) be a compact Einstein manifold of dimension m with a geodesic, bundle-like flow $\mathcal{F}$. Suppose that $\mathcal{F}$ is transversally Einstein and all the leaves of $\mathcal{F}$ are closed. Then*

(A) *$A = 0$, M is the locally Riemannian product,*

(B) *$A \neq 0$, the leaf space $M/\mathcal{F}$ admits a natural compact Kähler-Einstein Satake manifold structure.*

It is worthwhile to note that the transversal scalar curvature s_B for $\mathcal{F}$ is necessarily non-negative, and $s_B = 0$ if and only if $A = 0$.

Einstein-Weyl manifolds are Weyl (conformal) manifolds satisfying an analogue of the Einstein equations. They generalize the Einstein manifolds in the following sense : every Einstein manifold is considered as an Einstein-Weyl manifold, while there exist Einstein-Weyl manifolds which admit no Einstein metric. Materials on Einstein-Weyl manifolds may be found in several papers ([4] to [7], [13] to [16], [18]).

[a]The author wishes to acknowledge the financial support of the Korean Science and Engineering Foundation, 951-0106-015-1.

In [14], Pedersen and Swann extended the result of Bérard Bergery to Einstein-Weyl structures on principal S^1-bundles over compact Einstein bases. In the present paper we are interested in the canonical flow $\mathcal{F}_\omega$ determined from a given strict Gauduchon gauge (g, ω) on an Einstein-Weyl manifold. Our main purpose is to derive similar results for $\mathcal{F}_\omega$ as in [12], that is, to obtain sufficient conditions that $\mathcal{F}_\omega$ admits a transversal Kähler structure.

Theorem 1.2 (Main Theorem) *Let $(M, [g], D)$ be a compact Einstein-Weyl manifold of dimension m which admits the strict Gauduchon gauge (g, ω) and has constant conformal scalar curvature s_D. Suppose that the canonical flow $\mathcal{F}_\omega$ determined from (g, ω) is transversally Einstein. Then*

(A) *$s_D > 0$, $\mathcal{F}_\omega$ admits a transversal Kähler structure.*

(B) *$s_D = 0$, M is the locally Riemannian product which the first Betti number satisfies $b_1(M) = 1$.*

It should be remarked that Main Theorem extends the result of Narita ([9]) for Riemannian submersions on compact Einstein-Weyl manifolds with totally geodesic fibers of dimension one over Einstein manifolds.

The plan of this paper is as follows. Section 2 concerns with a brief review of Einstein-Weyl geometry. For our purpose, we introduce a new notion, so-called a strict Gauduchon gauge. The existence of a strict Gauguchon gauge on a given Weyl manifold is not, in general, assurred. In Section 3, we shall show that the sign of the conformal scalar curvature on an Einstein-Weyl manifold is closely related to the existence of a strict Gauduchon gauge. Section 4 is devoted to constructing certain examples of compact Einstein-Weyl manifolds which admit the strict Gauduchon gauge.

2 Strict Gauduchon Gauge and canonical flows

We start with the review of Einstein-Weyl geometry. A manifold M with conformal structure $[g]$ and torsion-free affine connection D compatible with the conformal structure is called *a Weyl manifold.* If we choose a metric g in the conformal structure, the compatiblity is equivalent to the existence of a 1-form ω on M such that

$$Dg = \omega \otimes g. \tag{1}$$

A Weyl manifold $(M, [g], D)$ is said to be *Einstein-Weyl* if the symmetrized Ricci tensor $S(r_D)$ of D is proportional to g pointwise, that is, there exists a function Λ on M such that

$$S(r_D)(E, F) := r_D(E, F) + r_D(F, E) = \Lambda g(E, F) \tag{2}$$

for all tangent vector fields $E, F \in \Gamma(TM)$.

It is noted that since D is not a metric connection, its Ricci curvature r_D is not necessarily symmetric and the multiplier Λ is not in general constant. If s_D denotes the trace of r_D with respect to g, then

$$S(r_D) = \frac{2s_D}{m} g, \tag{3}$$

where m is the dimension of M. s_D is called *the conformal scalar curvature.* We observe that s_D is constant if and only if Λ is constant.

If we fix our choice of metric g in the conformal structure and 1-form ω satisfying (1) and if ∇ is the Levi-Civita connection of g, then we may write

$$D_E F = \nabla_E F - \frac{1}{2}\{\omega(E)F + \omega(F)E - g(E,F)\omega^\sharp\} \tag{4}$$

for all $E, F \in \Gamma(TM)$. Here and hereafter, $\omega^\sharp$ is denoted by the dual vector field to ω with respect to g.

From this, it follows a formula for Ricci curvatures ([14])

$$\begin{aligned} r_D(E,F) = r(E,F) &+ \tfrac{1}{2}\{(m-1)(\nabla_E\omega)(F) - (\nabla_F\omega)(E)\} \\ &+ g(E,F)\{\tfrac{1}{2}d^*\omega - \tfrac{1}{4}(m-2)|\omega^\sharp|^2\} \\ &+ \tfrac{1}{4}(m-2)\omega(E)\omega(F) \end{aligned} \tag{5}$$

where $d^*\omega$ denotes the codifferential of ω and $|\omega^\sharp|$ the pointwise norm of $\omega^\sharp$ with respect to g.

Then the Einstein-Weyl equation (2), together with (3), reads to

$$r = \{\frac{s_D}{m} - \frac{1}{2}d^*\omega + \frac{1}{4}(m-2)|\omega^\sharp|^2\}g - \frac{1}{4}(m-2)(L_{\omega^\sharp}g + \omega\otimes\omega). \tag{6}$$

We are interested in the case where ω is nonsingular. In particular, we consider Einstein-Weyl manifolds admitting the special data (g,ω), so-called a strict Gauduchon gauge. We say that a Weyl manifold $(M,[g],D)$ admits *a strict Gauduchon gauge* if we can choose a metric g in the conformal structure and a nonsingular 1-form ω such that $\omega^\sharp$ is a Killing vector field of constant length with respect to g.

It may be natural to consider the problem whether every Weyl manifold admits a strict Gauduchon gauge and whether, if exists, it is unique (up to a constant). However, there are well-known facts related to this problem. The following result is found in [3].

Proposition 2.1 *Up to a constant, there is a unique metric g in the conformal structure of a compact Weyl manifold with respect to which the corresponding 1-form ω is co-closed.*

Such (g,ω) is called *the Gauduchon gauge.* Furthermore, Tod ([18]) demonstrated the following result.

Proposition 2.2 *If (g,ω) is the Gauduchon gauge of a compact Einstein-Weyl manifold, then $\omega^\sharp$ is a Killing vector field with respect to g.*

In the Gauduchon gauge, the Einstein-Weyl equation (6) is simplified as

$$r = \{\frac{s_D}{m} + \frac{1}{4}(m-2)|\omega^\sharp|^2\}g - \frac{1}{4}(m-2)\omega \otimes \omega. \tag{7}$$

Given a strict Gauduchon gauge (g,ω) on a Weyl manifold $(M,[g],D)$, we have the flow $\mathcal{F}_\omega$ whose leaves are the integral curves of $\omega^\sharp$, which is called *the canonical flow determined from* (g,ω). Note that $\mathcal{F}_\omega$ is bundle-like and geodesic on a Riemannian manifold (M,g). By normalizing length we may regard $\mathcal{F}_\omega$ as a flow generated by a unit Killing vector field $\omega^\sharp$ with respect to g.

If we consider a conformal change of metric $\tilde{g} = e^f g$, then the corresponding 1-form is $\tilde{\omega} = \omega + df$. Moreover, the new conformal scalar curvature is $\tilde{s}_D = e^{-f}s_D$, so the sign of the conformal scalar curvature is a conformal invariant.

3 Proof of Main Theorem

Our proof proceeds by dividing into three cases according to the sign of the conformal scalar curvature s_D. Recall that the sign of s_D is a conformal invariant.

To begin with, for the case where s_D is negative, or equivalently $S(r_D)$ is negative, we deduce the following two results.

Proposition 3.1 *Let $(M,[g],D)$ be an Einstein-Weyl manifold of dimension m with $s_D < 0$. Then M admits no any strict Gauduchon gauge (g,ω). Therefore, M admits no any canonical flow.*

Indeed, assume that $(M,[g],D)$ admits a strict Gauduchon gauge (g,ω). Then the canonical flow $\mathcal{F}_\omega$ is bundle-like and geodesic on (M,g). Thus a formula for the Ricci curvature of g ([1]) becomes

$$r(\omega^\sharp,\omega^\sharp) = |\omega^\sharp|^2|A|^2.$$

Here A denotes the O'Neill integrability tensor (for definition, see [11], [1]) for $\mathcal{F}_\omega$. It means that the A vanishes if and only if the horizontal bundle for $\mathcal{F}_\omega$ is integrable.

On the other hand, by using (7) we have $r(\omega^\sharp, \omega^\sharp) = \frac{s_D}{m}|\omega^\sharp|^2$, which yields that

$$s_D = m|A|^2 \geq 0, \tag{8}$$

which leads to a contradiction.

In addition, when M is compact we have a more stronger obstruction to the existence of the strict Gauduchon gauge, as we see Proposition 3.2 below ([16], [15]). Call a Gauduchon gauge (g, ω) with vanishing ω *a trivial Gauduchon gauge.*

Proposition 3.2 *Let $(M, [g], D)$ be a compact Einstein-Weyl manifold of dimension m with $s_D < 0$. Then the Gauduchon gauge (g, ω) must be trivial, that is, (M, g) must be Einstein.*

Remark 3.1 (A) The Bochner technique (refer to [2], [22]) yields that Proposition 3.2 is still valid in a more weaker condition that s_D is quasi-negative, i.e. $s_D \leq 0$ on M and $s_D(p_0) < 0$ for some point $p_0 \in M$. Such result is proved in [15] by somewhat another method.

(B) Ranjan ([17]) proved that if the Ricci curvature on a compact Riemannian manifold (M, g) is negative, then (M, g) cannot admit a bundle-like flow. We observe that for the Gauduchon gauge (g, ω) on a compact Einstein-Weyl manifold $(M, [g], D)$ the negativity of the Ricci curvature of g implies that of the conformal scalar curvature s_D. In fact, working in the Gauduchon gauge (7) yields

$$s = s_D + \frac{1}{4}(m-2)(m-1)|\omega^\sharp|^4. \tag{9}$$

(C) The hypothesis of compactness of M cannot be deleted in Proposition 3.2. Indeed, there exists a non-compact Einstein-Weyl manifold which has negative conformal scalar curvature and admits a Gauduchon gauge (g, ω) which is not trivial. The following example is found in [9].

Let H be the Heisenberg group

$$H := \left\{ \begin{pmatrix} 1 & s & u \\ 0 & 1 & t \\ 0 & 0 & 1 \end{pmatrix} \mid s, t, u \in \mathbf{R} \right\},$$

which is non-compact. Let K be the subgroup of H defined by

$$K := \left\{ \begin{pmatrix} 1 & 0 & u \\ 0 & 1 & 0 \\ 0 & 0 & 1 \end{pmatrix} \in H \right\}.$$

We take on H a Riemannian metric $g := ds^2 + dt^2 + (du - sdt)^2$. Then $\pi : H \longrightarrow H/K$ is a (non-trivial) Riemannian submersion with totally geodesic fibers and the curvature of H/K is zero. Moreover, H admits a Sasakian structure. Since the leaves of the Sasakian flow $\mathcal{F}_K$ determined from the structure vector field are exactly the fibers of π, $\mathcal{F}_K$ is transversally Einstein with vanishing transversal scalar curvature.

As will be seen later (refer to Proposition 4.1 and subsequent Remarks (B)), H admits an Einstein-Weyl structure with Gauduchon gauge (g, ω). In this case, ω is not identically zero. Since the scalar curvature of g is negative, (9) implies that its conformal scalar curvature is also negative. Indeed, it is explicitly of the form $-3(3 + 4\tan^2 u)/2$.

(*D*) It would be interesting to consider the following specialization of the existence problem of a strict Gauduchon gauge : does every compact Einstein-Weyl manifold with non-negative conformal scalar curvature admit the Gauduchon gauge which is not trivial?

Next we investigate on the case of vanishing the conformal scalar curvature.

Proposition 3.3 *Let $(M, [g], D)$ be an Einstein-Weyl manifold of dimension m. Then the following conditions are equivalent.*

(A) *M is locally conformal to Einstein,*

(B) *for a Gauduchon gauge (g, ω) $d\omega = 0$.*

Furthermore, if M is compact and admits the strict Gauduchon gauge (g, ω) then these conditions are equivalent to the following conditions

(C) $s_D = 0$,

(D) $A = 0$,

where A denotes the integrablity tensor for the canonical flow $\mathcal{F}_\omega$ determined from (g, ω). In this case, M is the locally Riemannian product which the first Betti number satisfies $b_1(M) = 1$.

Indeed, let (g, ω) be a Gauduchon gauge on M. If ω is closed, then $\omega = df$ locally. By the conformal change $\tilde{g} := e^{-f}g$, the corresponding 1-form is then $\tilde{\omega} = 0$, and vice versa. This means that $(A) \Leftrightarrow (B)$.

Let (g, ω) be the strict Gauduchon gauge on M. For M compact, it is well-known that $d\omega = 0$ if and only if $\Delta\omega = 0$, because $d^*\omega = 0$. Furhter, the Weitzenböck formula $\Delta\omega = \nabla^*\nabla\omega + r(\omega)$ combined with (7) yields

$$\Delta\omega = \frac{2s_D}{m}\omega,$$

which shows that $(B) \Leftrightarrow (C)$.

$(C) \Leftrightarrow (D)$ is immediate from the formula (8).

Now, If $d\omega = 0$, then 1-form ω defines a De Rham cohomology class $[\omega]$ on M. If $[\omega] = 0$ then $\omega = df$ globally. Then $\Delta f = 0$, which implies that the function f is constant. Hence $\omega = 0$. Therefore $b_1(M) \geq 1$. Moreover, it was shown in [15] that $b_1(M) = 1$.

Remark 3.2 In dimension 4, Einstein-Weyl manifolds with positive first Betti number have been studied intensively by Gauduchon ([4], [13]).

In order to complete the proof of Main Theorem, it remains the case of positivity of the conformal scalar curvature. We begin by recalling basic definitions of transversal Kähler geometry for foliations.

Let $\mathcal{F}$ be a bundle-like foliation on a Riemannian manifold (M, g) of codimension q. The metric g determines an orthogonal decomposion

$$TM = \mathcal{V} \oplus \mathcal{H},$$

where $\mathcal{V}$ is the tangent bundle of $\mathcal{F}$ and $\mathcal{H}$ is the horizontal bundle for $\mathcal{F}$. Correspondingly, $g = g_{\mathcal{V}} \oplus g_{\mathcal{H}}$. There exists a unique torsion-free metric connection ∇^B on $(\mathcal{H}, g_{\mathcal{H}})$, which is called *the transversal Levi-Civita connection.* We say (see [10]) that a bundle-like foliation $\mathcal{F}$ on a Riemannian manifold (M, g) is *transversal Kähler* if

(A) $\mathcal{F}$ admits a transversal almost complex structure J_B, with respect to which $g_{\mathcal{H}}$ is Hermitian, i.e., $g_{\mathcal{H}}(J_B X, J_B Y) = g_{\mathcal{H}}(X, Y)$ for $X, Y \in \Gamma(\mathcal{H})$,

(B) J_B is parallel with respect to the transversal Levi-Civita connection.

From now on, let $(M, [g], D)$ be a compact Einstein-Weyl manifold of dimension $m = 1 + q$ which admits the strict Gauduchon gauge (g, ω) and has positive constant conformal scalar curvature s_D. Furthermore, suppose that the canonical flow $\mathcal{F}_\omega$ determined from (g, ω) is transversally Einstein, that is the transversal Ricci curvature r_B of ∇^B satisfies

$$r_B = \frac{s_B}{q} g_{\mathcal{H}},$$

where s_B is the transversal scalar curvature of ∇^B. $\mathcal{F}_\omega$ may be regarded as the flow generated by a unit Killing vector field $\omega^\sharp$. Then by combining a formula for Ricci curvatures ([1])

$$r(X, Y) = r_B(X, Y) - 2g(A_X \omega^\sharp, A_Y \omega^\sharp) \quad \text{for } X, Y \in \Gamma(\mathcal{H})$$

with (7), we get

$$g(A_X\omega^\sharp, A_Y\omega^\sharp) = kg(X,Y), \tag{10}$$

where $k := \frac{1}{2}(\frac{s_B}{q} - \frac{s_D}{m} - \frac{m-2}{4})$. Here $\mathcal{H}$ denotes the horizontal bundle for $\mathcal{F}_\omega$. From Proposition 3.3, we deduce that k is a positive constant. Define an endomorphism J_B of $\mathcal{H}$ by

$$J_B X := \frac{1}{\sqrt{k}} A_X \omega^\sharp \quad \text{for } X \in \Gamma(\mathcal{H}). \tag{11}$$

It is easy to verify that J_B is a transversal almost complex structure with respect to which $g_\mathcal{H}$ is Hermitian. In particular, we can take q to be even.

The fundamental 2*-form* Φ of $\mathcal{F}$ defined by

$$\Phi(X,Y) := g_\mathcal{H}(X, J_B Y) \quad \text{for } X, Y \in \Gamma(\mathcal{H})$$

is closed, since

$$\Phi(X,Y) = \frac{1}{\sqrt{k}} g(X, A_Y\omega^\sharp) = -\frac{1}{\sqrt{k}} d\omega(X,Y). \tag{12}$$

Moreover, (9) implies that $s > 0$, so that

$$s_B = s + |A|^2 > 0. \tag{13}$$

With an argument similar to the proof of Sekigawa ([12], [8]), we have

$$\nabla^B J_B = 0. \tag{14}$$

4 Construction of Einstein-Weyl manifolds admitting strict Gauduchon Gauge

As was mentioned before, we have seen that every compact Einstein-Weyl manifold does not necessarily admit the strict Gauduchon gauge (Proposition 3.2), even though it admits the Gauduchon gauge (Proposition 2.2). Instead, on certain manifolds satisfying special conditions we shall construct Einstein-Weyl structures admitting strict Gauduchon gauge. The following result is related to the case where $s_D > 0$ in Main Theorem. The proof is greatly indebted to [9].

Proposition 4.1 *Let $(M, \phi, \bar{\xi}, \bar{\eta}, \bar{g})$ be a Sasakian manifold of dimension $m = 1 + 2n$. Suppose that the Sasakian flow $\mathcal{F}_{\bar{\eta}}$ generated by a unit Killing vector field $\bar{\xi}$ is transversally Einstein with positive transversal scalar curvature. Then M admits an Einstein-Weyl structure with strict Gauduchon gauge.*

Indeed, we begin with considering the conformal class $[\bar{g}]$ of $\bar{g}$. And then we choose a metric in the conformal structure as $g := \frac{1}{k}\bar{g}$ for a positive constant k. It is easy to see that $\xi = \sqrt{k}\bar{\xi}, \eta = \frac{1}{\sqrt{k}}\bar{\eta}$ and the curvature tensor of g holds

$$R_{E\xi}F = \sqrt{k}\{\eta(F)E - g(E, F)\xi\}. \tag{15}$$

We set $\omega := f\eta$ for a function f on M. Define a connection D by the formula (4), where ∇ is the Levi-Civita connection of g. Then $(M, [\bar{g}], D)$ is a Weyl manifold.

It is immediate that $\mathcal{F}_\eta$ is transversally Einstein, say $r_B = lg_{\mathcal{H}}$. Note that $l > 0$. Since $\mathcal{F}_\eta$ is a geodesic and bundle-like flow on (M, g), the Ricci curvature of g satisfies

$$\begin{aligned} r(\xi, \xi) &= 2nk, \\ r(\xi, X) &= 2nk\eta(X) = 0, \\ r(X, Y) &= r_B(X, Y) - 2kg(A_X\xi, A_Y\xi), \end{aligned} \tag{16}$$

for $X, Y \in \Gamma(\mathcal{H})$. Let us now to find conditions for constants k, l and the function f which allow (g, ω) to satisfy the Einstein-Weyl equation (6).

It follows from (16) that for unit $X \in \Gamma(\mathcal{H})$

$$\begin{aligned} S(r_D)(\xi, \xi) &= 4nk + \tfrac{1}{2}(m-2)(2\xi(f) + f^2) + d^*\omega - \tfrac{1}{2}(m-2)f^2, \\ S(r_D)(\xi, X) &= \tfrac{1}{2}(m-2)X(f), \\ S(r_D)(X, X) &= 2(l - 2k) + d^*\omega - \tfrac{1}{2}(m-2)f^2. \end{aligned} \tag{17}$$

Thus (g, ω) satisfies the Einstein-Weyl equation if and only if the system of differential equations

$$X(f) = 0, \quad 2\xi(f) + f^2 = \frac{4}{m-2}\{l - (m+1)k\} \tag{18}$$

has the global solution. In particular, this holds if f is constant on M. In our situation f is explicitly determined by the condition

$$f^2 := \frac{4}{m-2}(l - k(m+1)) \quad \text{where}\, 0 < k < \frac{l}{m+1}. \tag{19}$$

It is clear that $\omega^\sharp$ is a nonsingular Killing vector field on M of constant length with respect to g. Therefore we conclude that (g, ω) is a strict Gauduchon gauge on an Einstein-Weyl manifold $(M, [\bar{g}], D)$.

Remark 4.1 (*A*) With respect to the strict Gauduchon gauge (g, ω) obtained in Proposition 4.1, the canonical flow $\mathcal{F}_\omega$ is transversally Einstein. The conformal scalar curvature is given by $s_D = m(m-1)k$, which is a positive constant.

(B) In case that $\mathcal{F}_{\bar{\eta}}$ is transversally Einstein with vanishing transversal scalar curvature, we can not take f to be constant as a solution of (18). However, when M is the Heisenberg group H appeared in Section 3 the system (18) has a non-constant global solution f. Thus H admits an Einstein-Weyl structure.

(C) By applying the very similar method, we have the following construction which shows that our Main Theorem generalizes Theorem 1.1.

Proposition 4.2 *Let (M, g) be a compact Einstein manifold of dimension m with a geodesic, bundle-like flow $\mathcal{F}$ generated by a unit Killing vector field $V = \theta^\sharp$. Suppose that $\mathcal{F}$ is transversally Einstein with positive transversal scalar curvature. Then M admits an Einstein-Weyl structure with strict Gauduchon gauge.*

It is worth observing that both Proposition 4.1 and Proposition 4.2 may be also proved by applying the canonical variation method (refer to [1], [14] for definition) of the metric g instead of the homothetic change.

In order to find examples related to the case where $s_D = 0$, we examine a special class of locally conformal Kähler manifolds (for such manifolds, see [19] to [21]). A locally conformal Kähler manifold (M, g, J, Ω) has a unique torsion-free affine connection D satisfying $Dg = \omega \otimes g$, where ω denotes *the Lee form* of M, namely

$$d\Omega = \omega \wedge \Omega.$$

Thus $(M, [g], D)$ is a Weyl manifold. We note that the Lee form ω vanishes if and only if M is a Kähler manifold.

If, in addition, ω is parallel with respect to the Levi-Civita connection ∇ of g, M is called *a generalized Hopf manifold.* In this case, $\omega^\sharp$ is a Killing vector field of constant length with respect to g. Recalling the well-known result ([20]) that a compact locally conformal Kähler manifold which is not globally conformal Kähler cannot be a Kähler manifold, we have

Proposition 4.3 *Let (M, g, J, Ω) be a compact generalized Hopf manifold which is not globally conformal Kähler. Then M admits a Weyl structure with strict Gauduchon gauge (g, ω), where ω is the Lee form.*

Remark 4.2 From Proposition 4.3, we can easily find examples of compact Einstein-Weyl manifolds with strict Gauduchon gauge and vanishing conformal scalar curvature. In particular, if a generalized Hopf manifold M is locally conformal Kähler-flat, i.e. whose locally conformal Kähler structure consists of flat Kähler metrics, then the curvature tensor of the connection D vanishes.

Therefore, the strict Gauduchon gauge (g, ω) satisfies the Einstein-Weyl equation (2) with vanishing conformal scalar curvature. The Hopf manifold with the metric

$$g := \frac{1}{\sum_{i=1}^{m} z_i \bar{z}_i} \sum_{j=1}^{m} dz_j d\bar{z}_j \quad \text{for}\, (z_1, ..., z_m) \in \mathbf{C}^m - \{\mathbf{0}\}$$

is a sort of such manifolds.

References

1. A. L. Besse, *Einstein Manifolds, Springer, Berlin, Heidelberg, New York, 1987*
2. S. Bochner, *Vector fields and Ricci curvature,* Bull. Amer. Math. Soc. **52**, 776–797 (1946)
3. P. Gauduchon, *La 1-forme de torsion d'une variété hermitienne compacte,* Math. Ann. **267**, 495–518 (1984)
4. P. Gauduchon, *Structures de Weyl-Einstein espaces de twisteurs et variété de type* $S^1 \times S^3$ (to appear)
5. T. Higa, *Weyl manifolds and Einstein-Weyl manifolds,* Comm. Math. Univ. Saint Pauli **42**, 143–160 (1993)
6. J. Hitchin, *Complex manifolds and Einstein equations,* Lect Notes in Math. **970**, 79–99 (1982)
7. P. E. Jones and K. P. Tod, *Minitwister spaces and Einstein-Weyl spaces,* Classical Quantum Gravity **2**, 565–577 (1985)
8. S. Morimoto, *Almost complex foliations and its application to contact geometry,* Nat. Sci. Rep. Ochanomizu Univ. **43**, 11–23 (1992)
9. F. Narita, *Riemannian submersions and Riemannian manifolds with Einstein-Weyl structures* (to appear)
10. S. Nishikawa and Ph. Tondeur, *Transversal infinitesimal automorphisms for harmonic Kähler foliations,* Tohoku Math. J. **40**, 599–611 (1988)
11. B. O'Neill, *A The fundamental equations of a submersion,* Mich. Math. J. **13**, 459–469 (1966)
12. H. K. Pak, *On one-dimensional metric foliations in Einstein spaces,* Illinois J. Math. **36**, 594–599 (1992)
13. H. Pedersen, Y. S. Poon and A.Swann, *The Hitchin-Thrope inequality for Einstein-Weyl manifolds,* Bull. London Math. Soc. **26**, 191–194 (1994)
14. H. Pedersen and A.Swann, *Riemannian submersions, four manifolds and Einstein-Weyl geometry,* Proc. London Math. Soc. **66**, 381–399 (1993)

15. H. Pedersen and A.Swann, *Einstein-Weyl geometry, the bach tensor and conformal scalar curvature,* J. reine angew. Math. **441**, 99–113 (1993)
16. H. Pedersen and K. P. Tod, *Three-dimensional Einstein-Weyl geometry,* Adv. Math. **97**, 74–109 (1993)
17. A. Ranjan, *Structural equations and integral formula for foliated manifolds,* Geom. Dedicata **20**, 85–91 (1986)
18. K. P. Tod, *Compact 3−dimensional Einstein-Weyl structures,* J. London Math. Soc. **45**, 341-351 (1992)
19. I. Vaisman, *Locally conformal Kähler manifolds with parallel Lee form,* Rend. Mat. Roma **12**, 263–284 (1979)
20. I. Vaisman, *On locally and globally conformal Kähler manifolds,* Trans. Amer. Math. Soc. **262**, 533–542 (1980)
21. I. Vaisman, *Generalized Hopf manifolds,* Geom. Dedicata **13**, 231–255 (1982)
22. H. Wu, *The Bochner technique in differential geometry,* Mathematical Reports, Harwood Academic Publishers **3**, 298–542 (1988)

Jeffrey-Weitsman-Witten invariants of Seifert fibred manifolds

Jinsung Park
Department of Mathematics, Korea Institute for Advanced Study,
207-43 Cheongryangri-dong, Dongdaemun-gu,
Seoul 130-012, Republic of Korea
E-mail: jspark@math.snu.ac.kr

We introduce a result in [P] aboout Jeffrey-Weitsman-Witten invariants of Seifert fibred manifolds.

1 Definition of Jeffrey-Weitsman-Witten invariant

We review the Chern-Simons gauge theory to understand the definition of the Jeffrey-Weitsman-Witten invariant. For the detail of the Chern-Simons gauge theory, see [R,S,W], [J,W].

Let X be a 2-dimensional manifold and P be a principal $SU(2)$-bundle over X. Let $\mathcal{A}, \mathcal{A}_{\mathcal{F}}, \mathcal{G}$ be the affine space of connection one forms of P, the space of flat connections of P and the gauge transformation group of P respectively. Let $\mathcal{M}_2$ be the moduli space of the flat connections of P.

We consider a 3-dimensional manifold Y_1 with a boundary X. Moreover we assume that a neighborhood of X in Y_1 is diffeomorphic to $X \times [0,1)$. For $A \in \mathcal{A}, g \in \mathcal{G}$ we consider a $U(1)$-valued function $\mathcal{S}(A,g)$ defined by

$$\mathcal{S}(A,g) \equiv \exp(2\pi i(CS(\tilde{A}^{\tilde{g}}) - CS(\tilde{A}))) \tag{1}$$

where $\tilde{A}$ and $\tilde{g}$ are the extensions of A and g into Y_1, $\tilde{A}^{\tilde{g}}$ is the gauge transformation of $\tilde{A}$ by $\tilde{g}$ and the Chern-Simons invariant $CS(\tilde{A})$ is given by

$$CS(\tilde{A}) = \frac{1}{8\pi^2}\int_{Y_1} tr(d\tilde{A} \wedge \tilde{A} + \frac{2}{3}\tilde{A} \wedge \tilde{A} \wedge \tilde{A}). \tag{2}$$

Such an extension of $\tilde{g}$ always exists since $\pi_1(SU(2)) = \pi_2(SU(2)) = 0$. We choose the extensions so that $\tilde{A}$ and $\tilde{g}$ are pull-backs of A and g by the projection to X over $X \times [0,1)$ respectively. Then $\mathcal{S}$ is independent on the extension Y_1 and the extension $\tilde{A}$ and $\tilde{g}$. In fact the extensions $(\tilde{A}_1, \tilde{g}_1)$ and $(\tilde{A}_2, \tilde{g}_2)$ into Y_1 and Y_2 give a connection $\tilde{B}$ and a gauge transformation $\tilde{h}$ on $Y = Y_1 \cup Y_2$ so that

$$\begin{aligned}&\exp(2\pi i(CS(\tilde{A}_1^{\tilde{g}}) - CS(\tilde{A}_1)))\exp(2\pi i(CS(\tilde{A}_2^{\tilde{g}}) - CS(\tilde{A}_2)))^{-1}\\ &\qquad = \exp(2\pi i(CS(\tilde{B}^{\tilde{h}}) - CS(\tilde{B}))) = 1.\end{aligned} \tag{3}$$

The above function $\mathcal{S}$ over $\mathcal{A} \times \mathcal{G}$ is a cocycle since

$$\mathcal{S}(\tilde{A}, \tilde{g})\mathcal{S}(\tilde{A}^{\tilde{g}}, \tilde{h}) = \mathcal{S}(\tilde{A}, \tilde{g}\tilde{h}). \tag{4}$$

We can define a line bundle $\mathcal{L}$ over $\mathcal{M}_{\in}$ by

$$\mathcal{L} \equiv \mathcal{A}_{\mathcal{F}} \times_{\mathcal{S}} C \tag{5}$$

where the right side is the quotient space given by the equivalence relation

$$(A, z) \sim (A^g, \mathcal{S}(A, g)z) \tag{6}$$

for $A \in \mathcal{A}_{\mathcal{F}}, z \in C$.

We consider a 3-dimensional manifold Y and a principal $SU(2)$-bundle P_Y over Y. We decompose Y into two handle bodies Y_1 and Y_2. Let X be the intersection of Y_1 and Y_2. We apply above construction to 2-dimensional manifold X and $P_Y|_X = P$. We consider the restriction of line bundle $\mathcal{L}$ to Lagrangian submanifolds L_1, L_2 of $\mathcal{M}_{\in}$ where L_1, L_2 are made from the handle bodies Y_1, Y_2 of Y. Let $\mathcal{L}_{\rangle}$ be the restriction of the line bundle $\mathcal{L}$ to L_i. Then there is section $\mathcal{S}_{\rangle}(A)$ of $\mathcal{L}_{\rangle}$ over L_i defined by

$$\mathcal{S}_i(A) = \exp(2\pi i CS(A_{Y_i})) \tag{7}$$

for $[A] \in \mathcal{M}_{\in}$ and A_{Y_i} is an extension of A to Y_i.

Now we consider the intersection of the two Lagrangian submanifolds L_1, L_2 in $\mathcal{M}_2$. Then we can see that this intersection is the moduli space of flat connections of P_Y over Y which we denote by $\mathcal{M}_3$. By the correspondence between the flat connection A and the $SU(2)$ -representation ρ_Y of $\pi_1(Y)$, $\mathcal{M}_3$ can be identified with $R(Y, SU(2))$. There may occur singularities of $\mathcal{M}_2$ within $\mathcal{M}_3$. But the set of singularities is a measure zero set in $\mathcal{M}_3$. So we may not care these singularities in the following construction since we shall integrate over the dense subset $R(Y, SU(2))^-$ of $\mathcal{M}_3 = R(Y, SU(2))$.

We consider the k-tensor power of $\mathcal{L}$, $\mathcal{L}^{\otimes^k}$ over $\mathcal{M}_2$ and their restrictions to two Lagrangian submanifolds L_1, L_2. We denote these by $\mathcal{L}_1^{\otimes^k}, \mathcal{L}_2^{\otimes^k}$. Then we can pair two sections $\mathcal{S}_i^k = \mathcal{S}_i^{\otimes^k}$ of $\mathcal{L}_i^{\otimes^k}$ in $\mathcal{M}_3$ by the hermitian product of the complex line C. We denote this pairing by $\langle \mathcal{S}_1^k, \mathcal{S}_2^k \rangle$. Then this can be considered as an $U(1)$ valued function on $\mathcal{M}_3 = R(Y, SU(2))$. We can see easily that $\langle \mathcal{S}_1^k, \mathcal{S}_2^k \rangle$ at a connection A is the exponential of the Chern-Simons invariant of Y, that is, $\exp(2k\pi i CS(A))$.

We recall that the half density derived from the R-torsion can be considered as a measure of $R(Y, SU(2))^-$. The Jeffrey-Weitsman-Witten invariant is defined by integrating the pairing $\langle \mathcal{S}_1^k, \mathcal{S}_2^k \rangle$ with respect to half density $\tau(Y, Ad(\rho_Y))^{\frac{1}{2}}$ over $R(Y, SU(2))^-$ using the correspondence between the flat connection A of P and the $SU(2)$-representation ρ_Y of $\pi_1(Y)$. We formulate this construction as the following definition.

Definition 1.1. For integer k, the Jeffrey-Weitsman-Witten invariant $Z(Y, k)$ is defined by

$$Z(Y,k) = \int_{R(Y,SU(2))^-} \langle \mathcal{S}_1^k, \mathcal{S}_2^k \rangle \tau(Y, Ad(\rho_Y))^{\frac{1}{2}}. \tag{8}$$

This definition is given in [J,W]. This definition is motivated from the asymptotic expansion of the Witten invariant $Z_Y(k)$ of 3-dimensional manifold Y [W]. The asymptotic expansion of $Z_Y(k)$ is given by

$$\begin{aligned} Z_Y(k) \simeq \tfrac{1}{2} \textstyle\sum_i (\tau(Y, A_i))^{\frac{1}{2}} \exp(-\tfrac{3\pi i + 2\pi i SF(A_i)}{4}) \\ \exp(2(k+2)\pi i CS_Y(A_i)) \end{aligned} \tag{9}$$

where the sum is taken over the finite set of flat connections A_i, $\tau(Y, A_i)$ is the Reidemeister torsion for A_i of Y and $SF(A_i)$ is the spectral flow from trivial connection to the flat connection A_i. The above formula is given in [F,G]. We can see that if the moduli space of flat connections is a discrete set, then definition 5.1 is almost same as the leading term of the above asymptotic expansion since the the square root of the R-torsion becomes a point mass in this case.

2 Jeffrey-Weitsman-Witten invariants for Seifert fibred manifolds

Now we introduce our result in [P] for Jeffrey-Weitsman-Witten invariant $Z(M, k)$ of Seifert fibred manifold M with the non-discrete irreducible $SU(2)$-representation space $R(M, SU(2))^-$.

We denote the Seifert fibred manifold with the Seifert invariant $\{g; (\alpha_1, \beta_1), \ldots, (\alpha_m, \beta_m)\}$ by $M = M(g; (\alpha_1, \beta_1), \ldots, (\alpha_m, \beta_m))$. We have that $\pi_1(M)$ is

$$\{a_i, b_i, q_j, h : [h, a_i] = [h, b_i] = [h, q_j] = 1, q_j^{\alpha_j} h^{\beta_j} = 1, \prod_{j=1}^{m} q_j \prod_{i=1}^{g} [a_i, b_i] = 1\}.$$

We assume that g is larger than 2 or equal to 2 from now on. The irreducible $SU(2)$-representation of $\pi_1(M)$ is well known. We review some facts that we need.

Since h is central in $\pi_1(M)$, an irreducible representation ρ takes h to ± 1 in $SU(2)$. So the trace of $\rho(q_j)$ is $2\cos(\frac{\pi n_j}{\alpha_j})$. The set of numbers $\{n_1, \ldots, n_m\}$ which are called as 'rotation numbers' determines a connected component of the irreducible $SU(2)$-representation space $R(M, SU(2))^-$. The rotation number n_j is even(odd) if β_j is even(odd).

To compute $Z(Y,k)$, we must integrate the pairing $\langle \mathcal{S}_1^k, \mathcal{S}_2^k \rangle$ with respect to the half density of the R-torsion over $R(M, SU(2))^-$. We know that the value $\langle \mathcal{S}_1^k, \mathcal{S}_2^k \rangle$ at the flat connection A is simply the Chern-Simons invariant of A. This invariant is constant in each connected component $R(M,(n_j))$. By the result of [A], the value for the fixed connected component $R(M,(n_j))$ is given by

$$\exp\left[2k\pi i\left(-\sum_{j=1}^{m}\left(\frac{\beta_j^* n_j^2}{\alpha_j} + \frac{2\epsilon n_j}{\alpha_j}\right) + \epsilon^2 \sum_{j=1}^{m} \frac{\beta_j}{\alpha_j}\right)\right] \tag{10}$$

where $\epsilon = \frac{1}{2}, 1$ if $\rho_M(h) = -1, 1$ and $\beta_j \beta_j^* = 1$ (mod) α_j as above.

From the result in [P] for the half density volume of $R(M, SU(2))^-$ the value of the Jeffrey-Weitsman-Witten invariant $Z(Y,k)$ over the connected component $R(M,(n_i))$ which is determined by $\{n_1, \ldots, n_m\}$ is given by

$$\begin{aligned} &\int_{R(M,(n_j))} \langle \mathcal{S}_1^k, \mathcal{S}_2^k \rangle (\tau(M, Ad(\rho_M)))^{\frac{1}{2}} \\ &= 2^m \langle \mathcal{S}_1^k, \mathcal{S}_2^k \rangle \prod_{j=1}^{m} \frac{|\sin(\frac{\pi n_j \beta_j^*}{\alpha_j})|}{|\alpha_j|^{\frac{1}{2}}} \int_{R(M,(n_j))} D_{H_1(M, su(2)_{\rho_M})}. \end{aligned} \tag{11}$$

From (10), (11) we have the following theorem.

Theorem 2.1. *For the integer k, the Jeffrey-Weitsman-Witten invariant $Z(M,k)$ of the Seifert fibred manifold M with the Seifert invariant $(g, (\alpha_1, \beta_1, \cdots, \alpha_m, \beta_m))$ is given by*

$$\begin{aligned} c_M \sum_{\{(n_j)\}} \quad &\exp\left[2k\pi i\left(-\sum_{j=1}^{m}\left(\frac{\beta_j^* n_j^2}{\alpha_j} + \frac{2\epsilon n_j}{\alpha_j}\right) + \epsilon^2 \sum_{j=1}^{m} \frac{\beta_j}{\alpha_j}\right)\right] \\ &\times \prod_{j=1}^{m} \left[\frac{|\sin(\frac{\pi n_j \beta_j^*}{\alpha_j})|}{|\alpha_j|^{\frac{1}{2}}} \sum_{n=1}^{\infty}\left(\frac{1}{n^{2(g-1)+m}} \prod_{j=1}^{m} \frac{|\sin(\frac{\pi n n_j}{\alpha_j})|}{|\sin(\frac{\pi n_j}{\alpha_j})|}\right)\right] \end{aligned} \tag{12}$$

where the above sum is taken over the finite set of the rotation numbers $\{(n_j)\}$, $c_M = 2^{m-1}Vol(S^2)^m Vol(SU(2))^{g+m-1}$ *and* $\epsilon = \frac{1}{2}$ *or* 1 *if* $\rho_M(h) = -1$ *or* 1.

We can see that $Z(M,k)$ depends only on the manifold M via the Seifert invariant $(g;(\alpha_i,\beta_i))$ since the set of all the rotation numbers (n_j) is determined by $\pi_1(M)$.

References

1. D. R. Auckly, *Topological methods to compute Chern-Simons invariants,* Math. Proc. Camb. Phil. Soc. **115** (1994), 229-251.
2. D. S. Freed, R. E. Gompf, *Computer calculation of Witten's 3-manifold invariant,* Commun. Math. Phys. **141** (1991), 79-117.
3. L. C. Jeffrey, J. Weitsman, *Half density Quantization of the moduli space of flat connections and Witten's semiclassical manifold invariants,* Topology **32** (1993), 509-529.
4. Jinsung Park, *Half-density volumes of representation spaces of some 3-manifolds and their application,* Duke Math. Jour. **86** (1997), 493-515.
5. T. R. Ramadas, I. M. Singer and J. Weitsman, *Some Comments on Chern-Simons Gauge theory,* Commun. Math. Phys. **126** (1989), 409-420.
6. E. Witten, *Quantum Field Theory and the Jones Polynomial,* Commun. Math. Phys. **121** (1989), 351-399.

COMPLEX-ANALYTICITY OF HARMONIC MAPS

KEISUKE UENO
Department of Mathematical Sciences, Faculty of Science,
Yamagata University, Yamagata 990-8560, Japan
E-mail : ueno@sci.kj.yamagata-u.ac.jp

1 Introduction

The aim of this paper is to make a preliminary report of my joint work with Seiki Nishikawa on complex-analyticity of harmonic maps between Kähler manifolds.

It is well-known that any holomorphic map between Kähler manifolds is a harmonic map. The converse, that is, complex-analyticity of harmonic maps, also holds under some appropriate geometric conditions.

In fact, Siu [5] proved that any harmonic map between compact Kähler manifolds, without boundaries, is either holomorphic or anti-holomorphic, if it satisfies the rank condition and the curvature of the target manifold is strongly negative (Fact 1).

Moreover, Nishikawa and Shiga [3] studied that problem for relatively compact domains in complete Kähler manifolds, and obtained conditions which imply the biholomorphic equivalence of such domains (Fact 2).

A typical example of Kähler manifolds having strongly negative curvature, in the sense of Siu [5], is the complex hyperbolic space, that is, the complex space form of constant holomorphic sectional curvature -1. Moreover this has the Eberlein-O'Neill compactification by adding the sphere at infinity. Thus we can regard it as a manifold with boundary, and expect that complex-analyticity also holds for harmonic maps between complex hyperbolic spaces.

Different from the case of relatively compact domains, the metric tensor in this case is not well-defined up to the ideal boundary. This allows the harmonic maps under consideration may behave more wildly near the ideal boundary. However, we can prove that if a proper harmonic map, which has sufficiently high regularity up to the ideal boundary, between complex hyperbolic spaces is a nondegenerate CR-map at the boundary, then it must be a holomorphic map (Theorem).

We organize this paper as follows. In section 2, in order to illustrate the idea of our proof, we give an outline of the proof of Siu's result as well as Nishikawa-Shiga's one. Then, in section 3, we shall discuss the complex-analyticity of harmonic maps between complex hyperbolic spaces. We study this problem in connection with the Dirichlet problem at infinity for harmonic maps, and make some comments on it in the last section.

2 The case of compact manifolds

The first remarkable result on complex-analyticity of harmonic maps is due to Siu.

Fact 1. (Siu [5]) *Let (M,g) and (M',h) be compact Kähler manifolds without boundaries. If the curvature tensor of (M',h) is strongly negative, and if $u : M \to M'$ is a harmonic map satisfying $\mathrm{rank}_{\mathbf{R}} du_p \geq 4$ at some point $p \in M$, then u is either holomorphic or anti-holomorphic.*

Here we call the curvature tensor R' of (M',h) is strongly negative, if it holds that

$$\sum_{\alpha,\beta,\gamma,\delta} R'_{\alpha\bar{\beta}\gamma\bar{\delta}}(A^\alpha \overline{B^\beta} - C^\alpha \overline{D^\beta})\overline{(A^\delta \overline{B^\gamma} - C^\delta \overline{D^\gamma})} > 0$$

for arbitrary complex numbers $A^\alpha, B^\beta, C^\gamma, D^\delta$ which satisfy $A^\alpha \bar{B}^\beta - C^\alpha \bar{D}^\beta \neq 0$ for at least one pair of indices (α, β). For example, the curvature tensor of the invariant metric of the unit ball in $\mathbf{C}^m$ is strongly negative, hence so is its compact quotient. Note that if the curvature tensor of (M',h) is strongly negative, then its sectional curvature is negative.

We shall give a rough sketch of Siu's proof of Fact 1. The following Bochner type formula plays an important role.

Lemma 1. (Siu [5]) *Let (M,g) and (M',h) be Kähler manifolds. Then for a smooth map $u : M \to M'$, the following identity holds.*

$$\partial\bar{\partial}\langle h, \bar{\partial}u \wedge \partial\bar{u}\rangle = \langle R', \bar{\partial}u \wedge \partial\bar{u} \wedge \partial u \wedge \bar{\partial}\bar{u}\rangle - \langle h, D\bar{\partial}u \wedge \bar{D}\partial\bar{u}\rangle,$$

where R' denotes the curvature tensor of (M',h) and $\langle\ ,\ \rangle$ means contraction of tensors.

Note that this formula does not depend on the metric tensor of the source manifold. Thus we do not need to pose any assumption on the curvature of M.

Proof of Fact 1. Let ω be the Kähler form of (M, g). Then, from Siu's Bochner type formula, we have

$$\int_M \partial\bar{\partial}\langle h, \bar{\partial}u \wedge \partial\bar{u}\rangle \wedge \omega^{m-2}$$

$$= \int_M \langle R', \bar{\partial}u \wedge \partial\bar{u} \wedge \partial u \wedge \bar{\partial}\bar{u}\rangle \wedge \omega^{m-2} - \int_M \langle h, D\bar{\partial}u \wedge \bar{D}\partial\bar{u}\rangle \wedge \omega^{m-2}.$$

It follows from Stokes' theorem that the left hand side vanishes. On the other hand, since u is a harmonic map and (M', h) has negative sectional curvature, there exist a nonpositive function χ and a nonnegative function σ such that

$$\langle h, D\bar{\partial}u \wedge \bar{D}\partial\bar{u}\rangle \wedge \omega^{m-2} = \chi\omega^m,$$

$$\langle R', \bar{\partial}u \wedge \partial\bar{u} \wedge \partial u \wedge \bar{\partial}\bar{u}\rangle \wedge \omega^{m-2} = \sigma\omega^m.$$

Hence

$$0 = \int_M (\sigma\omega^m - \chi\omega^m) \geq 0,$$

which implies $\chi = \sigma = 0$ on M. Then the strong negativity of the curvature of M' and the condition for the rank of u imply that $\partial u = 0$ or $\bar{\partial}u = 0$ on a neighborhood of p. By virtue of the unique continuation property $\partial u = 0$ or $\bar{\partial}u = 0$ on M. □

On the other hand, if manifolds have boundaries, we need to pose some condition on the boundary map. In the following results, we assume that the boundary map is a CR-map, which enable us to determine the sign of the boundary integral.

Fact 2. (Nishikawa-Shiga [3]) Let (M, g) and (M', h) be complete Kähler manifolds of complex dimension $m \geq 2$, and $D_1(\subset M), D_2(\subset M')$ be relatively compact domains with smooth boundaries. Suppose that (M', h) has strongly negative curvature and ∂D_1 is pseudoconvexD If a CR-diffeomorphism $f : \partial D_1 \to \partial D_2$ extends to a homotopy equivalence of D_1 to D_2, then f extends to a biholomorphic diffeomorphism of D_1 to D_2. In particular, D_1 and D_2 are biholomorphically equivalent.

Although they assumed that manifolds have the same dimension, it is not essential if one just want to get complex-analyticity of harmonic maps.

Proposition 1. Let (M, g) and (M', h) be complete Kähler manifolds of complex dimensions $m, n \geq 2$, and $D_1(\subset M), D_2(\subset M')$ be relatively compact domains with smooth boundaries. Suppose (M', h) has strongly negative curvature and ∂D_1 is pseudoconvexD Let u be a map from $\overline{D_1}$ to $\overline{D_2}$. If $u : D_1 \to D_2$ is a harmonic map satisfying $\mathrm{rank}_{\mathbf{R}} du_p \geq 4$ at some point $p \in D_1$ and its boundary value is a CR-map from ∂D_1 into ∂D_2, then u must be a holomorphic map.

Outline of a proof of Proposition 1. Fix a point p arbitrarily and let ψ be a defining function of ∂D_1. Then for a neighborhood U of p, we have $D_1 \cap U = \{z \in U \mid \psi(z) < 0\}$. Taking a suitable local coordinate system $(z^1, \ldots, z^m)$ of U, we can show that u is a CR-map at $\partial D_1 \cap U$ if and only if the following holds.

$$\partial_i u^{\bar{\alpha}} = \partial_m u^{\bar{\alpha}} \cdot \frac{\partial_i \psi}{\partial_m \psi},$$

where $(u^1, \ldots, u^n)$ is a local coordinate system nearby $u(p), \partial_i u^{\bar{\alpha}} = \partial u^{\bar{\alpha}} / \partial z^i$ and so on. Hence, in terms of these local coordinates, we have

$$\bar{\partial}\langle h, \bar{\partial} u \wedge \partial \bar{u} \rangle \wedge \omega^{m-2}$$

$$= - \left(\sum_{i=1}^{m-1} \partial_{\bar{i}} \partial_i \psi \right) \left(\sum_{\alpha=1}^{n} |\partial_{\bar{m}} u^{\alpha}|^2 \right) \times \text{positive } (2m-1)\text{-form on } \partial D_1.$$

Now $\sum_{i=1}^{m-1} \partial_{\bar{i}} \partial_i \psi \geq 0$, since ∂D_1 is pseudoconvex. Thus

$$0 \geq \int_M \langle R', \bar{\partial} u \wedge \partial \bar{u} \wedge \partial u \wedge \bar{\partial} \bar{u} \rangle \wedge \omega^{m-2} - \int_M \langle h, D \bar{\partial} u \wedge \bar{D} \partial \bar{u} \rangle \wedge \omega^{m-2}.$$

Recalling that the right hand side is nonnegative and following the previous argument, we see that u is either holomorphic or anti-holomorphic. However, it must be a holomorphic map, since it is a CR-map at the boundary. □

We now refer to a result for complex-analyticity of pluriharmonic maps from complete Kähler manifold into the complex space form for later use.

Fact 3. (Udagawa[6]) Let (M, g) be a complete Kähler manifold of $\dim_{\mathbf{C}} M \geq 2$ and (M', h) a complex space form of constant holomorphic sectional curvature $c (\neq 0)$. If a pluriharmonic map $u : M \to M'$ satisfies $\mathrm{rank}_{\mathbf{R}}\, du_p \geq 3$ at some point $p \in M$, then u is either holomorphic or anti-holomorphic.

We do not present a proof. However, we note that he used pointwise identities instead of using the integral formula, which enable one to remove the assumption that the source manifold is compact. This is actually possible, since u is a pluriharmonic map.

3 The case of complex hyperbolic spaces

Let $M = (\mathbf{B}^m, g)$ be the ball model of complex hyperbolic space of dimension m. Namely, $\mathbf{B}^m = \{z \in \mathbf{C}^m \mid \|z\| < 1\}$ and g is the Bergman metric on $\mathbf{B}^m$. We denote by $\overline{M}$ the Eberlein-O'Neill compactification of M.

Then we can prove the following

Theorem. (Nishikawa-Ueno[4]) *Let $M = (\mathbf{B}^m, g)$ and $M' = (\mathbf{B}^n, h)$, where g and h are Bergman metrics and $m, n \geq 2$. If a proper harmonic map $u \in C^\infty(M, M') \cap C^5(\overline{M}, \overline{M'})$ is a nondegenerate CR-map at the ideal boundary of M, then u must be a holomorphic map.*

Let N be the nilpotent part of the Iwasawa decomposition of $\mathrm{Isom}_0(M)$ and A the abelian part of it. Then A is real one-dimensional and is identified with $\mathbf{R}_+ = \{t \in \mathbf{R} \mid t > 0\}$ via the exponential mapping. Hence M is diffeomorphic to the product manifold $\mathbf{R}_+ \times N$, which is the unbounded model of the complex hyperbolic space. Moreover, the Bergman metric g can be expressed as a doubly warped product metric

$$g = \frac{dy^2}{y^2} + \frac{1}{y^2} g_1 + \frac{1}{y^4} g_2,$$

where $y \in \mathbf{R}_+$, N is the classical Heisenberg group of real dimension $2m-1$ and $g_1 + g_2$ is a left invariant metric on N.

This expression is an analogue of the upper half plane model of Poincaré disc.

Let ∞ denote the point at infinity in the direction of $\mathbf{R}_+$ and $M(\infty)$ the ideal boundary of M. Then the boundary portion, $M(\infty)\backslash\{\infty\}$, is identified with $\{0\} \times N$. Thus our representation of M, $\mathbf{R}_+ \times N$, defines a local chart around the ideal boundary, and $M(\infty)\backslash\{\infty\}$ has the Lie group structure via N.

It is known that N is a 2-term nilpotent Lie group. Namely, if $\mathbf{n}$ denotes the Lie algebra of N, then the derived algebra $\mathbf{n}_2 = [\mathbf{n}, \mathbf{n}]$ is the center of $\mathbf{n}$ and $\dim_{\mathbf{R}} \mathbf{n}_2 = 1$. Consider the decomposition $\mathbf{n} = \mathbf{n}_1 \oplus \mathbf{n}_2$ and take an orthonormal basis $\{e_i\}_{i=1}^{2m-2}$ for $(\mathbf{n}_1, (g_1)_e)$ and $\{e_{2m-1}\}$ for $(\mathbf{n}_2, (g_2)_e)$. We denote their left invariant extensions on N by the same letters. Then $\left\{e_0 = \partial/\partial y, \{e_i\}_{i=1}^{2m-1}\right\}$ is a left invariant frame on TM.

The same discussion applies to M', that is, M' is realized as

$$M' = \left(\mathbf{R}_+ \times N', \frac{dY^2}{Y^2} + \frac{1}{Y^2}h_1 + \frac{1}{Y^4}h_2\right),$$

where $Y \in \mathbf{R}_+$ and $h_1 + h_2$ is a left invariant metric on N'. We decompose $\mathbf{n}' = \mathbf{n}'_1 \oplus \mathbf{n}'_2$ and take a frame field $\{f_0 = \partial/\partial Y, f_i\}_{i=1}^{2n-1}$ on TM' in the same way.

For a smooth map $u : M \to M'$ we set $v_i^\alpha = f_\alpha^*(du(e_i))$. Then Donnelly proved the following

Fact 4. (Donnelly [1]) *Let $u \in C^\infty(M, M') \cap C^1(\overline{M}, \overline{M}')$ be a proper harmonic map and set $\varphi = u|_{M(\infty)}$. Then it holds that*

$$(*) \qquad d\varphi_p((\mathbf{n}_1)_p) \subseteq (\mathbf{n}'_1)_{\varphi(p)}$$

for any point p at the ideal boundary of M. We denote by $(\mathbf{n}_i)_p$ the $\mathbf{n}_i$-component of the tangent space T_pN.

Definition. *Let u and φ be as above. We say that u is nondegenerate if*

$$d\varphi_p((\mathbf{n}_2)_p) \not\subset (\mathbf{n}'_1)_{\varphi(p)}$$

for any point p at the ideal boundary of M.

Here we note that u is nondegenerate at $p \in M(\infty)$ if and only if $v_{2m-1}^{2n-1}(p) \neq 0$.

Fact 5. (Donnelly [1]) *Let $u \in C^\infty(M, M') \cap C^2(\overline{M}, \overline{M}')$ be a proper harmonic map. If u is nondegenerate, then the following identities hold at the ideal boundary of M.*

$$(1) \quad v_j^{2n-1} = e_0 \cdot v_j^{2n-1} = 0 \qquad (0 \le j \le 2m-2),$$

$$(2) \quad v_0^\alpha = 0 \qquad (1 \le \alpha \le 2n-1),$$

$$(3) \quad 2m(v_0^0)^4 - \left\{\sum_{i=1}^{2m-2}\sum_{\alpha=1}^{2n-2}(v_i^\alpha)^2\right\}(v_0^0)^2 - 2(v_{2m-1}^{2n-1})^2 = 0.$$

In particular, $v_0^0 > 0$ at the ideal boundary of M.

For any fixed $n \in N$, let $c(t) = (t, n) \in \mathbf{R}_+ \times N$ and $p = c(0)$. Then the above fact means that $v_0^0(p)$ is the velocity of the curve $u(c(t))$ at $u(p)$, and hence it is perpendicular to $M(\infty)$ at $u(p)$. Moreover the identity (3) asserts that the velocity v_0^0 is completely determined by the first derivatives of the boundary map. This result plays an essential role in Donnelly's proof of uniqueness and existence results for Dirichlet problem at infinity for harmonic maps[1].

Now, for the convenience of calculation, we take complex frame fields $\{\xi_i\}, \{\eta_\alpha\}$ as follows.

$$\xi_0 = \frac{1}{2}(e_0 - \sqrt{-1}ye_{2m-1}), \quad \xi_i = \frac{1}{2}(e_i - \sqrt{-1}e_{i+m-1}) \ (1 \le i \le m-1),$$

$$\eta_0 = \frac{1}{2}(f_0 - \sqrt{-1}Yf_{2n-1}), \quad \eta_\alpha = \frac{1}{2}(f_\alpha - \sqrt{-1}f_{\alpha+n-1}) \ (1 \le \alpha \le n-1).$$

We set

$$u_P^A := \eta_A^*(du(\xi_P)),$$

$$D\bar{\partial}u = \sum_{i,j,\alpha} D_j u_{\bar{i}}^\alpha \xi_j^* \wedge \bar{\xi}_i^* \otimes \eta_\alpha.$$

Then investigating the asymptotic behavior of the harmonic map equation and making use of CR-condition, we can deduce the following identity and asymptotic estimates.

Proposition 2. *Under the same assumption of Theorem, we obtain the following.*

(1) $(v_0^0)^2 = v_{2m-1}^{2n-1} = \dfrac{1}{2m-2}\displaystyle\sum_{\alpha=1}^{2n-2}\sum_{i=1}^{2m-2}(v_i^\alpha)^2$ at $M(\infty)$.

(2) $u_{\bar{i}}^\alpha = O(y^3), \quad D_j u_{\bar{i}}^\alpha = O(y^2) \quad (y \to 0).$

We note that the first identity is equivalent to the following:

$$h(du(ye_0), du(ye_0)) = \frac{1}{2m-2}\sum_{i=1}^{2m-2} h(du(ye_i), du(ye_i))$$
$$= h(du(y^2 e_{2m-1}), Y^2 f_{2n-1}) = 1 \qquad \text{at } M(\infty).$$

Recall that $\{ye_0, \{ye_i\}_{i=1}^{2m-2}, y^2e_{2m-1}\}$ (resp. $\{Yf_0, \{Yf_\alpha\}_{\alpha=1}^{2n-2}, Y^2f_{2n-1}\}$) is an orthonormal basis of TM (resp. TN) with respect to g (resp. h). Hence this identity implies that norms of each components of the differential map du must closely relate each other, if a proper harmonic map is a nondegenerate CR-map at the ideal boundary.

Now we shall present the idea of our proof of Theorem briefly.

Let ω_0 be the canonical Kähler form on $\mathbf{C}^m$ and $B_r^m = \{z \in \mathbf{C}^m \mid \|z\| < r\}$ $(0 < r < 1)$. Then we have

$$\int_{B_r^m} \partial\bar{\partial}\langle h, \bar{\partial}u \wedge \partial\bar{u}\rangle \wedge \omega_0^{m-2}$$

$$= \int_{B_r^m} \langle R', \bar{\partial}u \wedge \partial\bar{u} \wedge \partial u \wedge \bar{\partial}\bar{u}\rangle \wedge \omega_0^{m-2} - \int_{B_r^m} \langle h, D\bar{\partial}u \wedge \bar{D}\partial\bar{u}\rangle \wedge \omega_0^{m-2}.$$

It follows from Stokes' theorem and the asymptotic estimates in Proposition 2 that the right hand side goes to 0 as $r \to 1$. Thus we have

$$\int_{\mathbf{B}^m} \langle R', \bar{\partial}u \wedge \partial\bar{u} \wedge \partial u \wedge \bar{\partial}\bar{u}\rangle \wedge \omega_0^{m-2} - \int_{\mathbf{B}^m} \langle h, D\bar{\partial}u \wedge \bar{D}\partial\bar{u}\rangle \wedge \omega_0^{m-2} = 0.$$

Each terms on the right hand side must vanish, thus u is a pluriharmonic map.

On the other hand, it holds from Proposition 2 and Fact 5 that u satisfies $\mathrm{rank}_{\mathbf{R}} du_p \geq 3$ at some point $p \in M$. Hence Udagawa's result (Fact 3) asserts that u must be a holomorphic map, since u is a CR-map at the ideal boundary of M. □

4 Comment

Donnelly [1] proved the following existence result.

Fact 6. (Donnelly [1]) *Let $\varphi \in C^{2,\varepsilon}(M(\infty), M'(\infty))$ be nondegenerate and satisfy the condition $(*)$. Then there exists a harmonic map $u : M \to M'$, which assumes the boundary value φ continuously.*

Since the distribution $\mathbf{n}_1$ defines the contact structure on the boundary, φ preserves that structure if it satisfies the condition $(*)$.

Thus, proper holomorphic maps relate the complex geometry of the complex hyperbolic space with the CR-geometry on the ideal boundary. On the other hand, via nondegenerate proper harmonic maps, the contact geometry on the ideal boundary may corresponds to the interior geometry, say symplectic geometry, of the complex hyperbolic space.

It is also natural, for other negative manifolds, to expect that proper harmonic maps relate a geometry on the ideal boundary with the interior one. Anyway, this problem is yet to be investigated.

References

1. H. Donnelly, *Dirichlet problem at infinity for harmonic maps: rank one symmetric spaces*, Trans. Amer. Math. Soc. **344** (1994), 713–735.
2. R. S. Hamilton, Harmonic maps of manifolds with boundary, Lecture Notes in Math. **471**, Springer Verlag, Berlin, Heidelberg, New York, 1975.
3. S. Nishikawa and K. Shiga, *On the holomorphic equivalence of bounded domains in complete Kähler manifolds of nonpositive curvature*, J. Math. Soc. Japan **35** (1983), 273–278.
4. S. Nishikawa and K. Ueno, *Complex-analyticity of proper harmonic maps between complex hyperbolic spaces*, in preparation.
5. Y. T. Siu, *The complex-analyticity of harmonic maps and the strong rigidity of compact Kähler manifolds*, Ann. of Math. **112** (1980), 73–111.
6. S. Udagawa, *Pluriharmonic maps and minimal isometric immersions of Kähler manifolds*, J. London Math. Soc. **37** (1988), 375–384.

Textbooks from International Press

www.intlpress.com
PO Box 38-2872, Cambridge, MA 02238-2872

Calculus: A Computer Algebra Approach,

by Iris L. Ashel and D. Goldfeld

Uniquely designed for use with computer algebra systems and sophisticated calculators, this course also works well with a computer laboratory. The students are encouraged to use technology for manual computation while they rapidly progress through the concepts of differential and integral calculus, mathematical modeling and optimization, ordinary differential equations, differential calculus for vector valued and multi-variable functions. The students will progress to vector geometry and coordinate systems, two and three dimensional graphical display, multiple integration, vector fields and line integrals, and on to Fourier series and the Fourier expansion theorem.

Basic Partial Differential Equations,

by D. Bleecker and G. Csordas

Using partial differential equations as a tool to predict systems based on underlying physical principles, this text is accessible to students with calculus background without presupposing courses in linear algebra or ordinary differential equations. The text takes an inductive approach, motivating students to see the relevance and applications of the general principles which govern the subject, but rigorous proofs are still available.

The problems included range from routine to very challenging. The problems are constructed to require genuine understanding of principles as well as procedures, and the solutions manual is available only to instructors—not to the general public. Suggestions from the author indicate how the book can be used in one semester, two quarter, or two semester courses.

A First Course in Differential Geometry,

by C.-C. Hsiung

This text treats the traditional topics of curves and surfaces in 3-dimensional Euclidean space with attention equally on local and global properties. Extension to higher dimensional spaces and more general surfaces is available. In addition to analytical approaches, the text gives geometric interpretation to its subjects and encourages the development of geometric intuition. The book uses vector analysis and exterior differential calculus but does not require tensor calculus.

Part 1 reviews point-set topology, advanced calculus, and linear algebra. Part 2 establishes a general local theory for curves in three dimensional Euclidean space including proving the uniqueness theorem for curves in E^3. Part 3 develops a Local theory for surfaces in E^3, and Part 4 covers orientation of surfaces, surfaces of constant Gaussian curvature, and global theorems for surfaces.

Mathematical Physics Books
from International Press

www.intlpress.com

PO Box 38-2872, Cambridge, MA 02238-2872

Quantum Groups: from Coalgebras to Drinfeld Algebras *by Steven Shnider and Shlomo Sternberg*

75 years of Radon Transform *by Simon Gindikin and Peter Melchior*

Perspectives in Mathematics Physics *by Robert Penner and S.T. Yau*

Proceedings of the International Conference on Mathematical Physics XIth volume edited by Daniel Iagolnitzer, XIIth volume edited by David DeWitt

Mirror Symmetry volume 1 edited by S.T.Yau, volume 2 edited by Brian Greene

Physics Series

Physics of the Electron Solid *by S.-T. Chui*

Proceedings of the Second International Conference on Computational Physics, *by D.H. Fenb and T.-Y Zhang*

Chen Ning Yang, a Great Physicist of the Twentieth Century, *by S.-T. Yau*

Yukawa Couplings and the Origins of Mass, *edited by P. Ramond*

Proceedings of the International Seminar on Singularities and Complex Geometry, *edited by Qi-Keng Lu, Stephen Yau and Anatoly Libgober*

Surveys in Differential Geometry, Volume I, Volume II, *edited by C.C. Hsiung and S.-T. Yau,* **Volume III,** *edited by Chuu Lian Terng and Karen Uhlenbeck.*

Topics in Symplectic 4-manifolds, volume 1, First International Press Lecture Series, *edited by Ronald Stern*

Tsing Hua Lectures on Geometry and Analysis, *edited by S.-T. Yau*

Current Developments in Mathematics

Proceedings of the annual conference sponsored by Harvard and MIT

1995 Proceedings: lectures by Henri Darmon, Fred Diamond, and Richard Taylor, Mikhail Lyubich, Ib Madsen, Curtis McMullen, and Gang Tian

1996 Proceedings: lectures by Richard Borcherds, Gerrit Heckman and Eric Opdam, Ehud Hrushovski,k and Yves Meyer

1997 Proceedings: lectures by A. Connes, L. C. Evans, P. Sarnak, and W. Soergel and open problems by A. Jaffe, D. Stroock, B. Mazur, C. Taubes, and D. Jerison.